Heat Pipes

Fifth Edition

Heat Pipes

Fifth Edition

D.A. Reay

David Reay & Associates, Whitley Bay, UK

P.A. Kew

Heriot-Watt University, Edinburgh, UK

AMSTERDAM • BOSTON • HEIDELBERG • LONDON • NEW YORK • OXFORD
PARIS • SAN DIEGO • SAN FRANCISCO • SINGAPORE • SYDNEY • TOKYO

Butterworth-Heinemann is an imprint of Elsevier

BH

Butterworth-Heinemann is an imprint of Elsevier
Linacre House, Jordan Hill, Oxford OX2 8DP
30 Corporate Drive, Suite 400, Burlington, MA 01803, USA

Fifth edition 2006

British Library Cataloguing in Publication Data
A catalogue record for this book is available from the British Library

Library of Congress Cataloguing in Publication Data
A catalogue record for this book is available from the Library of Congress

ISBN–13: 978-0-7506-6754-8
ISBN–10: 0-7506-6754-0

For information on all Butterworth-Heinemann publications visit our website at http://books.elsevier.com

Printed and bound in Great Britain
06 07 08 09 10   10 9 8 7 6 5 4 3 2 1

CONTENTS

CONTENTS

DEDICATION

It is well over 30 years since the first pens were put to paper (now an old-fashioned phrase) by Professor Peter Dunn, OBE, then a Professor at Reading University, and David Reay, a research engineer at International Research & Development Company (IRD) in Newcastle upon Tyne, to start writing 'Heat Pipes'.

When a Fifth Edition was first discussed 2 or 3 years ago, Professor Dunn indicated that he would step aside as co-author, and Dr Peter Kew at Heriot-Watt University took over his role. Dr Kew had also worked on heat pipes at IRD and continues to collaborate with David Reay on heat pipe consultancy work.

Professor Dunn has now turned his hand to Appropriate Technology, where he remains active, and his illustrious career is summarised below.

Professor Dunn was trained in Civil Engineering and worked on Permanent Way Design and Manufacture first in industry and later for the LMS Railway Research Department. He then changed his career interests and moved into the design of microwave valves and particle accelerators. He was responsible for the Radio Frequency Accelerating System of the 7 GeV Accelerator Nimrod. Later, as the head of a team at AERE, Harwell, he carried out work on the direct generation of heat to electricity from nuclear reactors. One of the methods studied was thermionic diodes, and it was at this time that he first met Dr George Grover who was responsible for a similar group at the Los Alamos Laboratory in USA. Professor Dunn is a founder member of the International Conference.

Dr Grover's new work on heat pipes was exciting and highly relevant and was the start of Professor Dunn's interest in the subject. He commenced a study of liquid metal heat pipes for reactor application.

Later, Professor Dunn moved to Reading University where he set up the, then, new Department of Engineering. With his colleague Dr Graham Rice, he carried on heat pipe work.

In recent years, Professor Dunn's interests have moved to Appropriate Technology and Third World Development, he has carried out projects, particularly in renewable energy, in many countries and for some time was chairman of Gamos, a small firm concerned with development work overseas.

It is with great pleasure and thanks that we dedicate this Edition of Heat Pipes to Professor Peter Dunn.

David Reay
Peter Kew
November 2005.

PREFACE TO FIFTH EDITION

Recent editions of 'Heat Pipes' have appeared at intervals of approximately a decade. It is perhaps the last 10 years that have seen a transformation in heat pipe technology and application, characterised by, in the first case, loop heat pipes, and secondly, the mass production of miniature heat pipes for thermal management in the ubiquitous desktop and laptop computers. Millions of units per month are now being made, particularly in the Far East.

The other major change has been in the co-authorship of 'Heat Pipes'. As readers will see from the Dedication, Professor Dunn has relinquished his role as co-author, and this has been taken over by Dr Peter Kew, who first started research on heat pipes in the 1980s. We hope that readers will judge Dr Kew to be a worthy successor to Professor Dunn.

The changes in technological emphasis have allowed us to make some more radical changes to the book, rather than the 'fine-tuning' characteristic of earlier updates. To this end, we must acknowledge the referees, who assessed the contents list and made many constructive suggestions regarding the content. We hope that they find the new edition satisfies their requirements.

Particular features of the Fifth Edition are a revamped theory chapter, 'Design Guide' (Chapter 4), substantial additions to Chapter 6 'Special Types of Heat Pipe' including sections on capillary pumped loops oscillating heat pipes and electrokinetic effects, and a chapter dedicated to electronics cooling applications. The growth in use of the World Wide Web has allowed us to replace the conventional 'List of Manufacturers' with a list of useful web sites. An Appendix also covers recent papers not cited in the main text on heat pipe applications – these illustrate the breadth and depth of uses to which heat pipes have been put, or where they are the subject of feasibility studies.

Where data remain relevant, although they may be in some cases 50 years old, they are retained, as are the original data sources. Theories, wick properties, working fluids and manufacturing technologies do not change rapidly – especially where proved techniques have been successful over extended periods. We make no apologies for keeping an archive of what we believe to be useful data within one publication.

We hope that readers find the updated version as useful as earlier editions.

D.A. Reay
P.A. Kew
November 2005.

PREFACE TO FIRST EDITION

Following the publication by G.M. Grover et al. of the paper entitled 'Structures of Very High Thermal Conductance' in 1964, interest in the heat pipe has grown considerably. There is now a very extensive amount of literature on the subject, and the heat pipe has become recognised as an important development in heat transfer technology.

This book is intended to provide the background required by those wishing to use or to design heat pipes. The development of the heat pipe is discussed and a wide range of applications described.

The presentation emphasises the simple physical principles underlying heat pipe operation in order to provide an understanding of the processes involved. Where necessary, a summary of the basic physics is included for those who may not be familiar with these particular topics.

Full design and manufacturing procedures are given and extensive data provided in Appendix form for the designer.

The book should also be of use to those intending to carry out research in the field.

ACKNOWLEDGEMENTS

Dr Cosimo Buffone of Thermacore Europe and Nelson J. Gernert of Thermacore, Inc. for substantial data on a range of heat pipes and case studies, in particular those related to thermal management of electronics systems.

Professor Yu. Maydanik for data on loop heat pipes and appropriate illustrations.

Professor Ali Akbarzadeh for illustrations and other data on his heat pipe turbine developments and electronics cooling concepts.

Professor Dumitru Fetcu, Transilvania University of Brasov, Romania and CTO Heat Pipe Division, iESi USA Inc., Canada, for illustration of sodium heat pipes and performance data.

Professor Leonard Vasiliev, Luikov Heat & Mass Transfer Institute, Minsk, for data and figures in Chapters 6 and 7.

Dr John Burns of Protensive Ltd, for data on spinning disc reactors.

Dr David Etheridge of Nottingham University for information on heat pipes linked to phase change materials in buildings.

Professor Graeme Maidment of London South Bank University for data on the thermal control of London Underground railway system, and supermarket chilled food cabinets (Chapter 7).

Professor Robert Critoph, University of Warwick, for data on the adsorption system that employs heat pipes (Chapter 7).

Peter C. Cologer, Director, Advanced Thermal Applications, Swales Aerospace.

Deschamps Technologies for permission to use Figure 7.21.

Dr C. James of the Food Refrigeration and Process Engineering Research Centre, University of Bristol, for data and illustrations in Section 7.8.2.

INTRODUCTION

Figure 4 reprinted from Applied Thermal Engineering, Vol. 25, Maydanik, Y.F., Loop heat pipes. Review article, pp 635–657, Copyright 2005, with permission from Elsevier.

CHAPTER 1

Figure 1.10 reprinted from Applied Thermal Engineering, Vol. 23, Swanson, T.D. and Birur, G.C. NASA thermal control technologies for robotic spacecraft, pp 1055–1066, Copyright 2003, with permission from Elsevier.

Figure 1.11 reprinted from Applied Thermal Engineering, Vol. 25, Bintoro, J.S., Akbarzadeh, A. and Mochizuki, M. A closed-loop electronics cooling by implementing single phase impinging jet and mini channels heat exchanger, pp 2740–2753, Copyright 2005, with permission from Elsevier.

CHAPTER 2

Figures 2.25 and 2.29 reprinted from Brautsch A., Heat transfer mechanisms during the evaporation process from mesh screen porous structures, Ph.D. Thesis, Copyright 2002.

Figures 2.27 and 2.28 reprinted from Proceedings of the 12th International Heat Transfer Conference, Grenoble, August 2002, Brautsch, A. and Kew, P.A. The effect of surface conditions on boiling heat transfer from mesh wicks, Copyright 2002, with permission from Elsevier.

Figure 2.44 reprinted from Applied Thermal Engineering, Vol. 20, Payakaruk, T. Terdtoon, P., Ritthidech, S. Correlations to predict heat transfer characteristics of an inclined closed two-phase thermosyphon at normal operating conditions, pp 781–790, Copyright 1999, with permission from Elsevier.

Table 2.13 reprinted from International Journal of Refrigeration, Vol. 20, Golobic, I. and Gaspersic, B. Corresponding states correlation for maximum heat flux in two-phase closed thermosyphon, pp 402–410, Copyright 1997, with permission from Elsevier.

CHAPTER 4

Figure 4.6 reprinted from International Journal of Heat and Mass Transfer, Vol. 42, El-Genk, M.S. and Saber, H.H. Determination of operation envelopes for closed two-phase thermosyphons. pp 889–903, Copyright 1999, with permission from Elsevier.

CHAPTER 5

Figure 5.8 reprinted from Applied Thermal Engineering, Vol. 24, Launay, S., Sartre, V. and Lallemand, M. Experimental study on silicon micro-heat pipe arrays, pp 233–243, Copyright 2004, with permission from Elsevier.

CHAPTER 6

Figure 6.7 reprinted from Applied Thermal Engineering, Vol. 23, Marcarino, P. and Merlone, A. Gas-controlled heat-pipes for accurate temperature measurements. Paper presented at 12th International Heat Pipe Conference, Moscow, pp 1145–1152, Copyright 2003, with permission from Elsevier.

Figure 6.9 reprinted from Nuclear Instruments and Methods in Physics Research A, Vol. 442, Chernenko, A., Kostenko, V., Loznikov, V., Semena, N., Konev, S., Rybkin, B., Paschin, A. and Prokopenko, I. Optimal cooling of HPGe spectrometers for space-born experiments, pp 404–407, Copyright 2000, with permission from Elsevier.

Figure 6.10 reprinted from Energy and Buildings, Vol. 34, Varga, S., Oliveira, A.C. and Afonso, C.F. Characterisation of thermal diode panels for use in the cooling season in buildings, pp 227–235, Copyright 2002, with permission from Elsevier.

Figures 6.13 and 6.14a reprinted from Applied Thermal Engineering, Vol. 23, Charoensawan, P., Khandekar, S., Groll, M. and Terdtoon, P. Closed loop pulsating heat pipes, Part A: parametric experimental investigations, pp 2009–2020, Copyright 2003, with permission from Elsevier.

Figure 6.14b reprinted from Electronics Cooling, Vogel, M. and Xu, G., 'Low Profile Heat Sink Cooling Technologies for Next Generation CPU Thermal Designs,' Figure 4, Electronics Cooling, Vol. 11, No. 1, February 2005, p. 16. Copyright 2005, with permission from Electronics Cooling.

Figures 6.15 and 6.16 reprinted from Applied Thermal Engineering, Vol. 23, Khandekar, S., Charoensawan, P., Groll, M. and Terdtoon, P. Closed loop pulsating heat pipes. Part B: visualization and semi-empirical modeling, pp 2021–2033, Copyright 2003, with permission from Elsevier.

Figure 6.17 reprinted from Applied Thermal Engineering, Vol. 23, Khandekar, S., Dollinger, N. and Groll, M. Understanding operational regimes of closed loop pulsating heat pipes: an experimental study, pp 707–719, Copyright 2003, with permission from Elsevier.

Figure 6.18 reprinted from Applied Thermal Engineering, Vol. 24, Sakulchangsatjatai, P., Terdtoon, P., Wongratanaphisan, T., Kamonpet, P. and Murakami, M. Operation modeling of closed-end and closed-loop oscillating heat pipes at normal operating condition, pp 995–1008, Copyright 2004, with permission from Elsevier.

Figure 6.19 reprinted from Applied Thermal Engineering, Vol. 25, Katpradit, T., Wongratanaphisan, T., Terdtoon, P., Kamonpet, P., Polchai, A. and Akbarzadeh, A. Correlation to predict heat transfer characteristics of a closed end oscillating heat pipe at critical state, pp 2138–2151, Copyright 2004, with permission from Elsevier.

Figures 6.20 and 6.22 reprinted from Applied Thermal Engineering, Vol. 25, Maydanik, Y.F. Loop heat pipes, pp 635–657, Copyright 2005, with permission from Elsevier.

Figures 6.24, 6.25 and 6.26 reprinted from Applied Thermal Engineering, Vol. 23, Bazzo, E. and Riehl, R.R. Operation characteristics of a small-scale capillary pumped loop, pp 687–705, Copyright 2003, with permission from Elsevier.

Figure 6.27 reprinted from Applied Thermal Engineering, Vol. 24, Khodabandeh, R. Thermal performance of a closed advanced two-phase thermo-syphon loop for cooling of radio base stations at different operating conditions, pp 2643–2655, Copyright 2004, with permission from Elsevier.

Figure 6.33 reprinted from Sensors and Actuators B, Vol. 106, Qian, S. and Bau, H.H. Magneto-hydrodynamic stirrer for stationary and moving fluids, pp 859–870, Copyright 2005, with permission from Elsevier.

Figures 6.46 and 6.47 reprinted from International Journal of Heat & Mass Transfer, Vol. 48, Vasiliev, L. and Vasiliev, L. Jr. Sorption heat pipe – a new thermal control device for space and ground application, pp 2464–2472, Copyright 2005, with permission from Elsevier.

CHAPTER 7

Figure 7.3 reprinted from Heat Recovery Systems & CHP, Vol. 12, Valisiev, L.L., Boldak, I.M., Domorod, L.S., Rabetsky, M.I. and Schirokov, E.I. Experimental device for the residential heating with heat pipe and electric heat storage blocks, pp 81–85, Copyright 1992, with permission from Elsevier.

Figures 7.4 and 7.5 reprinted from Applied Thermal Engineering, Vol. 24, Ampofo, F., Maidment, G. and Missenden, J. Underground railway environment in the UK. Part 3: methods of delivering cooling, pp 647–659, Copyright 2004, with permission from Elsevier.

Figure 7.6 reprinted from Progress in Nuclear Energy, Vol. 32, Ohashi, K., Hayakawa, H., Yamada, M., Hayashi, T. and Ishii, T. Preliminary study on the application of the heat pipe to the passive decay heat removal system of the modular HTR, pp 587–594, Copyright 1998, with permission from Elsevier.

Figures 7.8, 7.9 and 7.10 reprinted from Applied Thermal Engineering, Vol. 22, Critoph, R.E. Multiple bed regenerative adsorption cycle using the monolithic carbon–ammonia pair, pp 667–677, Copyright 2002, with permission from Elsevier.

Figure 7.11 reprinted from Applied Thermal Engineering, Vol. 23, Zhang, H and Zhuang, J. Research, development and industrial application of heat pipe technology. Proceedings of the 12th International Heat Pipe Conference, Moscow, 2002, pp 1067–1083, Copyright 2003, with permission from Elsevier.

Figure 7.12 reprinted from Renewable Energy, Vol. 14, Aghbalou, F. et al. A parabolic solar collector heat pipe heat exchanger reactor assembly for cyclohexane's dehydrogenation: a simulation study, pp 61–67, Copyright 1998, with permission from Elsevier.

CHAPTER 8

Figures 8.6, 8.7 and 8.8 reprinted from Microelectronics Reliability, Vol. 44, Moon, S.H. et al. Improving thermal performance of miniature heat pipe for notebook PC cooling, pp 135–140, Copyright 2002, with permission from Elsevier.

Figure 8.9 reprinted from Applied Thermal Engineering, Vol. 25, Murer, S. et al. Experimental and numerical analysis of the transient response of a miniature heat pipe, pp 2566–2577, Copyright 2005, with permission from Elsevier.

Figure 8.10 reprinted from Applied Thermal Engineering, Vol. 25, Maydanik, Y.F. Loop heat pipes, pp 635–657, Copyright 2005, with permission from Elsevier.

INTRODUCTION

The heat pipe is a device of very high thermal conductance. The idea of the heat pipe was first suggested by Gaugler [1] in 1942. It was not, however, until its independent invention by Grover [2, 3] in the early 1960s that the remarkable properties of the heat pipe became appreciated and serious development work took place.

The heat pipe is similar in some respects to the thermosyphon and it is helpful to describe the operation of the latter before discussing the heat pipe. The thermosyphon is shown in Fig. 1a. A small quantity of water is placed in a tube from which the air is then evacuated and the tube sealed. The lower end of the tube is heated causing the liquid to vapourise and the vapour to move to the cold end of the tube where it is condensed. The condensate is returned to the hot end by gravity. Since the latent heat of evaporation is large, considerable quantities of heat can be transported with a very small temperature difference from end to end. Thus, the structure will also have a high effective thermal conductance. The thermosyphon has been used for many years and various working fluids have been employed. (The history of the thermosyphon, in particular the version known as the Perkins Tube, is reviewed in Chapter 1.) One limitation of the basic thermosyphon is that in order for the condensate to be returned to the evaporator region by gravitational force, the latter must be situated at the lowest point.

The basic heat pipe differs from the thermosyphon in that a wick, constructed for example from a few layers of fine gauze, is fixed to the inside surface and capillary forces return the condensate to the evaporator. (See Fig. 1b.) In the heat pipe the evaporator position is not restricted and it may be used in any orientation. If, of course, the heat pipe evaporator happens to be in the lowest position, gravitational forces will assist the capillary forces. The term 'heat pipe' is also used to describe high thermal conductance devices in which the condensate return is achieved by other means, for example centripetal force, osmosis or electrohydrodynamics.

Several methods of condensate return are listed in Table 1. A review of techniques is given by Roberts [4], and others are discussed by Reay [13], Jeyadeven et al. [14] and Maydanik [15].

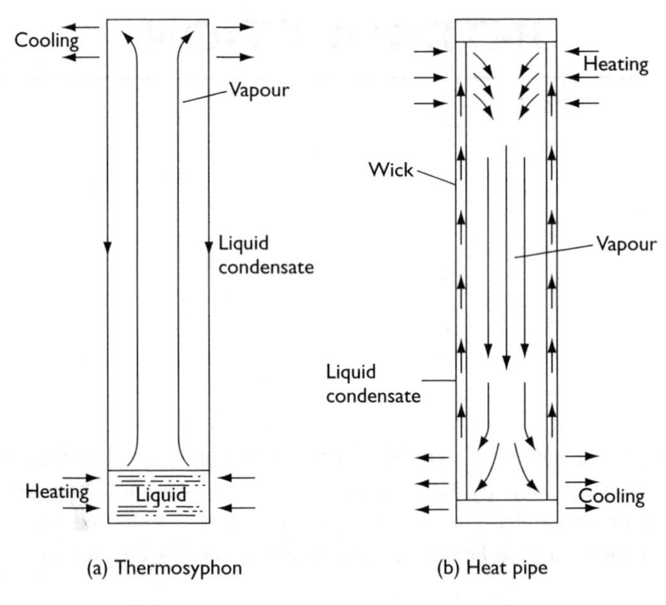

(a) Thermosyphon (b) Heat pipe

Fig. 1 The heat pipe and thermosyphon.

Table 1 Methods of condensate return

Gravity	Thermal syphon
Capillary force	Standard heat pipe
	Loop heat pipe
Centripetal force	Rotating heat pipe
Electrokinetic forces	Electrohydrodynamic heat pipe
	Electro-osmotic heat pipe
Magnetic forces	Magnetohydrodynamic heat pipe
	Magnetic fluid heat pipe
Osmotic forces	Osmotic heat pipe
Bubble pump	Inverse thermal syphon

1 THE HEAT PIPE–CONSTRUCTION, PERFORMANCE AND PROPERTIES

The main regions of the standard heat pipe are shown in Fig. 2. In the longitudinal direction (see Fig. 2a), the heat pipe is made up of an evaporator section and a condenser section. Should external geometrical requirements make this necessary, a further, adiabatic, section can be included to separate the evaporator and condenser. The cross-section of the heat pipe, Fig. 2b, consists of the container wall, the wick structure and the vapour space.

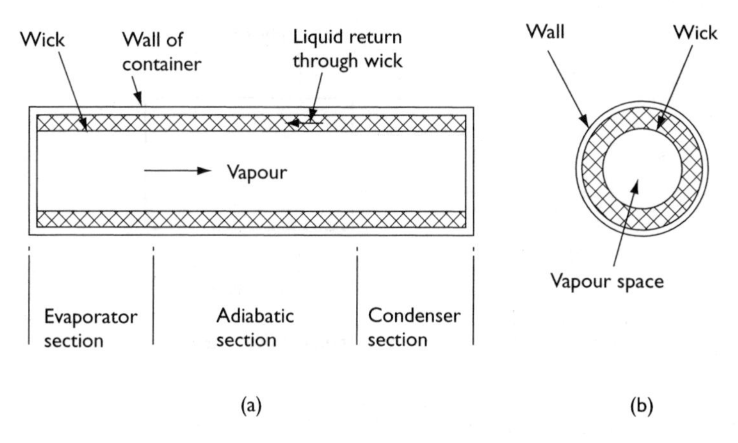

Fig. 2 The main regions of the heat pipe.

The performance of a heat pipe is often expressed in terms of 'equivalent thermal conductivity'. A tubular heat pipe of the type illustrated in Fig. 2, using water as the working fluid and operated at 150 °C would have a thermal conductivity several hundred times that of copper. The power handling capability of a heat pipe can be very high – pipes using lithium as the working fluid at a temperature of 1500 °C will carry an axial flux of 10–20 kW/cm^2. By suitable choice of working fluid and container materials, it is possible to construct heat pipes for use at temperatures ranging from 4 K to in excess of 2300 K.

For many applications, the cylindrical geometry heat pipe is suitable but other geometries can be adopted to meet special requirements.

The high thermal conductance of the heat pipe has already been mentioned; this is not the sole characteristic of the heat pipe.

The heat pipe is characterised by the following:

(i) Very high effective thermal conductance.
(ii) The ability to act as a thermal flux transformer. This is illustrated in Fig. 3.
(iii) An isothermal surface of low thermal impedance. The condenser surface of a heat pipe will tend to operate at uniform temperature. If a local heat load is applied, more vapour will condense at this point, tending to maintain the temperature at the original level.

Special forms of heat pipe can be designed having the following characteristics:

(iv) Variable thermal impedance.

A form of the heat pipe, known as the gas buffered heat pipe, will maintain the heat source temperature at an almost constant level over a wide range of heat input. This may be achieved by maintaining a constant pressure in the heat pipe but at the same time varying the condensing area in accordance with

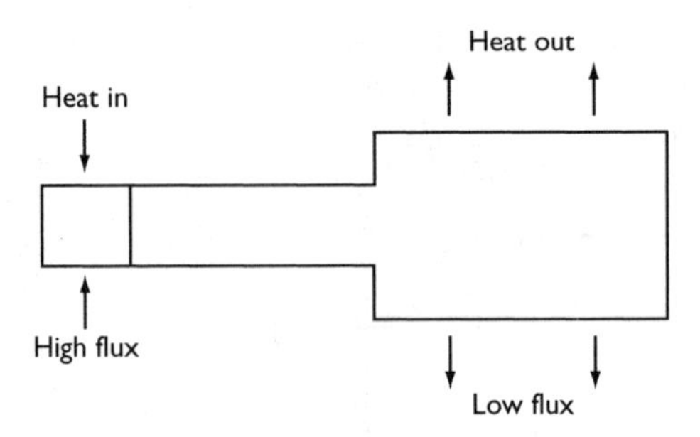

Fig. 3 The heat pipe as a thermal flux transformer.

the change in thermal input. A convenient method of achieving this variation of condensing area is that of 'gas buffering'. The heat pipe is connected to a reservoir having a volume much larger than that of the heat pipe. The reservoir is filled with an inert gas that is arranged to have a pressure corresponding to the saturation vapour pressure of the fluid in the heat pipe. In normal operation, a heat pipe vapour will tend to pump the inert gas back into the reservoir and the gas–vapour interface will be situated at some point along the condenser surface. The operation of the gas buffer is as follows.

Assume that the heat pipe is initially operating under steady state conditions. Now let the heat input increase by a small increment. The saturation vapour temperature will increase and with it the vapour pressure. The vapour pressure increases very rapidly for very small increases in temperature, for example the vapour pressure of sodium at 800 °C varies as the 10th power of the temperature. The small increase in vapour pressure will cause the inert gas interface to recede, thus exposing more condensing surface. Since the reservoir volume has been arranged to be large compared to the heat pipe volume, a small change in pressure will give a significant movement of the gas interface. Gas buffering is not limited to small changes in heat flux but can accommodate considerable heat flux changes.

It should be appreciated that the temperature, which is controlled in the more simple gas buffered heat pipes, as in other heat pipes, is that of the vapour in the pipe. Normal thermal drops will occur when heat passes through the wall of the evaporating surface and also through the wall of the condensing surface.

A further improvement is the use of an active feedback loop. The gas pressure in the reservoir is varied by a temperature-sensing element placed in the heat source.

(v) Loop heat pipes

The loop heat pipe (LHP), illustrated in Fig. 4, comprises an evaporator and a condenser, as in conventional heat pipes, but differs in having separate vapour and liquid lines, rather like the layout of the single-phase heat exchanger system used in buildings for heat recovery, the run-around coil. Those who recall the technical efforts made to overcome liquid–vapour entrainment in

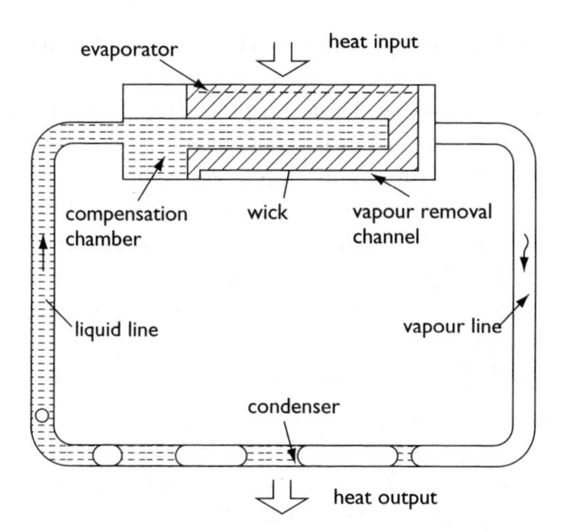

Fig. 4 Loop heat pipe (15).

heat pipes and, more importantly, in thermosyphons will know that isolation of the liquid path from the vapour flow (normally counter current) is beneficial. In the LHP, these flows are co-current in different parts of the tubing.

A unique feature of the LHP is the use of a compensation chamber. This two-phase reservoir helps to establish the LHP pressure and temperature, as well as maintain the inventory of the working fluid within the operating system. The LHP, described fully in Chapter 6, can achieve very high pumping powers, allowing heat to be transported over distances of several metres. This overcomes some of the limitations of other 'active' pumped systems that require external power sources.

(vi) Thermal diodes and switches

The former permit heat to flow in one direction only, while thermal switches enable the pipe to be switched off and on.

2 THE DEVELOPMENT OF THE HEAT PIPE

Initially Grover was interested in the development of high-temperature heat pipes, employing liquid metal working fluids, suitable for supplying heat to the emitters of thermionic electrical generators and removing heat from the collectors of these devices. This application is described in more detail in Chapter 7. Shortly after Grover's publication [3], work was started on liquid metal heat pipes by Dunn at Harwell and Neu and Busse at Ispra where both establishments were developing nuclear-powered thermionic generators. Interest in the heat pipe concept developed rapidly both for space and terrestrial applications. Work was carried out on many working fluids including metals, water, ammonia, acetone, alcohol, nitrogen and helium.

At the same time the theory of the heat pipe became better understood; the most important contribution to this theoretical understanding was by Cotter [5] in 1965. The manner in which heat pipe work expanded is seen from the growth in the number of publications, following Grover's first paper in 1964. In 1968, Cheung [6] lists 80 references; in 1971 Chisholm in his book [7] cites 149 references and by 1972 the NEL Heat Pipe Bibliography [8] contained 544 references. By the end of 1976, in excess of 1000 references to the topic were available and two International Heat Pipe Conferences had been held.

The III International Heat Pipe Conference, held in 1978 in Palo Alto, California, was sponsored by the American Institute of Aeronautics and Astronautics. Sixty-five papers were included in the proceedings [9], and it was noticeable that the term 'gravity-assisted heat pipe' was becoming popular as a description of units operating with the evaporator located below the condenser, while retaining some form of wick structure. Many so-called heat pipes are strictly thermosyphons, as they do not possess capillary or other means for transporting liquid internally.

Following the trend of approximately 3-year intervals, the IV International Heat Pipe Conference was held in 1981 in London. The proceedings [1] contain almost 70 papers, and of particular note is the contribution made to heat pipe technology during the past 3–4 years by Japan, particularly in applications technology in electronics and energy conservation (see Chapter 7).

In the first edition of this book, we stated that, with the staging of the first International Heat Pipe Conference in Stuttgart in 1973, the heat pipe had truly arrived. By 1977, it had become established as the most useful device in many mundane applications, as well as in retaining its more glamorous status in spacecraft temperature control [11].

The fourth edition was completed shortly after the VIII International Heat Pipe Conference, held in China in 1992, which was preceded by a successful VII Conference in Minsk, Byelorussia, in 1990 [12].

It was appropriate that the X International Heat Pipe Conference returned to Stuttgart in 1997, 24 years after the first Conference, and Prof. Manfred Groll, who initiated the series, presided over this anniversary event. The XII Conference, in Moscow (or to be strictly correct on a cruise along the Volga) in 2002, [16], and the subsequent Conference 2 years later in Shanghai, China, are indicative of the growth world-wide in research, development and application of heat pipes and some of the less conventional derivatives.

By the early twenty-first century, the heat pipe is a mass-produced item necessary in 'consumer electronics' products such as lap-top computers that are made by the tens of millions per annum. It also remains, judging by the literature appearing in scientific publications, a topic of substantial research activity.

Mention should also be made of the biannual series of Conferences hosted by Prof. Leonard Vasiliev in Minsk, which has successfully brought together the themes of heat pipes, heat pumps and refrigeration [17]. The most recent of these events, in September 2005, has yielded useful data for later Chapters. These topics complement one another rather well, as will be seen from some of the discussions on applications later in this book.

The most obvious pointer to the success of the heat pipe is the wide range of applications where its unique properties have proved beneficial. Some of these applications are discussed in detail in Chapters 7 and 8, but they include the following: electronics cooling, diecasting and injection moulding, heat recovery and other energy conserving uses, de-icing duties, cooking, control of manufacturing process temperatures, thermal management of spacecraft and in renewable energy systems.

3 THE CONTENTS OF THIS BOOK

Chapter 1 describes the development of the heat pipe in more detail. Chapter 2 gives heat transfer and fluid flow theory relevant to the operation of the classical wicked heat pipe, and details analytical techniques that are then applied to both heat pipes and thermosyphones. Chapter 3 discusses the main components of the heat pipe and the materials used, and includes compatibility data. Chapter 4 sets out design procedures and worked examples. Chapter 5 details how to make and test heat pipes, and Chapter 6 covers a range of special types, including loop heat pipes. Applications are fully discussed in Chapters 7 and 8.

A considerable amount of data are collected together in Appendices for reference purposes.

REFERENCES

[1] Gaugler, R.S. US Patent 2350348. Appl. 21 Dec, 1942. Published 6 June 1944.

[2] Grover, G.M. US Patent 3229759. Filed 1963.

[3] Grover, G.M., Cotter, T.P. and Erickson, G.F. Structures of very high thermal conductance. J. App. Phys., Vol. 35, p. 1990, 1964.

[4] Roberts, C.C. A review of heat pipe liquid delivery concepts. Advances in Heat Pipe Technology. Proceedings of IV International Heat Pipe Conference. Pergamon Press, Oxford, 1981.

[5] Cotter, T.P. Theory of heat pipes. Los Alamos Scientific Laboratory Report No. LA-3246-MS, 1965.

[6] Cheung, H. A critical review of heat pipe theory and application. UCRL 50453. 15 July 1968.

[7] Chisholm, D. The heat pipe. M & B Technical Library, TL/ME/2. Published by Mills and Boon Ltd., London, 1971.

[8] McKechnie, J. The heat pipe: a list of pertinent references. National Engineering Laboratory, East Kilbride. Applied Heat ST. BIB. 2-72, 1972.

[9] Anon. Proceedings of III International Heat Pipe Conference, Palo Alto, May 1978. AIAA Report CP 784, New York, 1978.

[10] Reay, D.A. (ed.). Advances in heat pipe technology. Proceedings of the IV International Heat Pipe Conference, Pergamon Press, Oxford, 1981.

[11] Anon. Proceedings of 2nd International Heat Pipe Conference, Bologna. ESA Report SP-112, Vols 1 & 2, European Space Agency, 1976.
[12] Anon. Proceedings of VII International Heat Pipe Conference, Minsk, Byelorussia (via Luikov Institute, Minsk), 1990.
[13] Reay, D.A. Microfluidics Overview. Paper presented at Microfluidics Seminar, East Midlands Airport, UK, April 2005. TUV-NEL, East Kilbride, 2005.
[14] Jeyadevan, B., Koganezawa, H. Nakatsuka, K. Performance evaluation of citric ion-stabilised magnetic fluid heat pipe. J. Magnetism and Magn. Mater., Vol. 289, pp 253–256, 2005.
[15] Maydanik, Yu. F. Loop heat pipes. Review article, Appl. Therm. Eng., Vol. 25, pp 635–657, 2005.
[16] Maydanik, Yu. F. (Guest Ed.). 12th International Heat Pipe Conference, 2002. Appl. Therm. Eng., Vol. 23, No. 9, June 2003.
[17] Heat Pipes, Heat Pumps, Refrigerators. Proceeding of the IV International Conference, Minsk, Belarus, 4–7 September 2000, CIS Association 'Heat Pipes', Minsk, 2000.

1

HISTORICAL DEVELOPMENT

The heat pipe differs from the thermosyphon by virtue of its ability to transport heat *against gravity* by an evaporation–condensation cycle. It is, however, important to realise that many heat pipe applications do not need to rely on this feature, and the Perkins tube, which predates the heat pipe by several decades and is basically a form of thermosyphon, is still used in heat transfer equipment. The Perkins tube must therefore be regarded as an essential part of the history of the heat pipe.

1.1 THE PERKINS TUBE

Angier March Perkins was born in Massachusetts, USA at the end of the eighteenth century, the son of Jacob Perkins, also an engineer [1]. In 1827, A.M. Perkins came to England, where he subsequently carried out much of his development work on boilers and other heat distribution systems. The work on the Perkins tube, which is a two-phase flow device, is attributed in the form of a patent to Ludlow Patton Perkins, the son of A.M. Perkins in the mid-nineteenth century. A.M. Perkins, however, also worked on single-phase heat distribution systems, with some considerable success, and although the chronological development has been somewhat difficult to follow from the papers available, the single-phase systems preceded the Perkins tube, and some historical notes on both systems seem appropriate.

The catalogue describing the products of A.M. Perkins & Sons Ltd, published in 1898, states that in 1831 A.M. Perkins took out his first patent for what is known as 'Perkins' system of heating by small bore wrought iron pipes. This system is basically a hermetic tube boiler in which water is circulated in tubes (in single phase and at high pressure) between the furnace and the steam drum, providing an indirect heating system. The boiler using hermetic tubes, described in UK Patent No. 6146, was produced for over 100 years on a commercial scale. The specification describes this closed cycle hot water heater as adapted for sugar making and refining evaporators, steam boilers and also for various processes requiring molten metals for alloying or working of other metals at high temperatures, suggesting that the tubes in the Perkins system operate with high-pressure hot water at temperatures well in excess of 150 °C.

The principle of the 'ever full' water boiler, as devised in the United States by Jacob Perkins to prevent the formation of a film of bubbles on the inner wall of the heat input section of the tubes, is applied as described in the above patent.

'As water expands about one-twentieth of its bulk being converted into steam, I provide about double that extra space in the "expansion tube" which is fitted with a removable air plug to allow the escape of air when the boiler is being filled. With this space for the expansion of the heated water the boiler is completely filled, and will at all times be kept in constant contact with the metal, however high the degree of heat such apparatus may be submitted to; and at the same time there will be no danger of bursting the apparatus with the provision of the sufficient space as named for the expansion of the water'.

In 1839 most of the well-known forms of A.M. Perkins hot water hermetic heating tubes were patented in UK Patent No. 8311, and in that year a new invention, a concentric tube boiler, was revealed. The hot water closed circuit heating tubes in the concentric tube system were fork-ended and dipped into two or more steam generation tubes. These resembled superheater elements as applied on steam locomotives and a large boiler operating on this principle would consist of many large firetubes, all sealed off at one end and traversed by the inner hot water tubes, connected up externally by U bends. This proved to be the most rapid producer of superheated steam manufactured by the Perkins Company and was even used as the basis for a steam actuated rapid firing machine gun, offered to the US Federal Government at the time of the Civil War. Although not used, they were 'guaranteed to equal the efficiency of the best Minie rifles of that day, but at a much lower cost for coal than for gun powder'. The system was, however, used in marine engines, '... it gives a surprising economy of fuel and a rapid generation, with lightness and compactness of form; and a uniform pressure of from 200 lbs to 800 lbs per sq. in., may be obtained by its use'.

Returning to the Perkins hermetic tube single-phase water circulating boiler, as illustrated in Fig. 1.1, some catalogues describe these units as operating at pressures up to 4000 psi and being pressure-tested in excess of 11 000 psi. Operators were quick to praise the cleanliness, both inside and outside, of the hermetic tubes after prolonged use.

The first use of the Perkins tube, i.e. one containing only a small quantity of water and operating on a two-phase cycle, is described in a patent by Jacob Perkins (UK Patent No. 7059, April 1936). The general description is as follows [2]:

'One end of each tube projects downwards into the fire or flue and the other part extends up into the water of the boiler; each tube is hermetically closed to prevent escape of steam. There will be no incrustation of the interior of the tubes and the heat from the furnace will be quickly transmitted upwards. The interior surfaces of the tubes will not be liable to scaleage or oxidation, which will, of course, tend

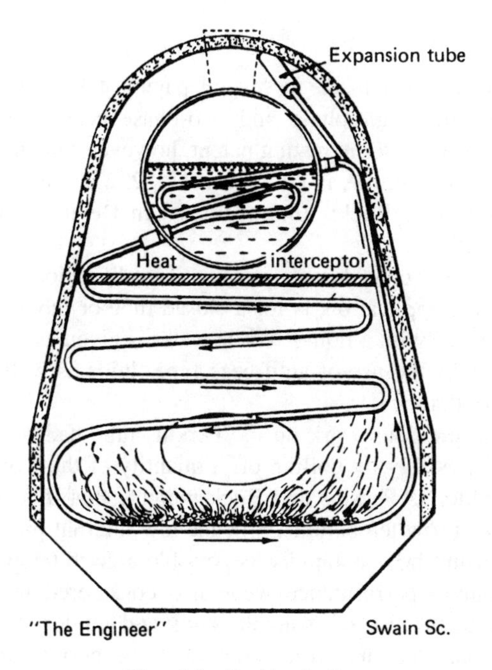

Fig. 1.1 Perkins boiler.

much to preserve the boiler so constructed.' The specification says 'These tubes are each one to have a small quantity of water depending upon the degree of pressure required by the engine; and I recommend that the density of the steam in the tubes should be somewhat more than that intended to be produced in the boiler, and, for steam and other boilers under the atmospheric pressure, that the quantity of water to be applied in each tube is to be about 1:1800 part of the capacity of the tube; for a pressure of 2 atm to be two 1:1800 parts; for 3 atm, three 1:1800 parts, and so on, for greater or less degrees of pressure, and by which means the tubes of the boiler when at work will be pervaded with steam, and any additional heat applied thereto will quickly rise to the upper parts of the tubes and be given off to the surrounding water contained in the boiler – for steam already saturated with heat requires no more (longer) to keep the atoms of water in their expanded state, consequently becomes a most useful means of transmitting heat from the furnace to the water of the boiler'.

The earliest applications for this type of tube were in locomotive boilers and in locomotive fire box superheaters (in France in 1863). Again, as with the single-phase sealed system, the cleanliness of the tubes was given prominent treatment in many papers on the subject. At the Institution of Civil Engineers in February 1837, Perkins stated that following a 7-month life test on such a boiler tube under representative operating conditions, there was no leakage or incrustation, no deposit of any kind occurring within the tube.

1.2 PATENTS

Reference has already been made to several patents taken by A.M. Perkins and J. Perkins on hermetic single-phase and two-phase heating tubes, normally for boiler applications. The most interesting patent, however, which relates to improvements in the basic Perkins tube, is UK Patent No. 22272, dated 1892, and granted to L.P. Perkins and W.E. Buck: 'Improvements in Devices for the Diffusion or Transference of Heat' [3].

The basic claim, with a considerable number of modifications and details referring to fluid inventory and application, is for a closed tube or tubes of suitable form or material, partially filled with a liquid. While water is given as one specific working fluid, the patent covers the use of antifreeze type fluids as well as those having a higher boiling point than water.

It is obvious that previous work on the Perkins tube had revealed that purging of the tubes of air, possibly by boiling off a quantity of the working fluid prior to sealing, was desirable, as Perkins and Buck indicate that this should be done for optimum operation at low temperatures (hence low internal pressures) and when it is necessary to transmit heat as rapidly as possible at high temperatures.

Safety and optimum performance were also considered in the patent, where reference is made to the use of 'suitable stops and guides' to ensure that tubes refitted after external cleaning were inserted at the correct angle and with the specified amount of evaporator section exposed to the heat source (normally an oil, coal or gas burner). Some form of entrainment had also probably occurred in the original straight Perkins tube, particularly when transferring heat over considerable distances. As a means of overcoming this limitation, the patent provides U bends so that the condensate return occurs in the lower portion of the tube, vapour flow to the heat sink taking place in the upper part.

Applications cited included heating of greenhouses, rooms, vehicles, dryers, and as a means of preventing condensation on shop windows, the tubes providing a warm convection current up the inner face of the window. Indirect heating of bulk tanks of liquid is also suggested. The use of the device as a heat removal system for cooling dairy products, chemicals and heat exchanging with the cooling water of gas engines is also proposed, as its use in waste heat recovery, the heat being recovered from the exhaust gases from blast furnaces and other similar apparatus, and used to preheat incoming air.

On this and other air/gas heating applications of the Perkins tube, the inventors have neglected to include the use of external finning on the tubes to improve the tube-to-gas heat transfer. Although not referring to the device as a Perkins tube, such modifications were proposed by F.W. Gay in US Patent No. 1725906, dated 27 August, 1929, in which a number of finned Perkins tubes or thermosyphons are arranged as in the conventional gas/gas heat pipe heat exchanger, with the evaporator sections located vertically below the condensers, a plate sealing the passage between the exhaust and inlet air ducts, as shown in Fig. 1.2. Working fluids proposed include methanol, water and mercury, depending upon the likely exhaust gas temperatures.

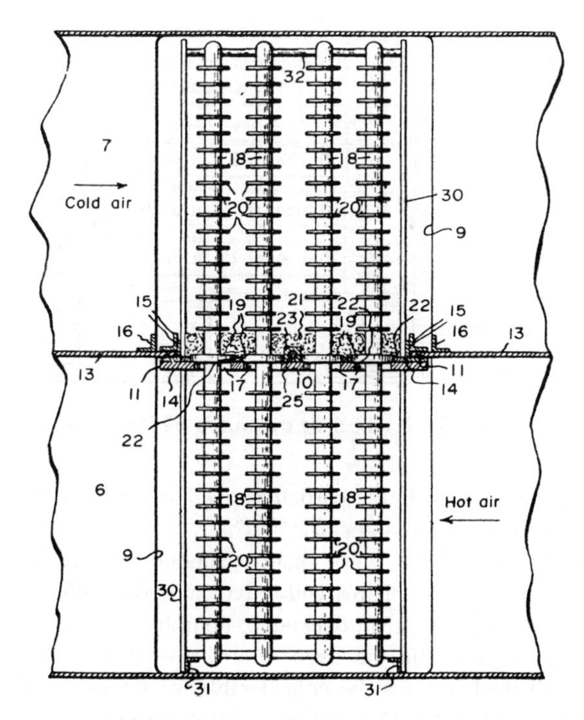

Fig. 1.2 Thermosyphon heat exchanger proposed by F.W. Gay.

1.3 THE BAKER'S OVEN

The main use of the Perkins tube was in baking ovens. One of the earliest forms of baking ovens to which the Perkins tube principle was applied was a portable bread oven supplied to the British army in the nineteenth century. In common with static ovens employing the Perkins tube, the firing was carried out remote from the baking chamber, the heat being transferred from the flames to the chamber by the vapour contained within the tubes. The oven operated at about 210 °C, and it was claimed that the fuel savings using this type of heating were such that only 25 per cent of the fuel typically consumed by conventional baking ovens was required [4].

A more detailed account of the baking oven is given in Ref. [5]. This paper, published in 1960 by the Institution of Mechanical Engineers, is particularly concerned with failures of the tubes used in these ovens, and the lack of safety controls which led to a considerable number of explosions in the tube bundles.

Gas or oil firing had replaced coal and coke in many of the installations, resulting in the form of an oven shown in Fig. 1.3, in which 80 tubes are heated at the evaporator end by individual gas flames. This illustration shows the simplest form of Perkins tube oven in which straight tubes are used. Other systems employ U tubes or a completely closed loop, as put forward by Perkins and Buck as a method

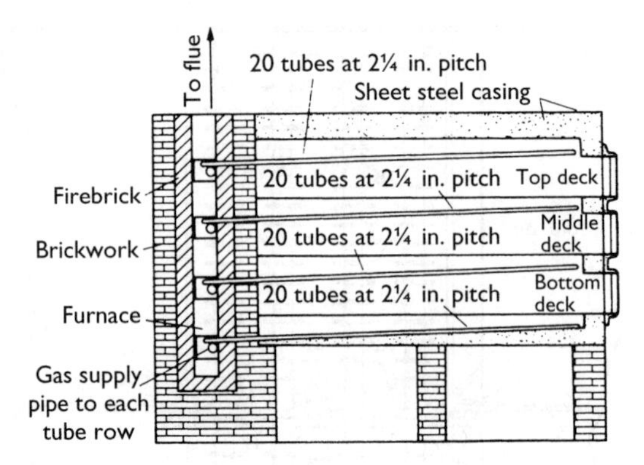

Fig. 1.3 Gas-fired baking oven using 80 Perkins tubes.

for overcoming entrainment. In the particular oven under consideration, maximum oven temperatures were of the order of 230 °C.

One feature of interest is the very small evaporator length, typically less than 5 cm, of many ovens. This compares with a typical overall tube length of 3 m and a condenser section of about 2.5 m depending upon the thickness of the insulating wall between the furnace and the oven. The diameter of the tubes is typically 3 cm and the wall thickness can be considerable, of the order of 5–6 mm. Solid drawn tubes were used on the last Perkins ovens constructed, with one end closed by swaging or forging prior to charging with the working fluid. Originally seam-welded tubes or wrought iron tubes were used. The fluid inventory in the tubes is typically about 32 per cent by volume, a very large proportion when compared with normal practice for heat pipes and thermosyphon, which generally have much longer evaporator sections, in addition. However, Perkins and Buck, in proposing a larger fluid inventory than that originally used in the Perkins tube, indicated that dryout had been a problem in earlier tubes, leading to overheating, caused by complete evaporation of the relatively small fluid inventory.

The inventory of 32 per cent by volume is calculated on the assumption that just before the critical temperature of saturated steam is reached, 374 °C, the water content would be exactly 50 per cent of the total tube volume. It can be shown by calculation that, up to the critical temperature, if the water inventory at room temperature is less than 32 per cent, it will all evaporate before the critical temperature is reached, resulting in overheating. If, however, the quantity of water is greater than 32 per cent by volume, the tube will be completely filled with water and unable to function before the critical temperature is reached.

An application of the thermosyphon revealed in the engineering literature early in the last century was the Critchley–Norris car radiator, which employed 110 thermosyphons for cooling water, the finned condenser sections projecting into the airstream. Ease of replacement of burst tubes was cited as an advantage. Note that the maximum speed of the vehicle to which they were fixed was 12 miles/h! The radiator is illustrated in Fig. 1.4.

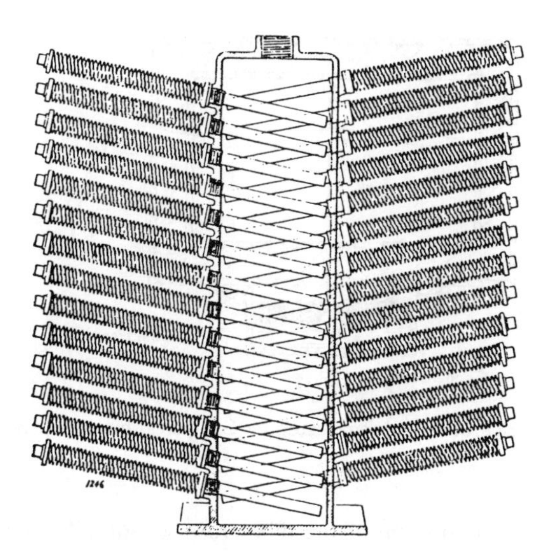

Fig. 1.4 The Critchley–Norris Radiator.

1.4 THE HEAT PIPE

As mentioned in the Introduction, the heat pipe concept was first put forward by R.S. Gaugler of the General Motors Corporation, Ohio, USA. In a patent application dated 21 December 1942 and published [5] as US Patent No. 2350348 on 6 June 1944, the heat pipe is described as applied to a refrigeration system.

According to Gaugler, the object of the invention was to '... cause absorption of heat, or in other words, the evaporation of the liquid to a point above the place where the condensation or the giving off of heat takes place without expending upon the liquid any additional work to lift the liquid to an elevation above the point at which condensation takes place'. A capillary structure was proposed as the means for returning the liquid from the condenser to the evaporator, and Gaugler suggested that one form of this structure might be a sintered iron wick. The wick geometries proposed by Gaugler are shown in Fig. 1.5. It is interesting to note the

Fig. 1.5 Gaugler's proposed heat pipe wick geometries.

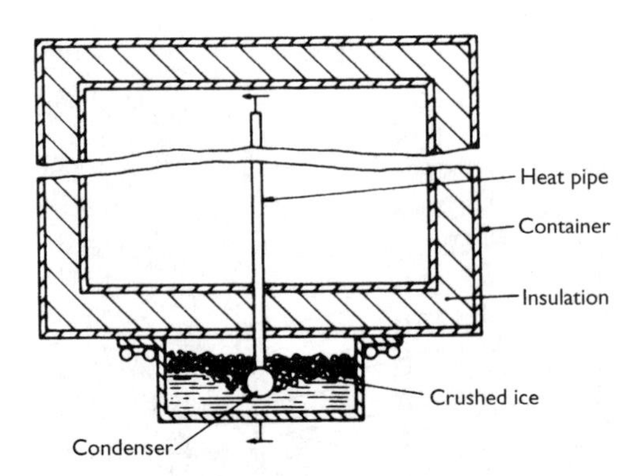

Fig. 1.6 The refrigeration unit suggested by Gaugler in his patent published in 1944.

comparatively small proportion of the tube cross-section allocated to vapour flow in all three of his designs.

One form of refrigeration unit suggested by Gaugler is shown in Fig. 1.6. The heat pipe is employed to transfer heat from the interior compartment of the refrigerator to a pan below the compartment containing crushed ice. In order to improve heat transfer from the heat pipe into the ice, a tubular vapour chamber with external fins is provided into which the heat pipe is fitted. This also acts as a reservoir for the heat pipe working fluid.

The heat pipe as proposed by Gaugler was not developed beyond the patent stage, as the other technology available at that time was applied to solve the particular thermal problem at General Motors Corporation.

Grover's patent [6] filed, on behalf of the US Atomic Energy Commission in 1963, coins the name 'heat pipe' to describe the devices essentially identical to that in the Gaugler patent. Grover, however, includes a limited theoretical analysis and presents results of experiments carried out on stainless steel heat pipes incorporating a wire mesh wick and sodium as the working fluid. Lithium and silver are also mentioned as working fluids.

An extensive programme was conducted on heat pipes at Los Alamos Laboratory, New Mexico, under Grover, and preliminary results were reported in the first publication on heat pipes [7]. Following this, the UK Atomic Energy Laboratory at Harwell started similar work on sodium and other heat pipes [8]. The Harwell interest was primarily the application to nuclear thermionic diode converters, a similar programme commenced at the Joint Nuclear Research Centre, Ispra, Italy under Neu and Busse. The work at Ispra built up rapidly and the laboratory become the most active centre for heat pipe research outside the United States [9, 10].

The work at Ispra was concerned with heat pipes for carrying heat to emitters and for dissipating waste heat from collectors. This application necessitated heat pipes operating in the temperature regions between about 1600 °C and 1800 °C (for emitters) and 1000 °C (for collectors). At Ispra the emphasis was on emitter heat pipes that posed more difficult problems concerning reliability over extended periods of operation.

The first commercial organisation to work on heat pipes was RCA [11, 12]. Most of their early support came from US Government contracts; during the 2-year period, mid-1964 to mid-1966, they made heat pipes using glass, copper, nickel, stainless steel, molybdenum and TZM molybdenum as wall materials. Working fluids included water, caesium, sodium, lithium and bismuth. Maximum operating temperatures of 1650 °C had been achieved.

Not all of the early studies on heat pipes involved high operating temperatures. Deverall and Kemme [13] developed a heat pipe for satellite use incorporating water as the working fluid, and the first proposals for a variable conductance heat pipe were made again for a satellite [14].

During 1967 and 1968, several articles appeared in the scientific press, most originating in the United States, indicating a broadening of the areas of application of the heat pipe to electronics cooling, air conditioning, engine cooling and others [15–17]. These revealed developments such as flexible and flat plate heat pipes. One point stressed was the vastly increased thermal conductivity of the heat pipe when compared with solid conductors such as copper, a water heat pipe with a simple wick having an effective conductivity several hundred times that of a copper rod of similar dimensions.

Work at Los Alamos Laboratory continued at a high level. Emphasis was still on satellite applications, and the first flights of heat pipes took place in 1967 [13]. In order to demonstrate that heat pipes would function normally in a space environment, a water/stainless steel heat pipe, electrically heated at one end, was launched into an earth orbit from Cape Kennedy on an Atlas–Agena vehicle. When the orbit had been achieved, the heat pipe was automatically turned on and telemetry data on its performance were successfully received at five tracking stations in a period lasting 14 orbits. The data suggested that the heat pipe operated successfully.

By now the theory of the heat pipe was well developed, based largely on the work of Cotter [18], also working at Los Alamos. So active were laboratories in the United States, and at Ispra that in his critical review of heat pipe theory and applications [19], Cheung was able to list over 80 technical papers on all aspects of heat pipe development. He was able to show that the reliability of liquid metal heat pipes under long-term operation (9000 h) at elevated temperatures (1500 °C) had been demonstrated. Heat pipes capable of transferring axial fluxes of $7\,\mathrm{kW/cm^2}$ had been demonstrated.

Cheung also referred to various forms of wick, including an arterial type illustrated in Fig. 1.7, which was developed by Katzoff [20]. Its operation was tested in a glass heat pipe using an alcohol as the working fluid. The function of the artery, which has become a common feature of heat pipes developed for satellite use, is to provide a low-pressure drop path for transporting liquid from the condenser to the evaporator, where it is redistributed around the heat pipe circumference using a fine pore wick provided around the heat pipe wall.

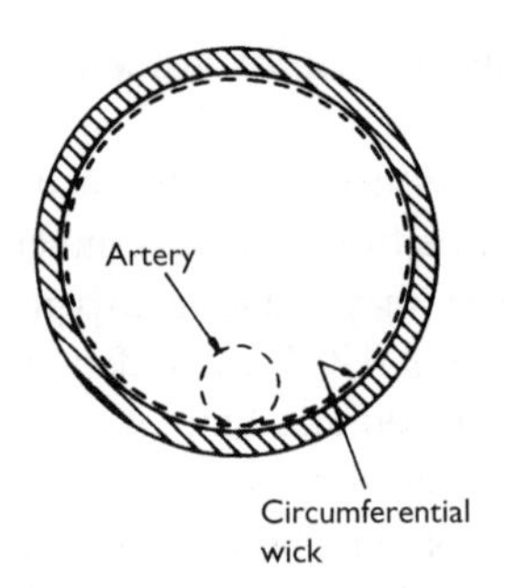

Fig. 1.7 An arterial wick form developed by Katzoff [20].

Following the first heat pipe test in space in 1967 [13], the first use of heat pipes for satellite thermal control was on GEOS-B, launched from Vandenburg Air Force Base in 1968 [21]. Two heat pipes were used, located as shown in Fig. 1.8. The heat pipes were constructed using 6061 T-6 aluminium alloy, with 120-mesh aluminium as the wick material. Freon 11 was used as the working fluid. The purpose of the heat pipes was to minimise the temperature differences between the various transponders in the satellite. Based on an operating period of

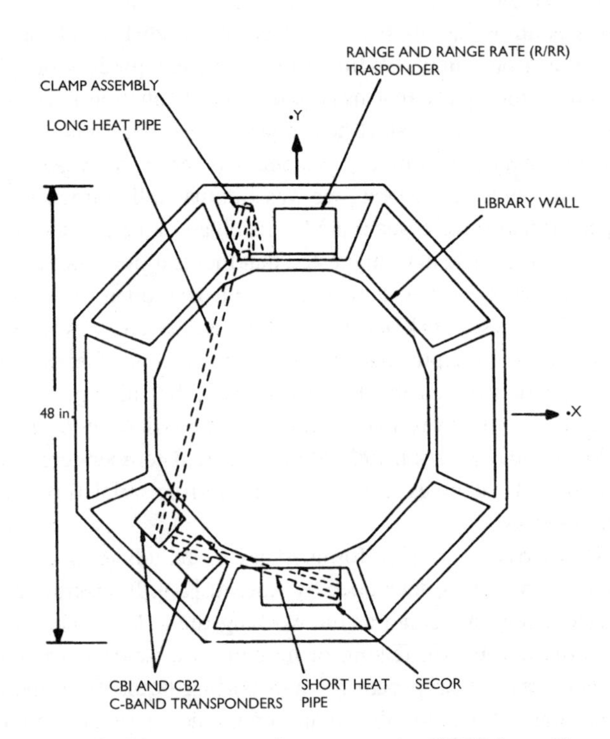

Fig. 1.8 Heat pipes used in space – the GEOS-B satellite.

145 days, the range between the maximum and minimum transponder temperatures was considerably smaller than that in a similar arrangement in GEOS-A, an earlier satellite not employing heat pipes. The heat pipes operated near-isothermally and performed well over the complete period of observation.

In 1968, Busse wrote a paper [22] that summarised the heat pipe activities in Europe at that time, and it is notable that the Ispra Laboratory of Euratom was still the focal point for European activities. Other laboratories were making contributions, however, including Brown Boveri, Karlsruhe Nuclear Research Centre, the Institüt für Kernenergetic, Stuttgart and Grenoble Nuclear Research Centre. The experimental programmes at the above mentioned laboratories were performed largely on heat pipes using liquid metals as working fluids and centred on life tests and measurements of the maximum axial and radial heat fluxes. Theoretical aspects of heat transport limitations were also studied. By now we were also seeing the results of basic studies on separate features of heat pipes, for example, wick development, factors affecting evaporator limiting heat flux and the influence of non condensable gases on performance.

In Japan, a limited experimental programme was conducted at the Kisha Seizo Kaisha Company [23]. Presenting a paper on this work in April 1968 to an audience of air conditioning and refrigeration engineers, Nozu et al. described an air heater utilising a bundle of finned heat pipes. This heat pipe heat exchanger is of considerable significance because of interest in energy conservation and environmental protection, as it can be used to recover heat from hot exhaust gases and can be applied in industrial and domestic air conditioning. Such heat exchangers are now available commercially and are referred to in Chapter 7.

During 1969, the published literature on heat pipes showed that establishments in the United Kingdom were increasingly aware of their potential including the British Aircraft Corporation (BAC) and the Royal Aircraft Establishment (RAE), Farnborough. RAE [24] was evaluating heat pipes and vapour chambers for the thermal control of satellites, and BAC had a similar interest.

It was during this year that Reay, then at IRD, commenced work on heat pipes, initially in the form of a survey of potential applications, followed by an experimental programme concerned with the manufacture of flat plate and tubular heat pipes. Work was also being carried out under Dunn, coauthor of the first four editions of *Heat Pipes*, at Reading University where several members of the staff had experience of the heat pipe activities at Harwell, described earlier. The National Engineering Laboratory at East Kilbride, now TUV-NEL, also entered the field.

Interest in the then USSR in heat pipes was evident from an article published in the Russian journal '*High Temperature*' [25], although much of the information described a summary of work published elsewhere.

The year 1969 saw reports of further work on variable conductance heat pipes, the principle contributions being made by Turner [26] at RCA and Bienert [27] at Dynatherm Corporation. Theoretical analyses were carried out on variable conductance heat pipes (VCHP's) to determine parameters such as reservoir size and practical aspects of reservoir construction and susceptibility to external thermal effects were considered.

A new type of heat pipe, in which the wick is omitted, was developed at this time by NASA [28]. The rotating heat pipe utilises centrifugal acceleration to transfer liquid from the condenser to the evaporator and can be used for cooling motor rotors and turbine blade rotors. Gray [28] also proposed an air-conditioning unit based on the rotating heat pipe, and this is illustrated in Fig. 1.9. (The rotating heat pipe is described fully in Chapter 5.) The rotating heat pipe does not of course suffer from the capillary pumping limitations that occur in conventional heat pipes, and its transport capability can be greatly superior to that of wicked pipes.

The application of heat pipes to electronics cooling in areas other than satellites was beginning to receive attention. Pipes of rectangular section were proposed by Sheppard [29] for cooling integrated circuit packages, and the design, development and fabrication of a heat pipe to cool a high-power airborne travelling wave tube is described by Calimbas and Hulett of the Philco-Ford Corporation [30].

Most of the work on heat pipes described so far has been associated with liquid metal working fluids, and for lower temperatures water, acetone, alcohols, etc. With the need for cooling detectors in satellite infra red scanning systems, to mention but one application, cryogenic heat pipes began to receive particular attention [31, 32]. The most common working fluid in these heat pipes was nitrogen, which was acceptable for temperature ranges between 77 and 100 K. Liquid oxygen was also used for this temperature range. The Rutherford High Energy Laboratory (RHEL) was the first organisation in the United Kingdom to operate cryogenic heat pipes [33], liquid hydrogen units being developed for cooling targets at the RHEL. Later RHEL developed a helium heat pipe operating at 4.2 K [34].

By 1970 a wide variety of heat pipes were commercially available from a number of companies in the United States. RCA, Thermo-Electron and Noren Products were among several firms marketing a range of 'standard' heat pipes, with the

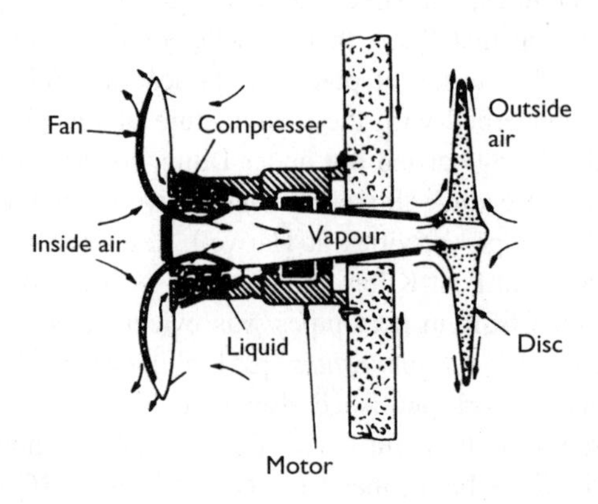

Fig. 1.9 A compact air-conditioning unit based on the wickless rotating heat pipe (Courtesy NASA).

ability to construct 'specials' for specific customer applications. During the next few years, several manufacturers were established in the United Kingdom (see Appendix 4), and a number of companies specialised in heat pipe heat recovery systems, based primarily on technology from the United States, entered what was becoming an increasingly competitive market [35].

The early 1970s saw a considerable growth in the application of heat pipes to solve terrestrial heat transfer problems, in addition to the continuing momentum in their development for spacecraft thermal control. The European Space Agency (ESA) commenced the funding of research and development work with companies and universities in Britain and Continental Europe, with a view to producing space-qualified heat pipes for ESA satellites. Other companies, notably Dornier, SABCA, Aerospatiale and Marconi, put considerable effort on development programmes, initially independent of ESA, in order to keep abreast of the technology. As a result of this, a considerable number of European companies were able to compete effectively in the field of spacecraft thermal control using heat pipes. In the 1990s, the role of European companies supplying heat pipes to national and pan-European space programmes was continuing apace, with Alcatel Space consolidating their early research on axially grooved heat pipes with new high-performance variants [38]. At the same time, Astrium was pioneering capillary pumped fluid loop systems for the French technological demonstrator spacecraft called STENTOR [40]. Capillary pumped loops are used by NASA on the Hubble Space Telescope, and the Space Shuttle in December 2001 carried an experiment to test a loop system with more than one evaporator in order to overcome a perceived limitation of single evaporators [42]. The experiment, known as CAPL-3, was photographed from the International Space Station and is visible in Fig. 1.10. (While the application of heat pipes in spacecraft may seem rather esoteric to the majority of potential users of

Fig. 1.10 CAPL-3 visible in the Space Shuttle Bay on STS-108 [42].

these devices, the 'technological fallout' has been considerable, both from this and applications in nuclear engineering, and has contributed significantly to the design procedures, reliability and performance improvements of the commercial products over the past 20 years.)

The effort expended in developing VCHPs, initially to meet the requirements of the US space programme, led to one of the most significant types of heat pipe, which is able to effect precise temperature control [36]. The number of techniques available for the control of heat pipes is large, and some are discussed in Chapter 6.

One of the major engineering projects of the 1970s, the construction and operation of the trans-Alaska oil pipeline, makes use of heat pipe technology. As described in Chapter 7, heat pipes are used in the pipeline supports in order to prevent melting of the permafrost, and the magnitude of the project necessitated McDonnell Douglas Astronautics Company producing 12 000 heat pipes per month, the pipes ranging in length between 9 and 23 m.

During this era [37], the Sony Corporation incorporated heat pipes in its tuner-amplifier products. Heat pipe heat sinks in this application proved to be 50 per cent lighter and 30 per cent more effective than conventional extruded aluminium units. The number of units delivered approached 1 million, and this was probably the first large-scale use of heat pipes in consumer electronic equipment.

While much development work was concentrated on 'conventional' heat pipes, the last few years have seen increasing interest in other systems for liquid transport. The proposed use of 'inverse' thermosyphons and an emphasis on the advantages (and possible limitations) of gravity-assisted heat pipes (see Chapter 2) have stood out as areas of considerable importance. The reason for the growing interest in these topics is not too difficult to find. Heat pipes in terrestrial applications have, probably in the majority of cases, proved particularly viable when gravity, in addition to capillary action, has aided condensate return to the evaporator. This is seen best of all in heat pipe heat recovery units, where slight changes in the inclination of the long heat pipes used in such heat exchangers can, as soon as reliance on the wick alone is effected, cut off heat transport completely.

The chemical heat pipe has so far not yet been commercially applied but 'conventional' very long heat pipes have been developed. 'Very long' in this case is 70–110 m, the units being tested at the Kyushu Institute of Technology in Japan [39]. These were expected to be used for under floor heating/cooling, snow melting or de-icing of roads. Long flexible heat pipes have also recently been used in Japan for cooling or isothermalising buried high-voltage cables.

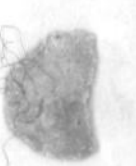

At the time of writing of this edition of *Heat Pipes*, the dominant application in terms of numbers of units produced is the cooling of chips in computers, in particular the central processors (CPUs). The demands of the computer hardware manufacturers dictate that thermal control systems are compact and are of low cost. The heat pipes used, of typically 2–3 mm diameter, may cost about US1 each, and mass production methods can be developed to achieve low costs, where millions of units are produced per month, in conjunction with manufacture in countries where labour costs are currently low.

However, with increasing powers applied in portable computers, the growing demand on cooling systems will lead to loads that may not be manageable with conventional heat pipe technology. Thus, one must, at this stage of their development, ask can heat pipes address our future thermal challenges?

1.5 CAN HEAT PIPES ADDRESS OUR FUTURE THERMAL CHALLENGES?

Some computer and chip manufacturers have indicated that by the time of publication of this book, an increase in heat dissipation to around 200 W for desktop and server computers, more specifically from the CPUs. The CPU size is little more than 100 mm^2. Thus, the equivalent heat flux is approaching 200 W/cm^2.

Workers in Australia and Japan are examining methods for overcoming the perceived limitations of conventional heat pipes where heat fluxes may be too high. Until heat pipe technology comes up with a passive solution, a pumped system such as that shown in Fig. 1.11 is being proposed. This uses a combination of impinging jet cooling, linked to a heat sink formed of a highly compact heat exchanger [41].

NASA has also highlighted the challenges of thermal management in a different environment, i.e. in robotic spacecraft [42]. Swanson and Birur, based at the

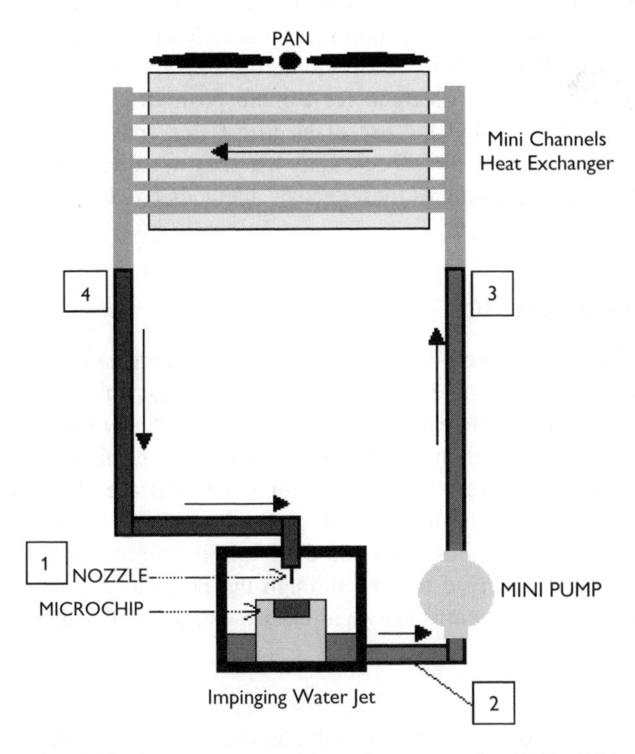

Fig. 1.11 Impinging water jet cooling system for CPUs [41].

Goddard Space Flight Center and the Jet Propulsion Laboratory, respectively, have indicated that loop heat pipes and capillary pumped loops will not satisfy future thermal control requirements. These include very low temperature operation (40 K) or operation in very high temperature environments, tight temperature control, high heat fluxes (>100 W/cm^2) and the need to minimise mass, e.g. for nanosatellites.

NASA sees challenges to heat-pipe-based two-phase thermal control systems coming from spray cooling (as in the example cited above), phase change thermal storage (that can be integrated with heat pipes – see Chapter 7) and mechanically pumped units.

If we are to address this fact, which has implications for the use of heat-pipe-based technologies in a large number of emerging systems, it is worth examining the opportunity to use unconventional liquid transport systems such as electrokinetics. (There is nothing new here, such systems were discussed in the first edition of *Heat Pipes*, but were deemed not to be necessary at the time). Electrokinetics is often discussed in the context of microfluidics – fluid flow at the microscale, for example, in heat pipe wick pores or in channels of around 100 μm or less.

Within two-phase systems (liquid–vapour), the influence of capillary action can be high, from a negative or positive viewpoint. Where the force is used to drive liquid, as in the microheat pipes illustrated in Fig. 1.12, there is a wealth of literature examining optimum configurations of channel cross-section, pore sizes, etc. However, the impression is that in many microfluidic devices, capillary action is a disadvantage that needs to be overcome by external 'active' liquid transport mechanisms. Perhaps there is room for a more positive examination of passive liquid transport designs, but the use of electric fields and other more advanced procedures can provide the control necessary that may not always be available using passive methods alone. Some of these 'active' methods are described below.[1]

1.6 ELECTROKINETICS

Electrokinetics is the name given to the electrical phenomena that accompany the relative movement of a liquid and a solid. These phenomena are ascribed to the presence of a potential difference at the interface between any two phases at which movements occur. Thus, if the potential is supposed to result from the existence of electrically charged layers of opposite sign at the interface, then the application of an electric field must result in the displacement of one layer with respect to the other. If the solid phase is fixed and the liquid is free to move, as in a porous material, the liquid will tend to flow through the pores as a consequence of the applied field. This movement is known as electro-osmosis and is discussed in detail in Chapter 6.

[1] A rotating heat pipe, see Chapter 6, is an active heat transfer device, in that external energy is used to rotate the system, so the borderline between 'active' and 'passive' can be blurred.

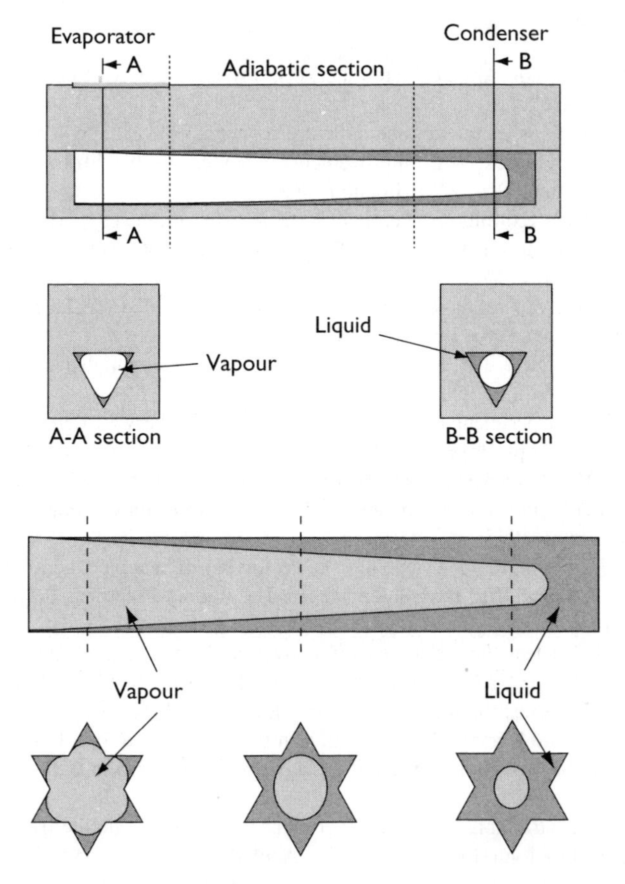

Fig. 1.12 Microheat pipe cross-sections showing capillary paths (Courtesy UCL).

Electro-osmosis using an alternating current (ac) has been used as the basis of microfluidic pumps, allowing normal batteries to be used at the modest applied voltages.

For those wishing to follow up the numerous techniques for influencing capillary forces, the paper by Le Berre et al. [43] on electrocapillary force actuation in microfluidic elements is worthy of study.

For the future, the heat pipe has a continuing role to play in many applications – its use in domestic equipment is yet to be exploited, its role in spacecraft is guaranteed by its unique features and the growing use of renewable energy systems, which can benefit from mass-produced passive heat transport systems, must auger well for the technology in the future. The research community will continue to support these applications by improving, perhaps incrementally, performance while reducing unit sizes and cost.

REFERENCES

[1] Anon. Memoirs of Angier March Perkins. Proc. Inst. Civ. Eng., Vol. 67, Pt. 1, pp 417–419, 1882.
[2] King, C.R. Perkins' hermetic tube boilers. Engineer, Vol. 152, pp 405–406, 1931.
[3] Perkins, L.P. and Buck, W.E. Improvements in devices for the diffusion or transference of heat. UK Patent No. 22272, London, 1892.
[4] Anon. The Paris Exhibition – Perkins' portable oven. Engineer, p 519, 1867.
[5] Gaugler, R.S. Heat transfer device. US Patent No. 2350348, Appl. 21 December 1942. Published 6 June 1944.
[6] Grover, G.M. Evaporation–condensation heat transfer device. US Patent No. 3229759. Appl. 2 December 1963. Published 18 January 1966.
[7] Grover, G.M., Cotter, T.P. and Erikson, G.F. Structure of very high thermal conductance. J. Appl. Phys., Vol. 35, No. 6, pp 1990–1991, 1964.
[8] Bainton, K.F. Experimental heat pipes. AERE-M1610, Harwell, Berks. Atomic Energy Establishment, Appl. Phys. Div., 1965.
[9] Grover, G.M., Bohdansky, J. and Busse, C.A. The use of a new heat removal system in space thermionic power supplies. EUR 2229e, Ispra, Italy, Euratom Joint Nuclear Research Centre, 1965.
[10] Busse, C.A., Caron, R. and Cappelletti, C. Prototypes of heat pipe thermionic converters for space reactors. IEE 1st Conference on Thermionic Electrical Power Generation, London, 1965.
[11] Leefer, B.I. Nuclear thermionic energy converter. Proceedings of 20th Annual Power Sources Conference, Atlantic City, NJ, 24–26 May 1966, pp 172–175, 1966.
[12] Judge, J.F. RCA test thermal energy pipe. Missiles Rockets, Vol. 18, pp 36–38, 1966.
[13] Deverall, J.E. and Kemme, J.E. Satellite heat pipe. USAEC Report LA-3278, Contract' W-7405-eng-36. Los Alamos Scientific Laboratory, University of California, September 1970.
[14] Wyatt, T. A controllable heat pipe experiment for the SE-4 satellite. JHU Tech. Memo APL-SDO-1134. John Hopkins University, Appl. Physics Lab., March 1965, AD 695 433.
[15] Feldman, K.T. and Whiting, G.H. The heat pipe and its potentialities. Eng. Dig., Vol. 28 No. 3, pp 86–86, 1967.
[16] Eastman, G.Y. The heat pipe. Sci. Am., Vol. 218, No. 5, pp 38–46, 1968.
[17] Feldman, K.T. and Whiting, G.H. Applications of the heat pipe. Mech. Eng., Vol. 90, pp 48–53, 1968.
[18] Cotter, T.P. Theory of heat pipes. USAEC Report LA-3246, Contract W7405-3ng-36. Los Alamos Scientific Laboratory, University of California, 1965.
[19] Cheung, H. A critical review of heat pipe theory and applications. USAEC Report UCRL-50453. Lawrence Radiation Laboratory, University of California, 1968.
[20] Katzoff, S. Notes on heat pipes and vapour chambers and their applications to thermal control of spacecraft. USAEC Report SC-M-66-623. Contract AT (29-1)-789. Proceedings of Joint Atomic Energy Commission/Sandia Laboratories Heat Pipe Conference, Vol. 1, Albuquerque, New Mexico, 1 June 1966, pp 69–89. Sandia Corporation, October 1966.
[21] Anand, D.K. Heat pipe application to a gravity gradient satellite. Proceedings of ASME Annual Aviation and Space Conference, Beverley Hills, California, 16–19 June 1968, pp 634–658.
[22] Busse, C.A. Heat pipe research in Europe. 2nd International Conference on Thermionic Electrical Power Generation, Stresa, Italy, May 1968. Report EUR 4210 f, e, 1969, pp 461–475.

[23] Nozu, S. et al. Studies related to the heat pipe. Trans. Soc. Mech. Eng. Jp. Vol. 35, No. 2, pp 392–401, 1969 (in Japanese).
[24] Savage, C.J. Heat pipes and vapour chambers for satellite thermal balance. RAE TR-69125, Royal Aircraft Establishment, Farnborough, Hant, June 1969.
[25] Moskvin, J.V. and Filinnov, J.N. Heat pipes. High Temp. Vol. 7, No. 6, pp 704–713, 1969.
[26] Turner, R.C. The constant temperature heat pipe. A unique device for the thermal control of spacecraft components. AIAA 4th Thermo-physics Conference Paper 69-632, San Francisco, 16–19 June 1969.
[27] Bienert, W. Heat pipes for temperature control. 4th Intersociety Energy Conversion Conference, Washington DC, September 1969, pp 1033–1041.
[28] Gray, V.H. The rotating heat pipe – a wickless hollow shaft for transferring high heat fluxes. ASME Paper 69-HT-19. American Society of Mechanical Engineers, New York, 1969.
[29] Sheppard, T.D., Jr. Heat pipes and their application to thermal control in electronic equipment. Proceedings of National Electronic Packaging and Production Conference, Anaheim, CA, 11–13 February 1969.
[30] Calimbas, A.T. and Hulett, R.H. An avionic heat pipe. ASME Paper 69-HT-16. American Society of Mechanical Engineers, New York, 1969.
[31] Eggers, P.E. and Serkiz, A.W. Development of cryogenic heat pipes. ASME 70-WA/Ener-1. American Society of Mechanical Engineers, New York, 1970.
[32] Joy, P. Optimum cryogenic heat pipe design. ASME Paper 70-HT/SpT-7. American Society of Mechanical Engineers, New York, 1970.
[33] Mortimer, A.R. The heat pipe. Engineering Note-Nimrod/NDG/70-34. Rutherford Laboratory, Nimrod Design Group, Harwell, October 1970.
[34] Lidbury, J.A. A helium heat pipe. Engineering Note NDG/72-11. Rutherford Laboratory, Nimrod Design Group, Harwell, April 1972.
[35] Reay, D.A. Industrial Energy Conservation: A Handbook for Engineers and Managers. Pergamon Press, Oxford, 1977.
[36] Groll, M. and Kirkpatrick, J.P. Heat pipes for spacecraft temperature control – an assessment of the state-of-the-art. Proceedings of 2nd International Heat Pipe Conference, Bologna, Italy, ESA Report SP112, Vol. 1, 1976.
[37] Osakabe, T. et al. Application of heat pipe to audio amplifier, in: Advances in Heat Pipe Technology. Proceedings of IV International Heat Pipe Conference. Pergamon Press, Oxford, 1981.
[38] Hoa, C., Demolder, B. and Alexandre, A. Roadmap for developing heat pipes for ALCATEL SPACE's satellites. Appl. Therm. Eng., Vol. 23, pp 1099–1108, 2003.
[39] Tanaka, O. et al. Heat transfer characteristics of super heat pipes. Proceedings of VII International Heat Pipe Conference, Paper B1P, Minsk, 1990.
[40] Figus, C., Ounougha, L., Bonzom, P., Supper, W. and Puillet, C. Capillary fluid loop development in Astrium. Appl. Therm. Eng., Vol. 23, pp 1085–1098, 2003.
[41] Bintoro, J.S., Akbarzadeh, A. and Mochizuki, M. A closed-loop electronics cooling by implementing single phase impinging jet and mini channels heat exchanger. Appl. Therm. Eng., Vol. 25, No. 17–18, pp 2740–2753, 2005.
[42] Swanson T.D. and Birur, G.C. NASA thermal control technologies for robotic spacecraft. Appl. Therm. Eng. Vol. 23, pp 1055-1065, 2003.
[43] Le Berre, M. et al. Electrocapillary force actuation of microfluidic elements. Microelectron. Eng., Vol. 78–79, pp 93–99, 2005.

2

HEAT TRANSFER AND FLUID FLOW THEORY

2.1 INTRODUCTION

As discussed in Chapter 6, there are many variants of heat pipe, but in all cases the working fluid must circulate when a temperature difference exists between the evaporator and the condenser. In this Chapter, the operation of the classical wicked heat pipe is discussed. Various analytical techniques are then outlined in greater detail, these techniques are then applied to the classical heat pipe and the gravity-assisted thermosyphon.

2.2 OPERATION OF HEAT PIPES

2.2.1 Wicked heat pipes

The overall thermal resistance of a heat pipe, defined by equation 2.1, should be low, providing that it functions correctly.

The operating limits for a wicked heat pipe, first described in [1], are illustrated in Fig. 2.1.

$$R = \frac{T_{\text{hot}} - T_{\text{cold}}}{\dot{Q}} \tag{2.1}$$

Each of these limits may be considered in isolation. In order for the heat pipe to operate the maximum capillary pumping pressure, $\Delta P_{\text{c,max}}$ must be greater than the total pressure drop in the pipe. This pressure drop is made up of three components.

(i) The pressure drop ΔP_{l} required to return the liquid from the condenser to the evaporator.
(ii) The pressure drop ΔP_{v} necessary to cause the vapour to flow from the evaporator to the condenser.

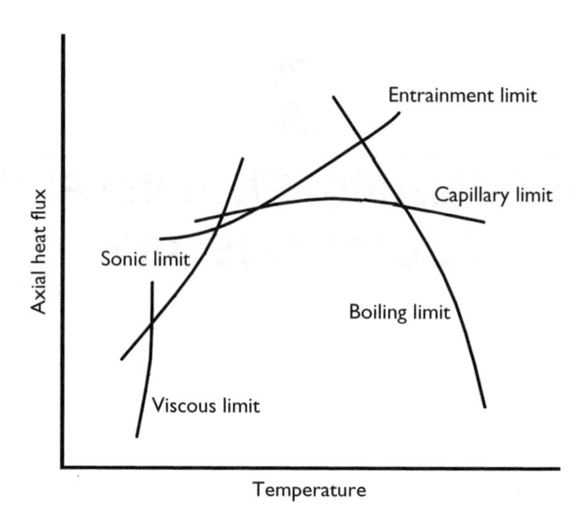

Fig. 2.1 Limitations to heat transport in a heat pipe.

(iii) The pressure due to the gravitational head, ΔP_g which may be zero, positive or negative, depending on the inclination of the heat pipe.

For correct operation,

$$\Delta P_{c,max} \geq \Delta P_l + \Delta P_v + \Delta P_g{}^1 \tag{2.2}$$

If this condition is not met, the wick will dry out in the evaporator region and the heat pipe will not operate. The maximum allowable heat flux for which equation 2.2 holds is referred to as the capillary limit. Methods of evaluating the four terms in equation 2.2. will be discussed in detail in this chapter. Typically, the capillary limit will determine the maximum heat flux over much of the operating range; however, the designer must check that a heat pipe is not required to function outside the envelope either at design conditions or at start-up.

During start-up and with certain high-temperature liquid metal heat pipes, the vapour velocity may reach sonic values. The sonic velocity sets a limit on the heat pipe performance. At velocities approaching sonic, compressibility effects must be taken into account in the calculation of the vapour pressure drop.

The viscous or vapour pressure limit is also generally the most important at start-up. At low temperature, the vapour pressure of the fluid in the evaporator is very low, and, since the condenser pressure cannot be less than zero, the maximum

[1] Note that this implies that the capillary pressure rise is considered as positive, while pressure drops due to liquid and vapour flow or gravity are considered positive.

difference in vapour pressure is insufficient to overcome viscous and gravitational forces, thus preventing satisfactory operation.

At high heat fluxes, the vapour velocity necessarily increases; if this velocity is sufficient to entrain liquid returning to the evaporator, then performance will decline, hence the existence of an entrainment limit.

The above limits relate to axial flow through the heat pipe. The final operating limit discussed will be the boiling limit. The radial heat flux in the evaporator is accompanied by a temperature difference that is relatively small until a critical value of heat flux is reached above which vapour blankets the evaporator surface resulting in an excessive temperature difference.

The position of the curves and shape of the operating envelope shown in Fig. 2.1 depends upon the wick material, working fluid and geometry of the heat pipe.

2.2.2 Thermosyphons

The two-phase thermosyphon is thermodynamically similar to the wicked heat pipe but relies on gravity to ensure liquid return from the condenser to the evaporator. A wick or wicks may be incorporated in at least part of the unit to reduce entrainment and improve liquid distribution within the evaporator.

2.2.3 Loop heat pipes and capillary pumped loops

As will be seen later, the characteristics that produce a wick having the capability to produce a large capillary pressure also lead to high values of pressure drop through a lengthy wick. This problem can be overcome by incorporating a short wick contained within the evaporator section and separating the vapour flow to the condenser from the liquid return.

2.3 THEORETICAL BACKGROUND

In this section, the theory underpinning the evaluation of the terms in equations 2.1 and 2.2 and the determination of the operating limits shown in Fig. 2.1 is discussed.

2.3.1 Gravitational head

The pressure difference, ΔP_g, due to the hydrostatic head of liquid may be positive, negative or zero, depending on the relative positions of the condenser and evaporator. The pressure difference may be determined from:

$$\Delta P_g = \rho_l g l \sin \phi \qquad (2.3)$$

where ρ_l is the liquid density (kg/m^3), g the acceleration due to gravity (9.81 m/s^2), l the heat pipe length (m), and ϕ the angle between the heat pipe and the horizontal (ϕ is positive when the condenser is lower than the evaporator).

2.3.2 Surface tension and capillarity

2.3.2.1 Introduction

Molecules in a liquid attract one another. A molecule in a liquid will be attracted by the molecules surrounding it and, on average, a molecule in the bulk of the fluid will not experience any resultant force. In the case of a molecule at or near the surface of a liquid, the forces of attraction will no longer balance out and the molecule will experience a resultant force inwards. Because of this effect, the liquid will tend to take up a shape having minimum surface area, in the case of a free falling drop in a vacuum this would be a sphere. Due to this spontaneous tendency to contract, a liquid surface behaves rather like a rubber membrane under tension. In order to increase the surface area, work must be done on the liquid. The energy associated with this work is known as the free surface energy, and the corresponding free surface energy per unit surface area is given the symbol σ_l. For example, if a soap film is set up on a wire support as in Fig. 2.2a and the area is increased by moving one side a distance dx the work done is equal to and the work done is equal to $F dx$, hence the increase in surface energy is $2\sigma_l l\,dx$.

The factor 2 arises since the film has two free surfaces. Hence, if T is the force per unit length for each of the two surfaces $2Tl dx = 2\sigma_l l dx$ or $T = \sigma_l$. This force per unit length is known as the surface tension. It is numerically equal to the surface energy per unit area measured in any consistent set of units, e.g. N/m.

Values for surface tension for a number of common working fluids are given in Appendix 1.

Since latent heat of vapourisation, L, is a measure of the forces of attraction between the molecules of a liquid, we might expect surface energy or surface tension σ_l to be related to L. This is found to be the case. Solids also will have a free surface energy, and in magnitude it is found to be similar to the value for the same material in the molten state.

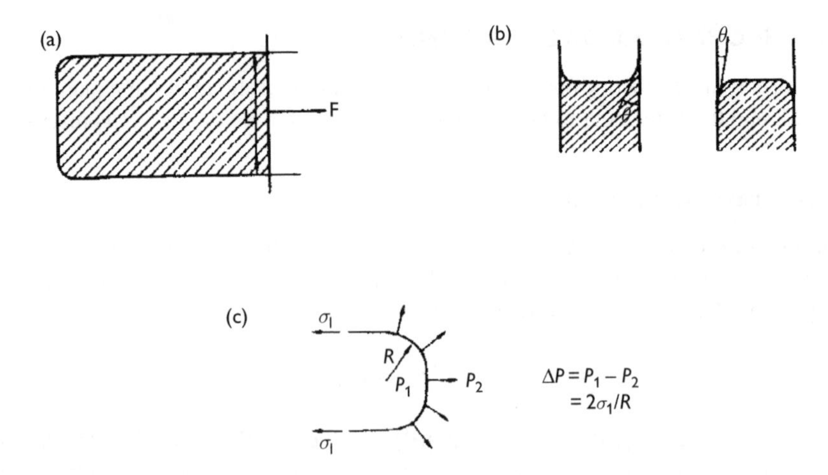

Fig. 2.2 Representation of surface tension and pressure difference across a curved surface.

When a liquid is in contact with a solid surface, molecules in the liquid adjacent to the solid will experience forces from the molecules of the solid in addition to the forces from other molecules in the liquid. Depending on whether these solid/liquid forces are attractive or repulsive, the liquid/solid surface will curve upwards or downwards, as indicated in Fig. 2.2b. The two best known examples of attractive and repulsive forces, respectively, are water and mercury. Where the forces are attractive, the liquid is said to 'wet' the solid. The angle of contact made by the liquid surface with the solid is known as the contact angle, θ. For wetting, θ will lie between 0 and $\pi/2$ rad and for nonwetting liquids, $\theta > \pi/2$.

The condition for wetting to occur is that the total surface energy is reduced by wetting.

$$\sigma_{sl} + \sigma_{lv} < \sigma_{sv}$$

where the subscripts, s, l and v refer to solid, liquid and vapour phases, respectively, as shown in Fig. 2.3a.

Wetting will not occur if $\sigma_{sl} + \sigma_{lv} > \sigma_{sv}$ as in Fig. 2.3c, while the intermediate condition of partial wetting $\sigma_{sl} + \sigma_{lv} = \sigma_{sv}$ is illustrated in Fig. 2.3b.

2.3.2.2 Pressure difference across a curved surface

A consequence of surface tension is that the pressure on the concave surface is greater than that on the convex surface. With reference to Figs. 2.2c and 2.4, this pressure difference ΔP is related to the surface energy σ_l and radius of curvature R of the surface.

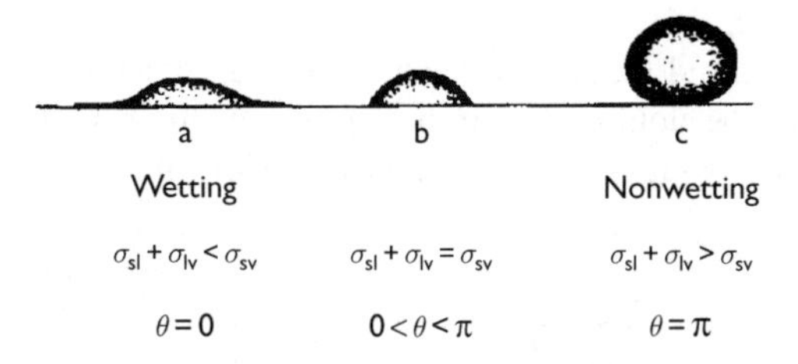

Fig. 2.3 Wetting and nonwetting contact.

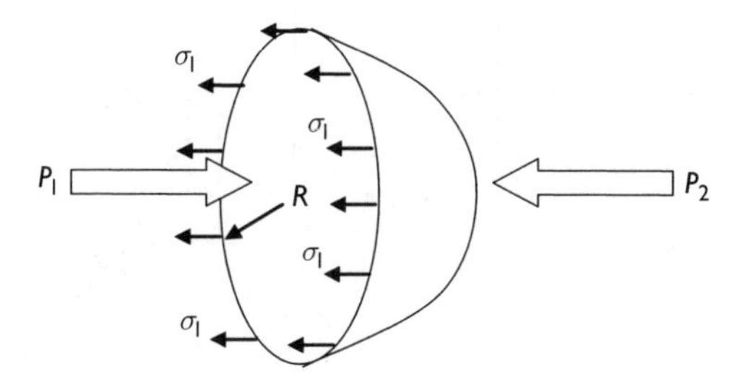

Fig. 2.4 Pressure difference across a curved liquid surface.

The hemispherical surface tension force acting around the circumference is given by $2\pi R\sigma_l$ and must be balanced by the net force on the surface due to the pressures which is $(P_1 - P_2)\,\pi R^2$ or $\Delta P\pi R^2$. Thus,

$$\Delta P = \frac{2\sigma_l}{R} \tag{2.4}$$

If the surface is not spherical, but can be described by two radii of curvature, R_1 and R_2 at right angles, then it can be shown that equation 2.4 becomes:

$$\Delta P = 2\sigma_l\left(\frac{1}{R_1} + \frac{1}{R_2}\right) \tag{2.5}$$

Due to this pressure difference, if a vertical tube of radius r is placed in a liquid that wets the material of the tube, the liquid will rise to a height above the plane surface of the liquid as shown in Fig. 2.5.

The pressure balance gives

$$(\rho_l - \rho_v)\,gh = \frac{2\sigma_l}{r}\cos\theta \approx \rho_l gh \tag{2.6}$$

For the case of a noncircular tube

$$\frac{1}{r} = \left(\frac{1}{R_1} + \frac{1}{R_2}\right) \tag{2.7}$$

where ρ_l is the liquid density, ρ_v the vapour density and θ the contact angle. This effect is known as capillary action or capillarity and is the basic driving force for the wicked heat pipe.

For nonwetting liquids, $\cos\theta$ is negative and the curved surface is depressed below the plane of the liquid level. In heat pipes, wetting liquids are always used, the capillary lift increasing with liquid surface tension and decreasing contact angle.

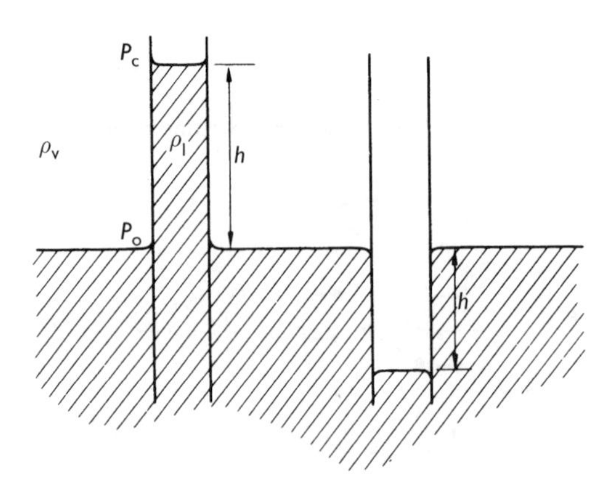

Fig. 2.5 Capillary rise in a tube.

2.3.2.3 Change in vapour pressure at a curved liquid surface

From Fig. 2.5, it can be seen that the vapour pressure at the concave surface is less than that at the plane liquid surface by an amount equal to the weight of a vapour column of unit area, length h.

$$P_c - P_o = g\rho_v h$$

assuming that ρ_v is constant.

Combining this with equation 2.6

$$P_c - P_o = \frac{2\sigma_l}{r}\frac{\rho_v}{(\rho_l - \rho_v)}\cos\theta \tag{2.8}$$

This pressure difference $P_c - P_o$ is small compared to the capillary head $(2\sigma_l/r)\cos\theta$ and may normally be neglected when considering the pressures within a heat pipe. It is, however, worth noting that the difference in vapour pressure between the vapour in a bubble and the surrounding liquid is an important phenomenon in boiling heat transfer.

2.3.2.4 Measurement of surface tension

There is a large number of methods for the measurement of surface tension of a liquid, and these are described in the standard texts [2, 3]. For our present purpose, we are interested in the combination of surface measure of the capillary force, $\sigma_l \cos\theta$. The simplest measurement is that of capillary rise h in a tube, which gives

$$\sigma_l \cos\theta \approx \frac{\rho_l g h r}{2} \tag{2.9}$$

In practical heat pipe design, it is also necessary to know r, the effective pore radius. This is by no means easy to estimate for a wick made up of a sintered porous structure or from several layers of gauze. By measuring the maximum height the working fluid will attain, it is possible to obtain information on the capillary head for fluid wick combinations. Several workers have reported measurements on maximum height for different structures and some results are given in Chapter 3. The results may differ for the same structure depending on whether the film was rising or falling; the reason for this is brought out in Fig. 2.6.

Another simple method, for the measurement of σ_l, due to Jäger, and shown schematically in Fig. 2.7 is sometimes employed. This involves the measurement of maximum bubble pressure. The pressure is progressively increased until the bubble breaks away and the pressure falls. When the bubble radius reaches that of the tube, pressure is a maximum P_{max} and at this point

$$P_{max} = \rho_l h g + \frac{2\sigma_l}{r} \tag{2.10}$$

This method has proved appropriate [4] for the measurement of the surface tension of liquid metals.

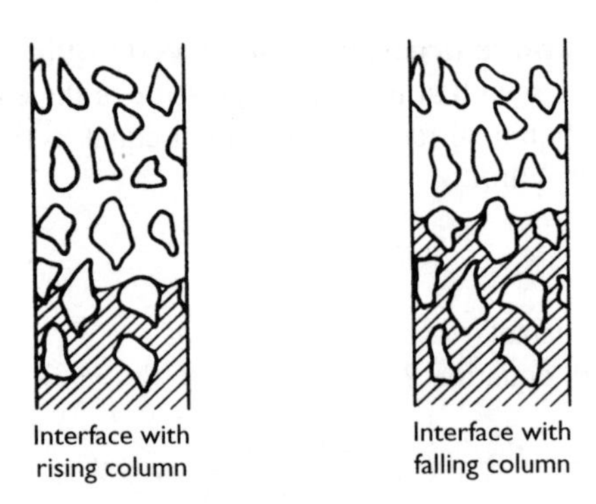

Fig. 2.6 Rising and falling column interface.

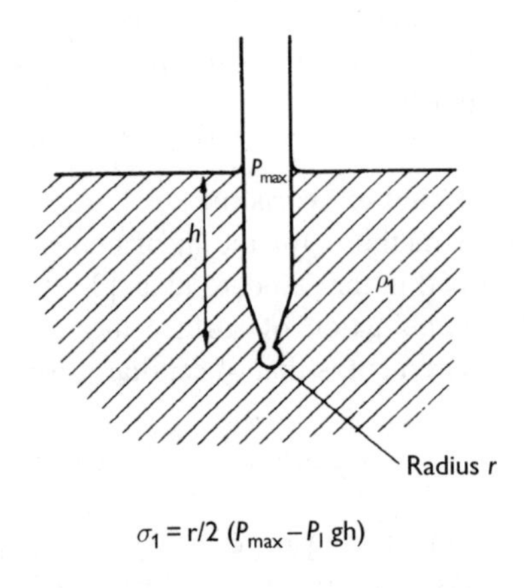

Fig. 2.7 Jäger's method for surface temperature measurement.

The surface tension of two liquids can be compared by comparing the mass of the droplets falling from a narrow vertical tube. If the mass of the droplets for liquids 1 and 2 are m_1 and m_2, respectively, then

$$\frac{m_1 \rho_{l1}}{m_2 \rho_{l2}} = \frac{\sigma_{f1}}{\sigma_{f2}} \tag{2.11}$$

2.3.2.5 Temperature dependence of surface tension

Surface tension decreases with increasing temperature, it is therefore important to take temperature effects into account when using the results of measurements at typical ambient temperatures.

Values of surface tension for over 2000 pure fluids have been tabulated by Jasper [5] and temperature corrections of the form $\sigma = a + bT$ are suggested.

For water, the following interpolating equation [6] gives good values of surface tension

$$\sigma_l = B(1 - T_r)^{\mu}(1 - b(1 - T_r)) \tag{2.12}$$

where T_r is the reduced temperature $= T/T_c$, $T_c = 647.096\,\text{K}$, $B = 235.8\,\text{m N/m}$, $b = 0.625$, $\mu = 1.256$.

Equation 2.12 is valid between the triple point (0.01 °C) and the critical temperature and is in agreement with the measured data within experimental uncertainty.

Eötvös proposed a relationship that was later modified by Ramsay and Shields to give

$$\sigma_l \left(\frac{M}{\rho_l}\right)^{2/3} = H(T_c - 6 - T) \tag{2.13}$$

where M is the molecular weight, T_c the critical temperature (K), T the temperature (K), H a constant, the value of which depends upon the nature of the liquid.

The Eötvös–Ramsay–Shields equation does not show agreement with the experimentally observed behaviour of liquid metal and molten salts. Bohdansky and Schins [7] have derived an equation that applies to the alkali metals. While Fink and Leibowitz [8] recommended

$$\sigma_l = \sigma_o \left(1 - \frac{T}{T_c}\right)^n \quad (\text{m N/m}) \tag{2.14}$$

For sodium, where $\sigma_o = 240.5\,\text{m N/m}$, $n = 1.126$, $T_c = 2503.7\,\text{K}$.

Alternatively, the surface tension of liquid metals may be estimated from the data provided by Iida and Guthrie [9] and summarised in Table 2.1. The value of surface tension may then be calculated

$$\sigma_l = \sigma_{lm} + (T - T_m)^{d\sigma/dT}$$

Table 2.1 Surface tension of selected liquid metals [9]

		M (kg/kmol)	Melting point T_m (K)	Boiling point (1bar) (K)	σ_{lm} at melting point(m N/m)	$d\sigma/dT$ (m N/mK)
Lithium	Li	7	452.2	1590	398	−0.14
Sodium	Na	23	371.1	1151	191	−0.1
Potassium	K	39	336.8	1035	115	−0.08
Caesium	Cs	133	301.65	1033	70	−0.06
Mercury	Hg	200	234.3	630	498	−0.2

A method of evaluating surface tension based upon the number and nature of chemical bonds was first suggested by Walden [10] and developed by Sugden [11] and Quale [12].

$$\sigma^{0.25} = \frac{\mathrm{P}}{M}(\rho_{\mathrm{l}} - \rho_{\mathrm{v}}) \tag{2.15}$$

where P was defined by Walden as a parachor. Values of the increments to be used in evaluating the parachor, adapted to give values of surface tension in N/m, are given in Ref. [13].

2.3.2.6 Capillary Pressure ΔP_c

Equation 2.4 shows that the pressure drop across a curved liquid interface is

$$\Delta P = \frac{2\sigma_{\mathrm{l}}}{R}$$

From Fig. 2.8 we can see that $R\cos\theta = r$ where r is the effective radius of the wick pores and θ the contact angle. Hence, the capillary head at the evaporator $\Delta P'_e$ is

$$\Delta P'_{\mathrm{e}} = 2\sigma_{\mathrm{l}}\frac{\cos\theta_{\mathrm{e}}}{r_{\mathrm{e}}} \tag{2.16a}$$

Similarly, for the condenser

$$\Delta P'_{c} = 2\sigma_{\mathrm{l}}\frac{\cos\theta_{\mathrm{c}}}{r_{\mathrm{c}}} \tag{2.16b}$$

and the capillary driving pressure, ΔP_{c}, is given by $\Delta P'_{\mathrm{e}} - \Delta P'_{\mathrm{c}}$.

It is worth noting that ΔP_{c} is a function only of the conditions where a meniscus exists. It does not depend on the length of the adiabatic section of the wick. This is particularly important in the design of loop heat pipes.

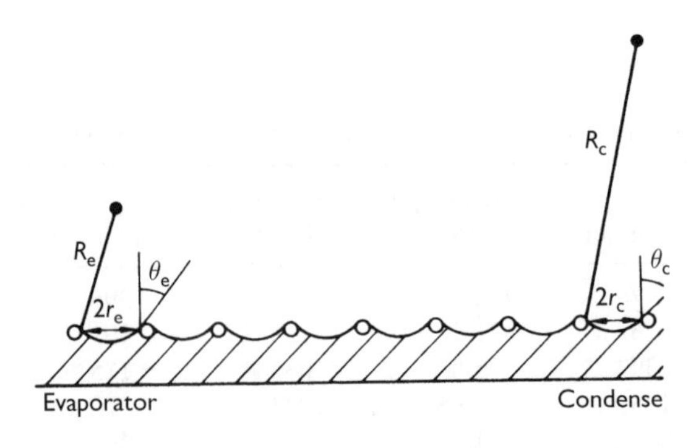

Fig. 2.8 Wick and pore parameters in evaporator and condenser.

2.3.3 Pressure difference due to friction forces

In this section, we will consider the pressure differences caused by frictional forces in liquids and vapours flowing in a heat pipe. Firstly, it is convenient to define some of the terms that will be used later in the chapter.

2.3.3.1 Laminar and turbulent flow

If one imagines a deck of playing cards or a sheaf of paper, initially stacked to produce a rectangle, to be sheared as shown in Fig. 2.9, it can be seen that the individual cards, or lamina, slide over each other. There is no movement of material perpendicular to the shear direction.

Similarly, in laminar fluid flow, there is no mixing of the fluid, and the fluid can be regarded as a series of layers sliding past each other.

Consideration of a simple laminar flow allows us to define viscosity. Fig. 2.10 illustrates the velocity profile for a laminar flow of a fluid over a flat plate.

The absolute or dynamic viscosity of a fluid, μ, is defined by

$$\tau = \mu \frac{\mathrm{d}v}{\mathrm{d}y}$$

where τ is the shear stress. At the wall, the velocity of the fluid must be zero, and the wall shear stress is given by

$$\tau_{\mathrm{w}} = \mu \left(\frac{\mathrm{d}v}{\mathrm{d}y} \right)_{\mathrm{w}}$$

Fig. 2.9 Shear applied to parallel sheets.

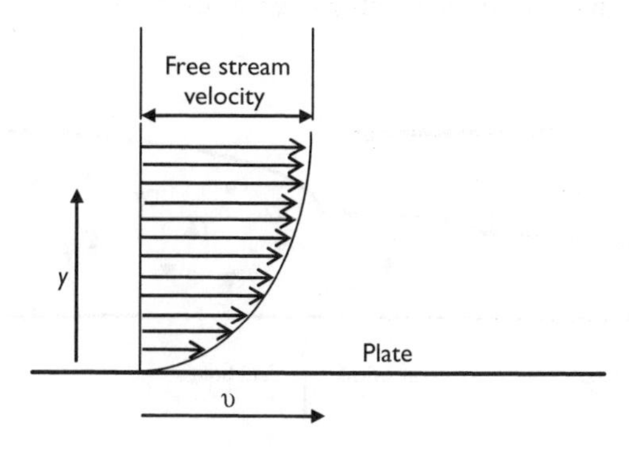

Fig. 2.10 Velocity profile in laminar flow over a flat plate.

In practice, laminar flow is observed at low speeds, in small tubes or channels, with highly viscous fluids and very close to solid walls. It is the flow normally observed when liquid flows through the wick of a heat pipe.

If the fluid layers seen in laminar flow break up and fluid mixes between the layers, then the flow is said to be turbulent. The turbulent mixing of fluid perpendicular to the flow direction leads to a more effective transfer of momentum and internal energy between the wall and the bulk of the fluid. This is illustrated in Fig. 2.11.

The heat transfer and pressure drop characteristics of laminar and turbulent flows are very different. In forced convection, the magnitude of the Reynolds number, defined below, provides an indication of whether the flow is likely to be laminar or turbulent.

$$\mathrm{Re} = \frac{\rho v d_{\mathrm{rep}}}{\mu}$$

where d_{rep} is a representative linear dimension. If Reynold's number is written

$$\mathrm{Re} = \frac{\rho v^2}{\mu v / d}$$

it can be seen to be a measure of the relative importance of inertial and viscous forces acting on the fluid.

For flow over a flat plate, as shown in Fig. 2.11, we may determine whether the flow in the boundary layer is likely to be laminar or turbulent by applying the following conditions:

$$\mathrm{Re}_x \left(= \frac{\rho V_\infty x}{\mu} \right) < 10^5 \quad \text{laminar flow}$$

$$\mathrm{Re}_x \left(= \frac{\rho V_\infty x}{\mu} \right) > 10^5 \quad \text{turbulent flow}$$

where x is the distance from the leading edge of the plate.

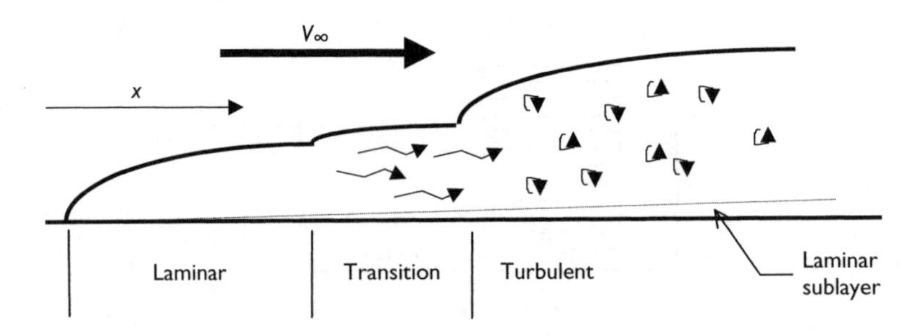

Fig. 2.11 Development of the boundary layer over a flat plate.

For values of Reynolds number between 10^5 and 10^6, the situation is complicated by two factors. Firstly, the transition is not sharp, it occurs over a finite length of plate. In the transition region, the flow may intermittently take on turbulent and laminar characteristics. Secondly, the position of the transition zone depends not only upon the Reynolds number but also influenced by the nature of the flow in the free stream and the nature of the surface. Surface roughness or protuberances on the surface tend to trip the boundary layer from laminar to turbulent.

For flow in pipes, channels or ducts, the situation is similar to that for a flat plate in the entry region, but in long channels the boundary layers from all walls meet and fully developed temperature and velocity profiles are established.

For fully developed flow in pipes or channels, the transition from laminar to turbulent flow occurs at a Reynolds number $\mathrm{Re}_d = \rho v d_e/\mu$ of approximately 2100. The dimension d_e is the on the channel equivalent or hydraulic diameter

$$d_e = \frac{4 \times \text{cross-sectional area}}{\text{wetted perimeter}} \tag{2.17a}$$

As expected, for a circular duct or pipe, diameter d, this is given by

$$d_e = \frac{4\pi d^2/4}{\pi d} = d \tag{2.17b}$$

For a square duct of side length x

$$d_e = \frac{4x^2}{4x} = x \tag{2.17c}$$

and for a rectangular duct of width a and depth b

$$d_e = \frac{4ab}{2(a+b)} \tag{2.17d}$$

For flow through an annulus having inner and outer diameters d_1 and d_2, respectively, the hydraulic diameter may be calculated

$$d_e = \frac{4\pi(d_2^2 - d_1^2)/4}{\pi(d_2 + d_1)} = \frac{4\pi(d_2 - d_1)(d_2 + d_1)/4}{\pi(d_2 + d_1)} = (d_2 - d_1) \tag{2.17e}$$

which is equal to twice the thickness of the annular gap.

The velocity profile in laminar flow in a tube is parabolic, while in turbulent flow the velocity gradient close to the wall is much steeper in turbulent flow, as shown in Fig. 2.12.

2.3.3.2 Laminar flow – the Hagen–Poiseuille equation

The steady-state laminar flow of an incompressible fluid of constant viscosity μ, through a tube of circular cross section, radius a, is described by the Hagen–Poiseuille equation.

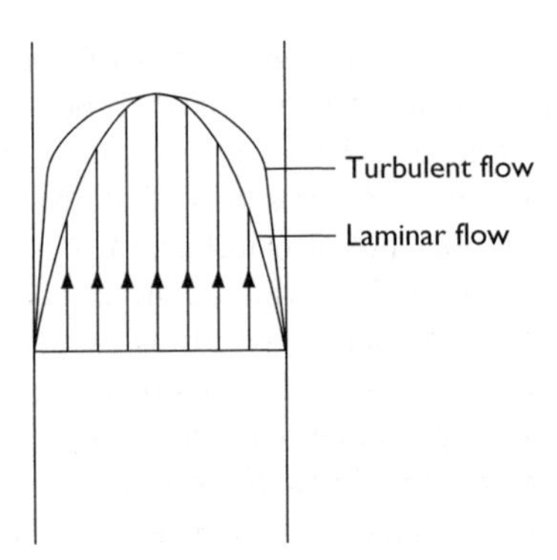

Fig. 2.12 Velocity distribution in a circular tube for laminar and turbulent flow.

This equation relates the velocity, v_r of the fluid at radius r, to the pressure gradient, $\mathrm{d}p/\mathrm{d}l$ along the tube.

$$v_r = \frac{a^2}{4\mu}\left[1 - \left(\frac{r}{a}\right)^2\right]\left(-\frac{\mathrm{d}p}{\mathrm{d}l}\right) \tag{2.18}$$

The velocity profile is parabolic, varying from the maximum value, v_m, given by

$$v_m = \frac{a^2}{4\mu}\left(-\frac{\mathrm{d}p}{\mathrm{d}l}\right) \tag{2.19}$$

on the axis of the tube to zero adjacent to the wall.

The average velocity, v, is given by

$$v = \frac{a^2}{8\mu}\left(-\frac{\mathrm{d}p}{\mathrm{d}l}\right) \tag{2.20}$$

or rearranging

$$\frac{\mathrm{d}p}{\mathrm{d}l} = -\frac{8\mu v}{a^2}$$

In a one-dimensional treatment, the average velocity may be used throughout. The volume flowing per second, S, is

$$S = \pi a^2 v = -\frac{\pi a^4}{8\mu}\left(-\frac{\mathrm{d}p}{\mathrm{d}l}\right) \tag{2.21}$$

and if ρ is the fluid density the mass flow $\dot{m}$ is given by

$$\dot{m} = \rho \pi a^2 v = -\rho\frac{\pi a^4}{8\mu}\left(-\frac{\mathrm{d}p}{\mathrm{d}l}\right) \tag{2.22}$$

For incompressible, fully developed flow, the pressure gradient is constant, so the term $(-\mathrm{d}p/\mathrm{d}l)$ in equations 2.18–2.22 may be replaced by $P_1 - P_2/l$ where $P_1 - P_2$ is the pressure drop ΔP_f over a length l of the channel.

The kinetic head, or flow energy, may be compared to the energy lost due to viscous friction over the channel length l. Both may be expressed in terms of the effective pressure difference, ΔP.

The kinetic energy term and the viscous term are given by

$$\Delta P_{\mathrm{KE}} = \frac{1}{2}\rho v^2$$

$$\Delta P_{\mathrm{F}} = \frac{8\mu v}{a^2}$$

respectively.

$$\frac{\Delta P_{\mathrm{KE}}}{\Delta P_{\mathrm{F}}} = \frac{\rho v a^2}{16\mu l} = \mathrm{Re}\frac{a}{32l} = \frac{\mathrm{Re}}{64}\frac{d}{l}$$

Thus, assuming the flow is laminar the kinetic and viscous terms are equal when

$$l = \frac{\mathrm{Re}}{64}d \tag{2.23}$$

For high l/d ratios, viscous pressure drop dominates.

2.3.3.3 Turbulent flow – the Fanning equation

The frictional pressure drop for turbulent flow is usually related to the average fluid velocity by the Fanning equation

$$\left(-\frac{\mathrm{d}p}{\mathrm{d}l}\right) = \frac{4}{d}f\frac{1}{2}\rho v^2 \tag{2.24a}$$

$$\frac{P_1 - P_2}{l} = \frac{4}{d}f\frac{1}{2}\rho v^2 \tag{2.24b}$$

where f is the Fanning friction factor.

f is related to the Reynolds number in the turbulent region and a commonly used relationship is the Blasius equation.

$$f = \frac{0.0791}{\mathrm{Re}^{0.25}}, \quad 2100 < \mathrm{Re} < 10^5 \tag{2.25}$$

The Fanning equation may be applied to laminar flow if

$$f = \frac{16}{\mathrm{Re}}, \quad \mathrm{Re} < 2100 \tag{2.26}$$

2.3.4 Flow in wicks

2.3.4.1 Pressure difference in the liquid phase

The flow regime in the liquid phase is almost always laminar. Since the liquid channels will not in general be straight nor of circular cross section and will often be interconnected, the Hagen–Poiseuille equation must be modified to take account of these differences.

Since mass flow will vary in both the evaporator and the condenser region, an effective length rather than the geometrical length must be used for these regions. If the mass change per unit length is constant, the total mass flow will increase or decrease, linearly along the regions. We can therefore replace the lengths of the evaporator l_e and the condenser l_c by $l_e/2$ and $l_c/2$. The total effective length for fluid flow will then be l_{eff} where

$$l_{eff} = l_a + \frac{l_e + l_c}{2} \tag{2.27}$$

Tortuosity within the capillary structure must be taken into account separately and is discussed below.

There are three principal capillary geometries

(i) Wick structures consisting of a porous structure made up of interconnecting pores. Gauzes, felts and sintered wicks come under this, these are frequently referred to as homogeneous wicks.
(ii) Open grooves.
(iii) Covered channels consisting of an area for liquid flow closed by a finer mesh capillary structure. Grooved heat pipes with gauze covering the groove and arterial wicks are included in this category. These wicks are sometimes described as composite wicks.

Some typical wick sections are shown in Fig. 2.13 and expressions for pressure drop within particular structures are discussed.

2.3.4.2 Homogeneous wicks

If ε is the fractional void of the wick, that is, the fraction of the cross section available for the fluid flow, then the total flow cross-sectional area A_f is given by

$$A_f = A\varepsilon = \pi\left(r_w^2 - r_v^2\right)\varepsilon \tag{2.28}$$

where r_w and r_v are the outer and inner radius of the wick, respectively.

If r_c is the effective pore radius, then the Hagen–Poiseuille equation (Equation 2.22) may be written

$$\dot{m} = \frac{\pi\left(r_w^2 - r_v^2\right)\varepsilon\, r_c^2 \rho_l}{8\mu_l}\frac{\Delta P_l}{l_{eff}} \tag{2.29}$$

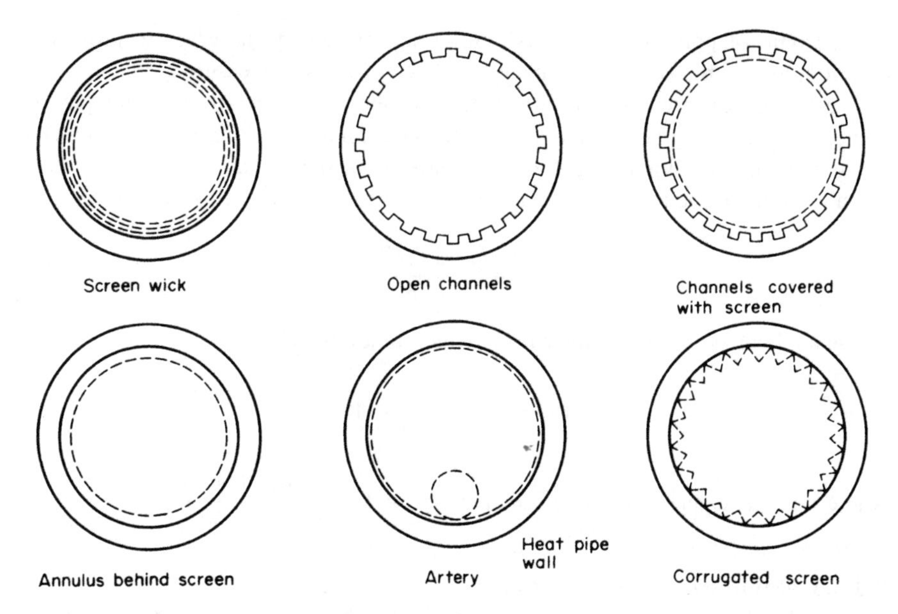

Fig. 2.13 A selection of wick sections.

or, relating the heat and mass flows, $\dot{Q} = \dot{m}L$, where L is the latent heat of vapourisation and rearranging

$$\Delta P_1 = \frac{8\mu_1 \dot{Q} l_{\text{eff}}}{\pi (r_w^2 - r_v^2)\, \varepsilon\, r_c^2 \rho_1 L} \tag{2.30}$$

For porous media, equation 2.30 is usually written

$$\Delta P_1 = \frac{b\mu_1 \dot{Q} l_{\text{eff}}}{\pi (r_w^2 - r_v^2)\, \varepsilon\, r_c^2 \rho_1 L} \tag{2.31}$$

The number 8, derived for round tubes, is replaced by the dimensionless constant, b, typically $10 < b < 20$ to include a correction for tortuosity.

While this relation can be useful for a theoretical treatment, it contains three constants, b, ε and r_c, which are all difficult to measure in practice. It is therefore useful to relate the pressure drop and flow rate for a wick structure by using a form of Darcy's law

$$\Delta P_1 = \frac{\mu_1 l_{\text{eff}} \dot{m}}{\rho_1 K A} \tag{2.32}$$

where K is the wick permeability.

Comparison of equation 2.32 with equation 2.31 shows that Darcy's law is the Hagen–Poiseuille equation with correction terms included with the constant K to take account of pore size, pore distribution and tortuosity. It serves to provide a definition of permeability, K, a quantity that can be easily measured.

The Blake–Kozeny equation is sometimes used in the literature. This equation relates the pressure gradient across a porous body, made up from spheres diameter D, to the flow of liquid. Like Darcy's law, it is merely the Hagen–Poiseuille equation with correction factors. The Blake–Koseny equation may be written

$$\Delta P_1 = \frac{150\mu_1 (1-\varepsilon')^2 l_{\mathrm{eff}} v}{D^2 \varepsilon'^3} \tag{2.33}$$

and is applicable only to laminar flow, which requires that

$$\mathrm{Re}' = \frac{\rho v D}{\mu (1-\varepsilon')} < 10$$

where v is the superficial velocity $\dot{m}/\rho_1 A$ and $\varepsilon' = \dfrac{\text{volume of voids}}{\text{volume of body}}$

2.3.4.3 Nonhomogeneous wicks

Longitudinal groove wick. For grooved wicks, the pressure drop in the liquid is given by

$$\Delta P_1 = \frac{8\mu_1 \dot{Q} l}{\pi \left(\frac{1}{2} d_{\mathrm{e}}\right)^4 N \rho_1 L} \tag{2.34}$$

where N is the number of grooves and d_{e} is the effective diameter defined by equation 2.17.

At high vapour velocities, shear forces will tend to impede the liquid flow in open grooves. This may be avoided by using a fine pore screen to form a composite wick structure.

Composite wicks. Such a system as arterial or composite wicks require an auxiliary capillary structure to distribute the liquid over the evaporator and condenser surfaces.

The pressure drop in wicks constructed by an inner porous screen separated from the heat pipe wall to give an annular gap for the liquid flow may be obtained from the Hagen–Poiseuille equation applied to parallel surface, provided that the annular width w is small compared to the radius of the pipe vapour space r_{v}

In this case

$$\Delta P_1 = \frac{6\mu_1 \dot{Q} l}{\pi r_{\mathrm{v}} w^3 \rho_1 L} \tag{2.35}$$

This wick structure is particularly suitable for liquid metal heat pipes. Variants are also used in lower temperature high-performance heat pipes for spacecraft. Crescent annuli may be used, in which it is assumed that the screen is moved down to touch the bottom of the heat pipe wall leaving a gap $2w$ at the top. In this case

$$\Delta P_{\mathrm{l}} = \frac{6\mu_{\mathrm{l}}\dot{Q}l}{\pi r_{\mathrm{v}} w^3 \rho_{\mathrm{l}} L} \tag{2.36}$$

In equations 2.35 and 2.36, the length should be taken as the effective length defined in equation 2.27.

2.3.5 Vapour phase pressure difference, ΔP_{v}

2.3.5.1 Introduction

The total vapour phase difference in pressure will be the sum of the pressure drops in the three regions of a heat pipe, namely the evaporator drop ΔP_{ve}, the adiabatic section drop ΔP_{va} and the pressure drop in the condensing region ΔP_{vc}. The problem of calculating the vapour pressure drop is complicated in the evaporating and condensing regions by radial flow due to evaporation or condensation. It is convenient to define a further Reynolds number, the radial Reynolds number

$$\mathrm{Re}_{\mathrm{r}} = \frac{\rho_{\mathrm{v}} v_{\mathrm{r}} r_{\mathrm{v}}}{\mu_{\mathrm{v}}} \tag{2.37}$$

to take account of the radial velocity component v_{r} at the wick where $r = r_{\mathrm{v}}$.

By convention, the vapour space radius r_{v} is used rather than the vapour space diameter that is customary in the definition of axial Reynolds number. Re_{r} is positive in the evaporator section and negative in the condensing section. In most practical heat pipes, Re_{r} lies in the range 0.1–100.

Re_{r} is related to the radial rate of mass injection or removal per unit length $\frac{\mathrm{d}\dot{m}}{\mathrm{d}z}$ as follows:

$$\mathrm{Re}_{\mathrm{r}} = \frac{1}{2\pi\mu_{\mathrm{v}}} \frac{\mathrm{d}\dot{m}}{\mathrm{d}z} \tag{2.38}$$

The radial and axial Reynolds numbers are related to uniform evaporation or condensation rates by the equation

$$\mathrm{Re}_{\mathrm{r}} = \frac{\mathrm{Re}}{4} \frac{r_{\mathrm{v}}}{z} \tag{2.39}$$

where z is the distance from either the end of the evaporator section or the end of the condenser section.

In Section 2.3.3.2, we showed in equation 2.23 that provided the flow is laminar, the pressure drop due to viscous forces in a length l is equal to the kinetic head when

$$l = \frac{\mathrm{Re}d}{64} = \frac{\mathrm{Re}r_{\mathrm{v}}}{32} \tag{2.40}$$

If we substitute

$$\mathrm{Re} = \frac{4\mathrm{Re}_r l}{r_v}$$

for the evaporator or condenser region we find that the condition reduces to

$$\mathrm{Re}_r = 8 \tag{2.41}$$

Fig. 2.14, taken from Busse [14], shows Re_r as a function of power per unit length for various liquid metal working fluids. Clearly, the kinetic head may be a significant component of the vapour pressure drop in the evaporator and result in an appreciable pressure rise in the condenser.

2.3.5.2 Incompressible flow: (simple one dimensional theory)

In the following treatment, we will regard the vapour as an incompressible fluid. This assumption implies that the flow velocity v is small compared to the velocity of sound c, in the vapour, i.e. the Mach number

$$\frac{v}{c} < 0.3$$

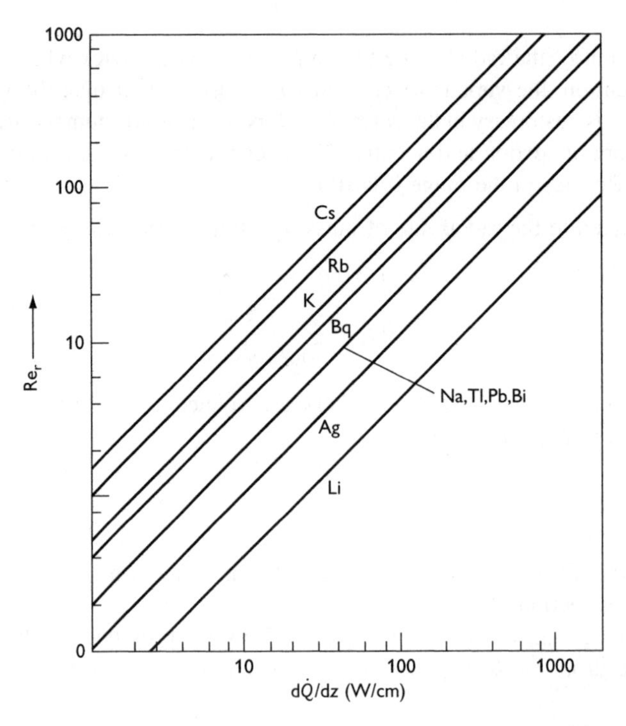

Fig. 2.14 Radial Reynolds number versus heat input per unit length of heat pipe (liquid metal working fluids) [14].

Alternatively, this condition implies that the treatment is valid for heat pipes in which ΔP_v is small compared with P_v, the average vapour pressure in the pipe. This assumption is not necessarily valid during start-up nor is it always true in the case of high-temperature liquid metal heat pipes. The effect of compressibility of the vapour will be considered in Section 2.3.5.6.

In the evaporator region, the vapour pressure gradient will be necessary to carry out two functions.

(i) To accelerate the vapour entering the evaporator section up to the axial velocity v. Since, on entering the evaporator, this vapour will have radial velocity but no axial velocity. The necessary pressure gradient is called the inertial term $\Delta P'_v$

(ii) To overcome frictional drag forces at the surface $r = r_v$ at the wick. This is the viscous term $\Delta P''_v$

We can estimate the magnitude of the inertial term as follows. If the mass flow per unit area of cross section at the evaporator is ρv, then the corresponding momentum flux per unit will be given by $\rho v \times v$ or ρv^2. This momentum flux in the axial direction must be provided by the inertial term of the pressure gradient.

Hence

$$\Delta P'_v = \rho v^2 \tag{2.42}$$

Note that $\Delta P'_v$ is independent of the length of the evaporator section. The way in which $\Delta P'_v$ varies along the length of the evaporator is shown in Fig. 2.15a. If we assume laminar flow, we can estimate the viscous contribution to the total evaporator pressure loss by integrating the Hagen–Poiseuille equation. If the rate of mass entering the evaporator per unit length $d\dot{m}/dz$ is constant, we find by integrating equation 2.22 along the length of the evaporator section

$$\Delta P''_v = \frac{8\mu_v \dot{m}}{\rho_v \pi r_v^4} \frac{l_e}{2} \tag{2.43}$$

Thus, the total pressure drop in the evaporator region ΔP_{ve} will be given by the sum of the two terms

$$\begin{aligned} \Delta P_{ve} &= \Delta P'_v + \Delta P''_v \\ \Delta P_{ve} &= \rho v^2 + \frac{8\mu_v \dot{m}}{\rho \pi r_v^4} \frac{l_e}{2} \end{aligned} \tag{2.44}$$

The condenser region may be treated in a similar manner, but in this case axial momentum lost as the vapour stream is brought to rest, so the inertial term will be negative, that is, there will be pressure recovery. For the simple theory, the two inertial terms will cancel and the total pressure drop in the vapour phase will be

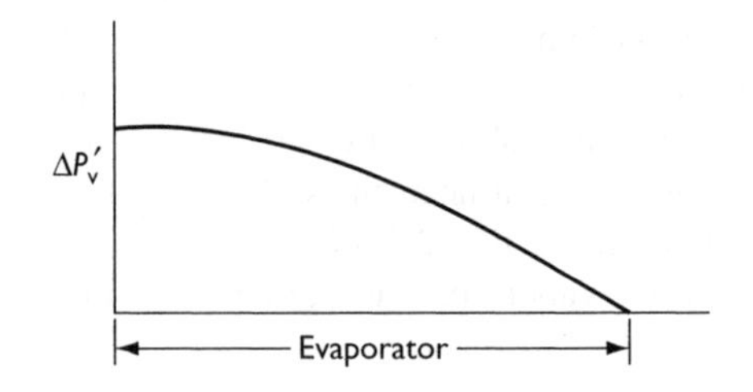

(a) Pressure variation along the evaporator

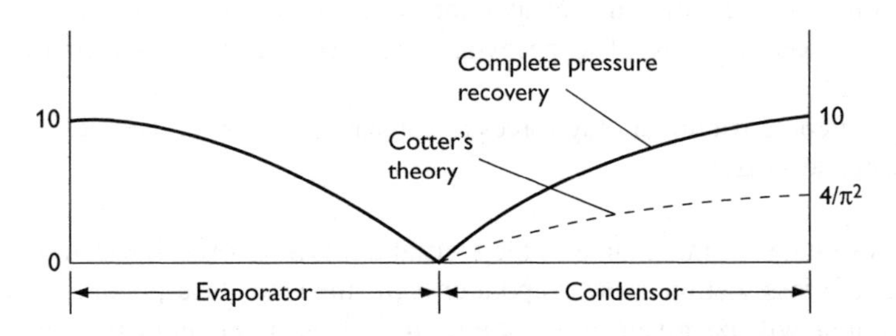

(b) Pressure recovery of inertial pressure term (Adiabatic section omitted)

Fig. 2.15 Vapour pressure change due to inertial effects.

due entirely to the viscous terms. It is shown later that it is not always possible to recover the initial pressure term in the condensing region.

In the adiabatic section, the pressure difference will contain only the viscous term which will be given either by the Hagen–Poiseuille equation or by the Fanning equation depending on whether the flow is laminar or turbulent.

For laminar flow

$$\Delta P_{\mathrm{va}} = \frac{8\mu_{\mathrm{v}}\dot{m}}{\rho\pi r_{\mathrm{v}}^4} l_{\mathrm{a}} \qquad \mathrm{Re} < 2100 \tag{2.45}$$

For turbulent flow

$$\Delta P_{\mathrm{va}} = \frac{2}{r_{\mathrm{v}}} f \frac{1}{2}\rho_{\mathrm{v}} v^2 l_{\mathrm{a}} \quad \mathrm{Re} > 2100 \tag{2.46}$$

where $f = \dfrac{0.0791}{\mathrm{Re}^{0.25}}$, $2100 < \mathrm{Re} < 10^5$ (equation 2.25)

Hence, the total vapour pressure drop, ΔP_{v} is given by

$$\begin{aligned}\Delta P_{\mathrm{v}} &= \Delta P_{\mathrm{ve}} + \Delta P_{\mathrm{vc}} + \Delta P_{\mathrm{va}} \\ &= \rho v^2 + \frac{8\mu_{\mathrm{v}}\dot{m}}{\rho\pi r_{\mathrm{v}}^4}\left[\frac{l_{\mathrm{e}} + l_{\mathrm{c}}}{2} + l_{\mathrm{a}}\right]\end{aligned} \tag{2.47}$$

for laminar flow with no pressure recovery and

$$\Delta P_v = \Delta P_{ve} + \Delta P_{vc} + \Delta P_{va} = \frac{8\mu_v \dot{m}}{\rho \pi r_v^4}\left[\frac{l_e + l_c}{2} + l_a\right] \tag{2.48}$$

for laminar flow with full pressure recovery.

Equations 2.47 and 2.48 enable the calculation of vapour pressure drops in simple heat pipe design and are used extensively.

2.3.5.3 Incompressible flow one-dimensional theories of Cotter and Busse

In addition to the assumption of incompressibility, the above treatment assumes a fully developed flow velocity profile and complete pressure recovery. It does, however, broadly correct results. A considerable number of papers have been published giving a more precise treatment of the problem. Some of these will be summarised in this and the following section.

The earliest theoretical treatment of the heat pipe was proposed by Cotter [1] For $Re_r << 1$, Cotter used the following result obtained by Yuan and Finkelstein for laminar incompressible flow in a cylindrical duct with uniform injection or suction through a porous wall.

$$\frac{dP_v}{dz} = \frac{8\mu_v \dot{m}}{\pi \rho_v r_v^4}\left[1 + \frac{3}{4}Re_r - \frac{11}{270}Re_r^2\right]$$

He obtained the following expression

$$\Delta P_{ve} = \frac{4\mu_v l_e \dot{Q}}{\pi \rho_v r_v^4 L} \tag{2.49}$$

which is equivalent to equation 2.43.

For $Re_r >> 1$, Cotter used the pressure gradient obtained by Knight and McInteer for flow with injection or suction through porous parallel plane walls. The resulting expression for pressure gradient is

$$\Delta P_{ve} = \frac{\dot{m}^2}{8\rho_v r_v^4}$$

which may be rewritten

$$\Delta P_{ve} = \frac{(\rho_v \pi r_v^2 v)^2}{8\rho_v r_v^4} = \frac{\pi^2}{8}\rho_v v^2 \approx \rho_v v^2 \tag{2.50}$$

This is equivalent to equation 2.42, derived previously, suggesting that inertia effects dominate the pressure drop in the evaporator.

A different velocity profile was used by Cotter in the condenser region, this gave pressure recovery in the condenser of

$$\Delta P_{\mathrm{vc}} = -\frac{4}{\pi^2}\frac{\dot{m}^2}{8\rho_{\mathrm{v}} r_{\mathrm{v}}^4} \tag{2.51}$$

which is $4/\pi^2$, or 0.405 of ΔP_{ve}, giving only partial pressure recovery.

In the adiabatic region, Cotter assumed fully developed laminar flow, hence equation 2.46

$$\Delta P_{\mathrm{va}} = \frac{8\mu_{\mathrm{v}}\dot{m}}{\rho\pi r_{\mathrm{v}}^4} l_{\mathrm{a}}$$

was used.

The full expression for the vapour pressure drop combined equations 2.46, 2.50 and 2.51 to give

$$\Delta P_{\mathrm{v}} = \left(1 - \frac{4}{\pi^2}\right)\frac{\dot{m}}{8\rho_{\mathrm{v}} r_{\mathrm{v}}^4} + \frac{8\mu_{\mathrm{v}}\dot{m} l_{\mathrm{a}}}{\pi\rho_{\mathrm{v}} r_{\mathrm{v}}^4} \tag{2.52}$$

Busse also considered the one-dimensional case, assuming a modified Hagen–Poiseuille velocity profile and using this to obtain a solution of the Navier–Stokes equation[2] [15] for a long heat pipe and obtained similar results.

2.3.5.4 Pressure recovery

We have seen that the pressure drop in the evaporator and condenser regions consists of two terms, an inertial term and a term due to viscous forces. Simple theory suggests that the inertial term will have the opposite sign in the condenser region and should cancel out that of the evaporator (Fig. 2.15b). There is experimental evidence for this pressure recovery. Grover et. al. [16] provided an impressive demonstration with a sodium heat pipe. In these experiments, they achieved 60 per cent pressure recovery. The radial Reynold number was greater than 10. For simplicity in Fig. 2.15b, the viscous component of the pressure drop has been omitted. The liquid pressure drop is also shown in Fig. 2.16a–c. Ernst [17] has pointed out that if the pressure recovery in the condenser region is greater than the liquid pressure drop (Fig. 2.16b), then the meniscus in the wick will be convex. While this is possible in principle, under normal heat pipe operation there is excess in the condenser region so that this condition cannot occur. For this reason, if $|\Delta P_{vc}| > |\Delta P_{lc}|$ it is usual to neglect pressure recovery and assume that there is no resultant pressure drop or gain in the condenser region, this is indicated in Fig. 2.16c.

[2] The Navier–Stokes equation, together with the continuity equation, is a relationship that relates the pressure, viscous and inertial forces in a three-dimensional, time-varying flow. A derivation and statement of the Navier–Stokes equation is given in Ref. [15].

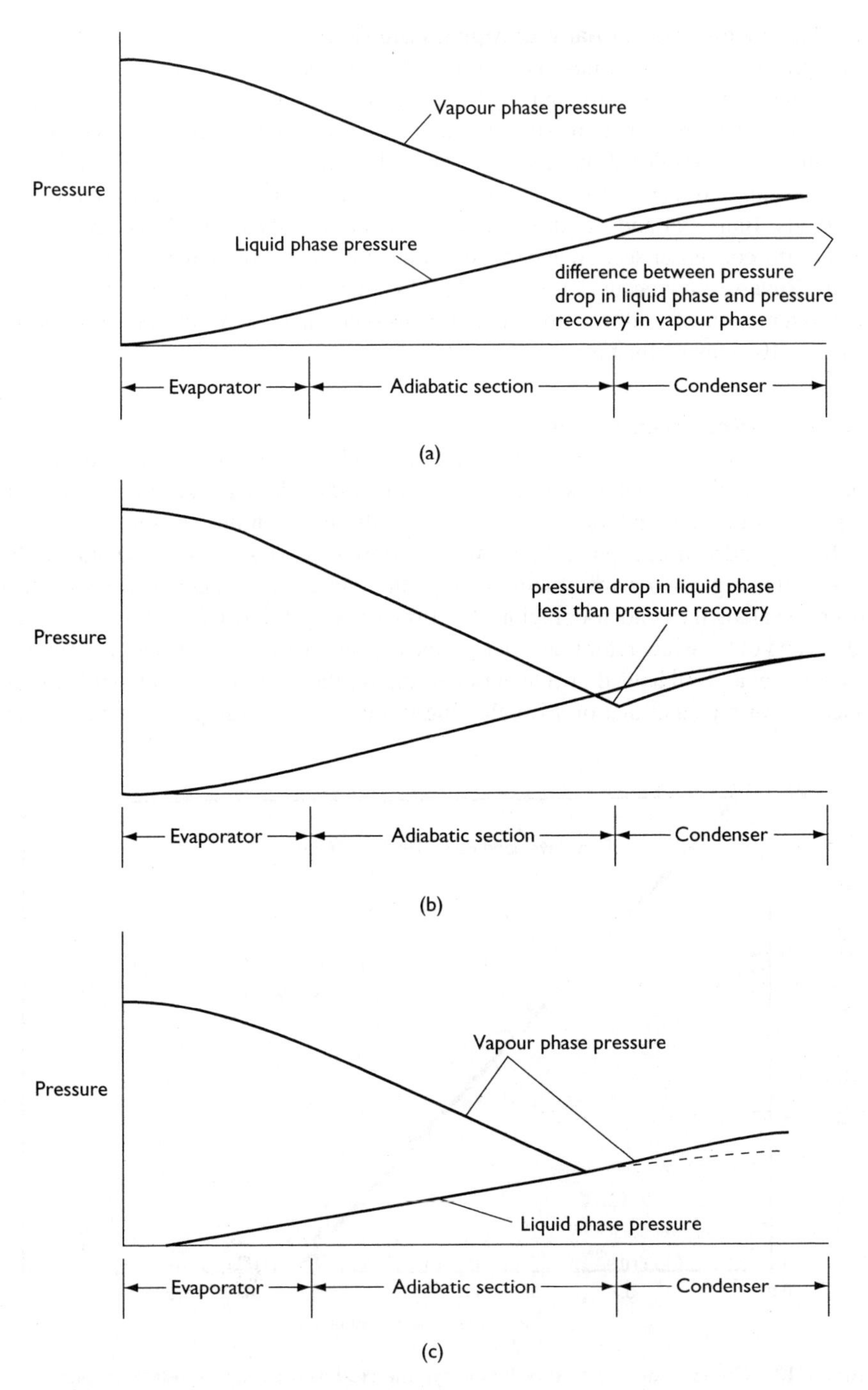

Fig. 2.16 Pressure profiles showing pressure recovery in a heat pipe [17].

2.3.5.5 Two-dimensional incompressible flow

The previous discussion has been restricted to one-dimensional flow. In practical heat pipes, the temperature and pressure are not constant across the cross section and this variation is particularly important in the condenser region. A number of authors have considered this two-dimensional problem. Bankston and Smith [18] and Rohani and Tien [19] have solved the Navier–Stokes equation by numerical methods. Bankston and Smith showed that axial velocity reversal occurred at the end of the condenser section under conditions of high evaporation and condensation rates. Reverse flow occurs for $|Re_r| > 2.3$. In spite of this extreme divergence from the assumption of uniform flow, one-dimensional analyses yield good results for $|Re_r| < 10$. This is illustrated in Fig. 2.17.

2.3.5.6 Compressible flow

So far we have neglected the effect of compressibility of the vapour on the operation of the heat pipe. Compressibility can be important during start-up and also in high-temperature liquid metal heat pipes; it is discussed in this section.

In a cylindrical heat pipe, the axial mass flow increases along the length of the evaporator region to a maximum value at the end of the evaporator; it will then decrease along the condenser region. The flow velocity will rise to a maximum value at the end of the evaporator region where the pressure will have fallen to a minimum. Deverall et al. [20] have drawn attention to the similarity in flow behaviour between such a heat pipe and that of a gas flowing through a converging–diverging nozzle.

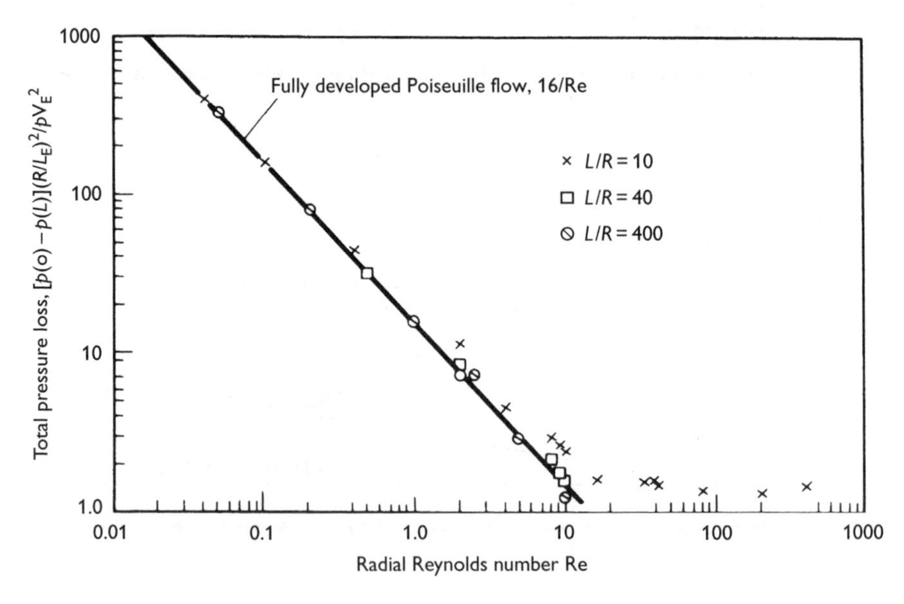

Fig. 2.17 Comparison of pressure loss in symmetrical heat pumps with that predicted for Poiseuille flow [18].

In the former, the area remains constant but the mass flow varies and in the latter the mass flow is constant but the cross-sectional area is changed. It is helpful to examine the behaviour of the convergent–divergent nozzle in more detail before returning to the heat pipe. Let the pressure of the gas at the entry to the nozzle be kept constant and consider the effect of reducing the pressure at the outlet. With reference to the curves in Fig. 2.18, we can see the effect of increasing the flow through a nozzle. For the flow of curve A, the pressure difference between inlet and outlet is small. The gas velocity will increase to a maximum value in the position of minimum cross section, or throat, falling again in the divergent region. The velocity will not reach the sonic value. The pressure passes through a minimum in the throat. If the outlet pressure is now reduced, the flow will increase and the situation shown in curve B can be attained. Here the velocity will increase through the convergent region, rising to the sonic velocity in the throat. As before, the velocity will reduce during travel through the diverging section and there will be some pressure recovery. If the outlet pressure is further reduced the flow rate will remain constant and the pressure profile will follow curve C. The gas will continue to accelerate after entering the divergent region and will become supersonic. Pressure recovery will occur after a shock front. Curve D shows that for a certain exit pressure, the gas can be caused to accelerate throughout the diverging region. Further pressure reduction will not affect the flow pattern in the nozzle region. It should be noted that after curve C, pressure reduction does not affect the flow pattern in the converging section, hence the mass flow did not increase after the throat velocity has attained the sonic value. This condition is referred to as choked flow.

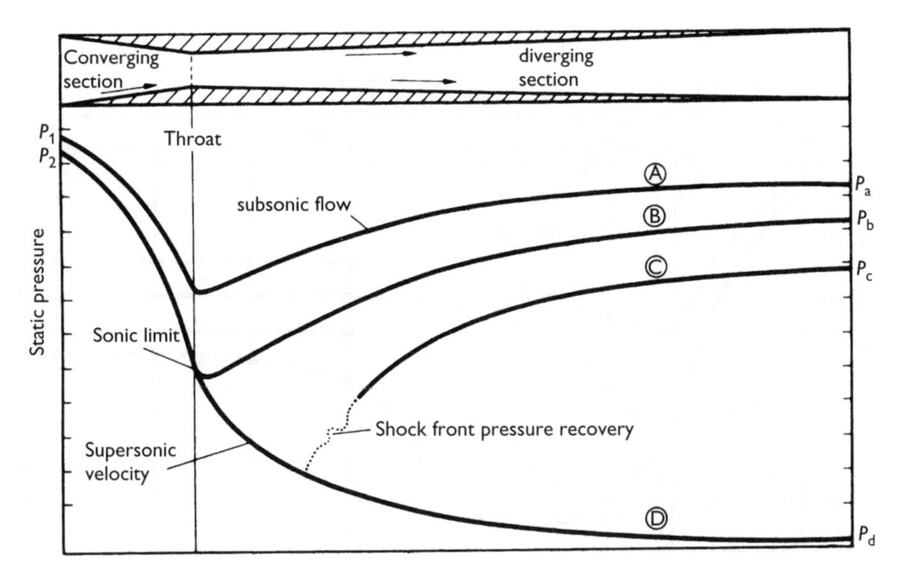

Fig. 2.18 Pressure profiles in a converging–diverging nozzle.

Kemme [21] has shown very clearly that a heat pipe can operate in a very similar manner to the diverging nozzle. His experimental arrangement is shown in Fig. 2.19. Kemme used sodium as the working fluid and maintained a constant heat input of 6.4 kW. He measured the axial temperature variation, but since this is related directly to pressure, his temperature profile can be considered to be the same as the pressure profile.

Kemme arranged to vary the heat rejection at the condenser by means of a gas gap, the thermal resistance of which could be altered by varying the argon–helium ratio of the gas. Kemme's results are shown in Fig. 2.19. Curve A demonstrates subsonic flow with pressure recovery; curve B, obtained by lowering the condenser temperature, achieved sonic velocity at the end of the evaporator and hence operated under choked flow conditions. Further decrease in the thermal resistance between the condenser and the heat sink simply reduced the condenser region temperature but did not increase the heat flow that was limited by the choked flow condition and fixed axial temperature drop in the evaporator [22]. It should be noted that under these conditions of sonic limitation considerable axial temperature and pressure changes will exist and the heat pipe operation will be far from isothermal.

Deverall et al. [20] have shown that a simple one-dimensional model provides a good description of the compressible flow behaviour. Consider the evaporator section using the nomenclature of Fig. 2.19.

The pressure $P_1 = P_o$, where P_o is the stagnation pressure of the fluid. The pressure drop along the evaporator is given by equation 2.42.

$$P_2 - P_o = \rho v^2 \tag{2.53}$$

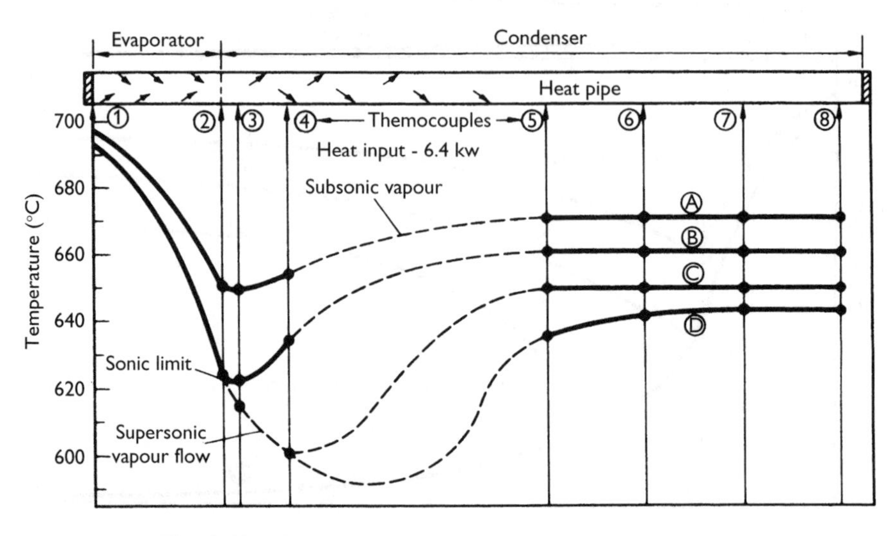

Fig. 2.19 Temperature profiles in a sodium heat pipe [21].

The equation state for a gas at low pressure may be written

$$P = \rho_v RT \tag{2.54}$$

and the sonic velocity, C, may be expressed

$$C = \sqrt{\gamma RT} \tag{2.55}$$

The Mach number, M, is defined

$$M = \frac{v}{C}$$

Hence, substituting into equation 2.53

$$\frac{P_o}{P_2} - 1 = \frac{\rho_v v^2}{P_2} = \frac{M^2 \gamma RT_2}{RT_2} = \gamma M^2 \tag{2.56}$$

Defining the total temperature T_o such that

$$\dot{m} c_p T_o = \dot{m}\left(c_p T_2 + \frac{v^2}{2}\right) \tag{2.57}$$

and remembering that $c_p = \dfrac{\gamma}{\gamma - 1}$

$$\frac{T_o}{T_2} = 1 + M^2 \frac{\gamma - 1}{2} \tag{2.58}$$

The density ratio may be expressed

$$\frac{\rho_o}{\rho_2} = \frac{P_o T_2}{P_2 T_o} = \frac{1 + \gamma M^2}{1 + (\gamma - 1)/\gamma M^2} \tag{2.59}$$

Finally, the energy balance for the evaporator section gives

$$\begin{aligned} \dot{Q} &= \rho_v A v L \\ &= \rho_v AMCL \end{aligned} \tag{2.60}$$

where C is the sonic velocity at T_2. The heat flow may be expressed in terms of the sonic velocity C_o corresponding to the stagnation temperature T_o. In this case

$$\dot{Q} = \frac{\rho_v AMC_o L}{\sqrt{2(\gamma + 1)}} \tag{2.61}$$

The pressure, temperature and density ratio for choked flow may be obtained by substituting $M = 1$ into equations 2.59–2.61, respectively, the values are presented in Table 2.2.

Table 2.2 Effect of γ on compressibility parameters

	Monatomic gas	Diatomic gas	Triatomic gas
γ	1.66	1.4	1.3
$\frac{P_o}{P_{2,c}} = 1 + \gamma$	2.66	2.4	2.3
$\frac{T_o}{T_{2,c}} = \frac{1+\gamma}{2}$	1.33	1.2	1.15
$\frac{\rho_o}{\rho_{2,c}} = 2$	2	2	2

2.3.5.7 Summary of vapour flow

The equations presented in this section permit the designer to predict the pressure drop due to the vapour flow in a heat pipe. These relatively simple equations give acceptable results for most situations encountered within heat pipes. An excellent summary of the analysis of vapour side pressure drop is presented in [23]. For applications that present exceptional problems, the designer should consider using one of the many computational fluid dynamics (CFD) computer packages on the market that can deal with three-dimensional and compressible flows.

2.3.6 Entrainment

In a heat pipe, the vapour flows from the evaporator to the condenser and the liquid is returned by the wick structure. At the interface between the wick surface and the vapour, the latter will exert a shear force on the liquid in the wick. The magnitude of the shear force will depend on the vapour properties and velocity and its action will be to entrain droplets of liquid and transport them to the condenser end. This tendency to entrain is resisted by the surface tension in the liquid. Entrainment will prevent the heat pipe operating and represents one limit to performance. Kemme observed entrainment in a sodium heat pipe and reports that the noise of droplets impinging on the condenser end could be heard.

The Weber number, We, which is representative of the ratio between inertial vapour forces and liquid surface tension forces provides a convenient measure of the likelihood of entrainment.

The Weber number is defined

$$\mathrm{We} = \frac{\rho_v v^2 z}{2\pi\sigma_l} \tag{2.62}$$

where ρ_v is the vapour density, v the vapour density, σ_l the surface tension, z a dimension characterising the vapour–liquid surface.

In a wicked heat pipe, z is related to the wick spacing.

It should be noted that some authors, including Kim and Peterson [24] define the Weber number as We′

$$\mathrm{We}' = \frac{\rho_v v^2 z}{\sigma_l} = 2\pi\,\mathrm{We} \tag{2.63}$$

A review of 11 models of entrainment by Kim and Peterson reported critical Weber numbers (as defined by equation 2.63) principally in the range 0.2–10.

It may be assumed that entrainment may occur when We is of the order 1. The limiting vapour velocity, v_c, is thus given by

$$v_c = \sqrt{\frac{2\pi\sigma_l}{\sigma_v z}} \tag{2.64}$$

and, relating the axial heat flux to the vapour velocity using

$$\dot{q} = \rho_v L v$$

The entrainment limited axial flux is given by

$$\dot{q} = \sqrt{\frac{2\pi\rho_v L^2 \sigma_l}{z}} \tag{2.65}$$

Extracting the fluid properties from equation 2.65 suggests $\rho_v L^2 \sigma_l$ as a suitable figure of merit for working fluids from the point of view of entrainment.

Cheung [25] has plotted this figure of merit against temperature for a number of liquid metals and this is reproduced as Fig. 2.20.

A number of authors report experimental results on the entrainment limit. Typically, they select a value of z which fits their results and show that the temperature dependence of the limit is as predicted by equation 2.65.

Kim and Peterson [24] defined the Weber number

$$\mathrm{We}' = \frac{\rho_v v^2 z}{\sigma_l} \tag{2.66}$$

and the viscosity number

$$N_{vl} = \frac{\mu_l}{\sqrt{\rho_l \sigma_l \lambda_c / 2\pi}} \tag{2.67}$$

where λ_c is the critical wavelength derived from the Rayleigh–Taylor instability [26]. N_{vi} represents the stability of a liquid interface. They derived a correlation, equation 2.68, from their experimental measurements with water on mesh wicks for the critical Weber number

$$\mathrm{We}_{vc} = 10^{-1.163} N_{vl}^{-0.744} \left(\frac{\lambda_c}{d_1}\right)^{-0.509} \left(\frac{D_h}{d_2}\right)^{0.276}. \tag{2.68}$$

where d_1 and d_2 are the mesh wire spacing and thickness, respectively, and D_h is the equivalent diameter of the vapour space.

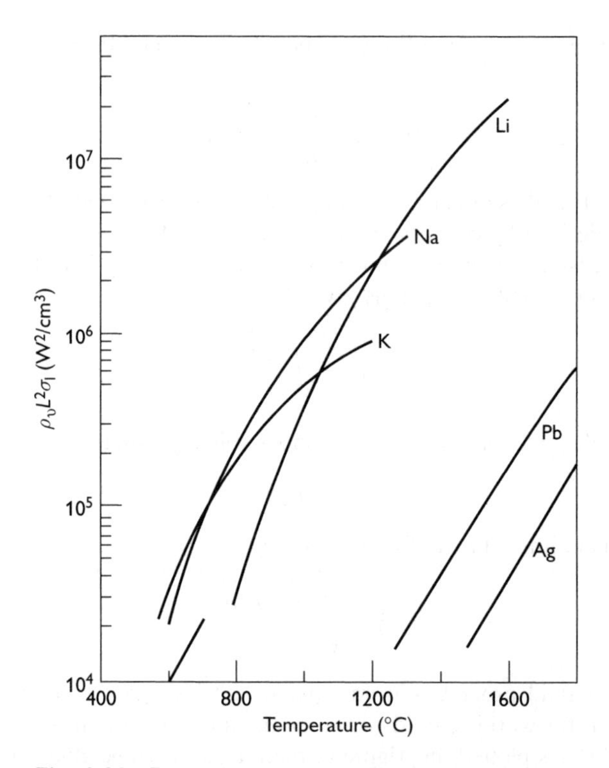

Fig. 2.20 Entrainment merit number for liquid metals.

2.3.7 Heat transfer and temperature difference

2.3.7.1 Introduction

In this section, we consider the transfer of heat and the associated temperature drops in a heat pipe. The latter can be represented by thermal resistances and an equivalent circuit is shown in Fig. 2.21.

Heat can both enter and leave the heat pipe by conduction from or to a heat source/sink by convection or by thermal radiation. The pipe may also be heated by eddy currents or by electron bombardment and cooled by electron emission. Further temperature drops will occur by thermal conduction through the heat pipe walls at both the evaporator and condenser regions. The temperature drops through the wicks arise in several ways and are discussed in detail later in this section. It is found that a thermal resistance exists at the two vapour–liquid surfaces and also in the vapour column. The processes of evaporation and condensation are examined in some detail both in order to determine the effective thermal resistances and also to identify the maximum heat transfer limits in the evaporator and condenser regions. Finally, the results for thermal resistance and heat transfer limits are summarised for the designer.

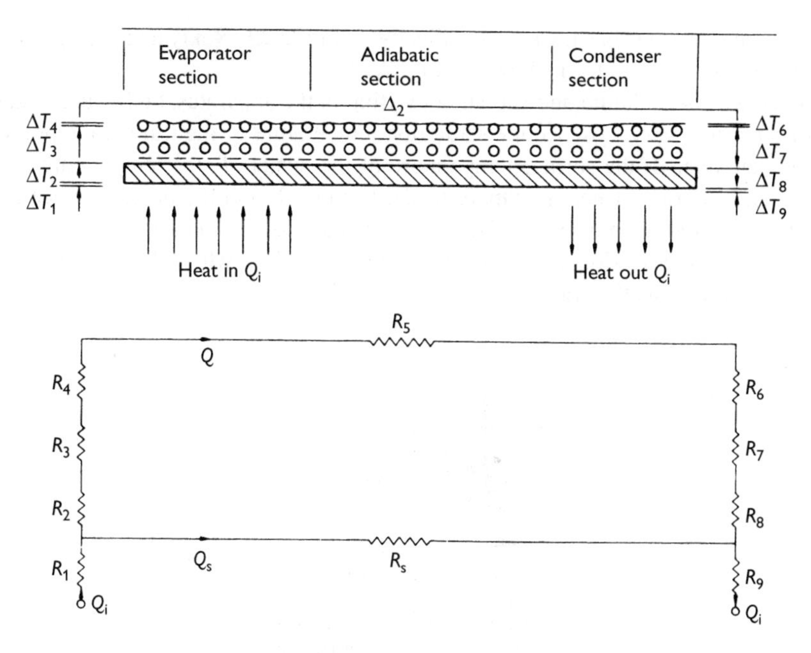

Fig. 2.21 Temperature drops and equivalent thermal resistances in a heat pipe.

2.3.7.2 Heat transfer in the evaporator region

For low values of heat flux, the heat will be transported to the liquid surface partly by conduction through the wick and liquid and partly by natural convection. Evaporation will be from the liquid surface. As the heat flux is increased, the liquid in contact with the wall will become progressively superheated and bubbles will form at nucleation sites. These bubbles will transport some energy to the surface by latent heat of vapourisation and will also greatly increase convective heat transfer. With further increase of flux, a critical value will be reached, burnout, at which the wick will dry out and the heat pipe will cease to operate. Before discussing the case of wicked surfaces, the data on heat transfer from plane, unwicked surfaces will be summarised. Experiments on wicked surfaces are then described and correlations given to enable the temperature drop through the wick and the burnout flux to be estimated. The subject is complex and further work is necessary to provide an understanding of the processes in detail.

2.3.7.3 Boiling heat transfer from plane surfaces

In 1934, Nukiyama [27] performed a pool boiling experiment, passing an electric current through a platinum wire immersed in water. The heat flux was controlled by the current through and voltage across the wire and the temperature of the wire was determined from its resistance. Similar results are obtained when boiling from a plane surface or from the surface of a cylinder. An apparatus for measuring pool boiling heat transfer is shown schematically in Fig. 2.22. Nukiyama then proposed a boiling curve of the form shown in Fig. 2.23.

Since we have a liquid and vapour coexisting in the apparatus, both must be at (or during boiling, very close to) the saturation temperature of the fluid at the pressure in the container. If the surface temperature of the heater, T, the temperature of the fluid, T_{sat}, the rate of energy supply to the heater, $\dot{Q}$, and the heater surface area, A, are measured, then a series of tests may be carried out and a graph of $\log \dot{Q}$, or more usually $\log \dot{q} = \log(\dot{Q}/A)$ plotted against $\log \Delta T_{\text{sat}}$, where $\Delta T_{\text{sat}} = (T - T_{sat})$, often referred to as the wall superheat.

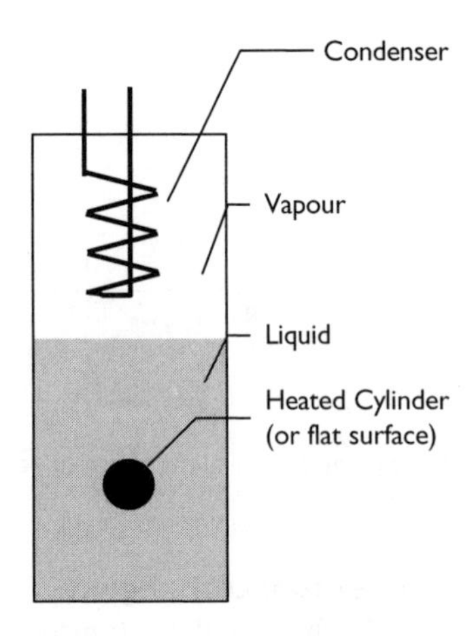

Fig. 2.22 Schematic diagram of pool boiling experiment.

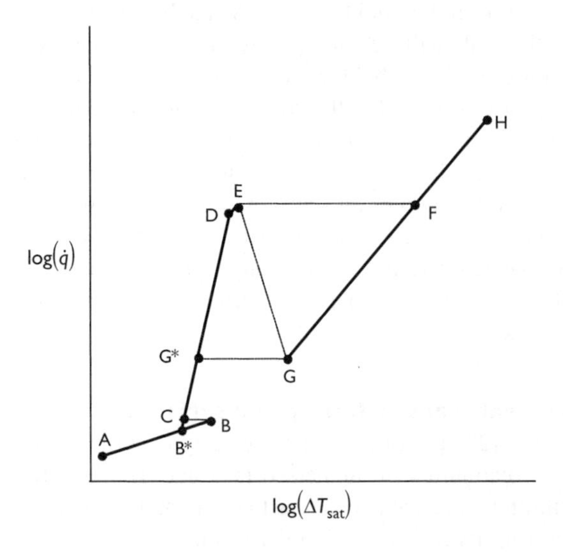

Fig. 2.23 Boiling curve for pool boiling from plane surface, cylinder or wire.

For the case of controlled heat flux (e.g. electric heating), the various regimes may be described.

For increasing heat flux, in the region A–B heat transfer from the heater surface is purely by single-phase natural convection. Superheated liquid rises to the surface of the reservoir and evaporation takes place at this surface. As the heat flux is increased beyond the value at B, bubbles begin to form on the surface of the heater, depart from the heater surface and rise through the liquid, this process is referred to as nucleate boiling. At this stage, a reduction of heater surface temperature to C may be observed. Reducing the heat flux would now result in the heat flux temperature difference relationship following the curve C–B*. This type of phenomenon, for which the relationship between a dependent and independent variable is different for increasing and decreasing values of the independent variable, is known as hysterisis.

After the commencement of nucleate boiling, further increase in heat flux leads to increased heater surface temperature to point D. Further increase beyond the value at D leads to vapour generation at such a rate that it impedes the flow of liquid back to the surface and transition boiling occurs between D and E. At E a stable vapour film forms over the surface of the heater and this has the effect of an insulating layer on the heater resulting in a rapid increase in temperature from E to F. The heat flux at E is known as the critical heat flux. The large temperature increase that occurs if an attempt is made to maintain the heat flux above the level of the critical heat flux is frequently referred to as burnout. However, if physical burnout does not occur it is possible to maintain boiling at point F and then adjust the heat flux, the heat flux temperature difference relationship will then follow the line G–H. This region on the boiling curve corresponds to the stable film boiling regime. Reduction of the heat flux below the value at G causes a return to the nucleate boiling regime at G*.

The factors that influence the shape of the boiling curve for a particular fluid include fluid properties, heated surface characteristics and physical dimensions and orientation of the heater. The previous history of the system also influences the behaviour, particularly at low heat flux.

Clearly several relationships, defining both the extent of each region and the appropriate shape of the curve for that region are required to describe the entire curve. It is the nucleate boiling region, C–D which is of greatest importance in most engineering applications. However, it is clearly important that the designer ensures that the critical heat flux is not inadvertently exceeded, and there are some systems that operate in the film boiling regime. Many correlations describing each region of the boiling curve have been published. Additionally, the temperature difference at which nucleation first occurs, i.e. the temperature at C influences the boiling regime during flow boiling and the hysteresis.

If the temperature of the heater, rather than the heat flux, was to be controlled then increasing temperature above that corresponding to the critical heat flux would result in a decrease in heat flux with increasing temperature from E to G, followed by an increase along the line G–H. The point G is sometimes referred to as the Liedenfrost Point. Temperature-controlled heating of a surface is found in many heat exchangers and boilers – the temperature of the wall being necessarily below the temperature of the other fluid in the heat exchanger. Experimentally, it is difficult to maintain surface temperatures over a wide range with the corresponding range of heat fluxes. To obtain boiling curves for varying ΔT_{sat}, it is usual to plunge an ingot of high conductivity material into a bath of the relevant fluid. The surface temperature is measured directly and the heat flux can then be calculated from the geometry of the ingot and the rate of change of temperature.

This suggests that some feature of the surface encourages bubble nucleation. The explanation for the importance of surface finish lies in the mechanism of bubble formation. It has been observed that nucleation occurs in cavities within the surface, these cavities contain minute bubbles of trapped gas or vapour which act as starting points for bubble growth. This is illustrated schematically in Fig. 2.24. When the bubble leaves the site, a small bubble remains in the cavity which acts as the start for the next bubble.

Consideration of idealised nucleation sites allows some indication of their necessary size if they are to play a part in boiling, with reference to an idealised conical cavity as shown in Fig. 2.24.

The pressure, P_B, inside a bubble is somewhat higher than the pressure in the surrounding liquid

$$P_B = P + \frac{2\sigma_l}{R} \tag{2.69}$$

where P is the liquid pressure, r the radius of curvature of the bubble and σ_l the surface tension of the liquid. The radius of curvature is a maximum when the bubble forms a hemispherical cap over the cavity, i.e. $r = R$, the radius of the mouth of the cavity. This is the condition for P_B to be a maximum. If the bubble is to grow, then the wall temperature must be sufficiently high to vapourise the liquid at a pressure P_B.

In order for the bubble to grow

$$T_W > T_{sat} + \frac{dT}{dP}(P_B - P) \tag{2.70}$$

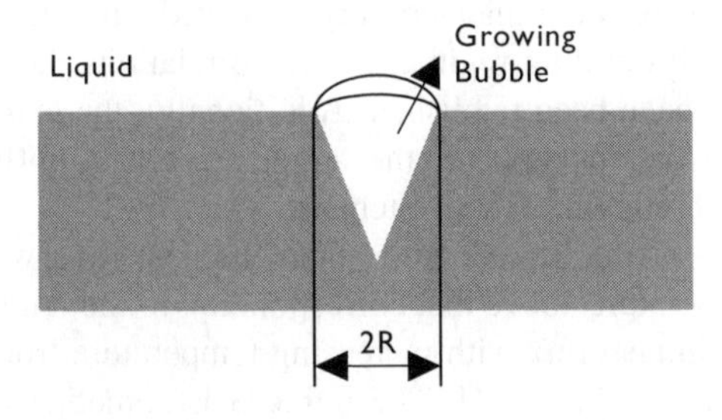

Fig. 2.24 Idealised cavity acting as a nucleation site.

The Clapeyron equation states that the slope of the vapour pressure curve is given by

$$\frac{\mathrm{d}P}{\mathrm{d}T} = \frac{L}{(v_\mathrm{v} - v_\mathrm{l})\,T_\mathrm{sat}} \tag{2.71}$$

if v_v is very much greater than v_l we can simplify this

$$\frac{\mathrm{d}T}{\mathrm{d}P} = \frac{v_\mathrm{v} T_\mathrm{sat}}{L} \tag{2.72}$$

Hence, for the bubble to grow

$$T_\mathrm{W} > T_\mathrm{sat} + \frac{2\sigma_\mathrm{l}}{R}\frac{v_\mathrm{v} T_\mathrm{sat}}{L} = \frac{2\sigma_\mathrm{l}}{R}\frac{T_\mathrm{sat}}{\rho_\mathrm{v} L} \tag{2.73}$$

The radius of the cavity and the superheat, ΔT_sat, at which nucleation from the cavity starts can be related

$$R = \frac{2\sigma_\mathrm{l} T_\mathrm{sat}}{L\rho_\mathrm{v}\Delta T_\mathrm{sat}} \tag{2.74}$$

For water boiling at 1 bar ΔT_sat is commonly of the order of 5 K. Substitution of values for the properties of water gives a value for the smallest active cavity to be approximately 6.5×10^{-6} m radius. This demonstrates that typical active cavities are of the order of 1–10 μm.

Hsu [28] has conducted a more rigorous analysis and showed that

$$\Delta T = \frac{3.06\sigma_\mathrm{l} T_\mathrm{sat}}{\rho_\mathrm{v} L\delta} \tag{2.75}$$

where δ is the thermal boundary layer thickness, as a first approximation this may be taken as the average diameter of cavities on the surface. For typical heat pipe evaporators, this is approximately 25 μm. Taking $\delta = 25$ μm allows the superheat for nucleation to be calculated for various fluids, values of this superheat for various fluids at atmospheric temperature are shown in Table 2.3.

Table 2.3 Superheat for the initiation of nucleate boiling at atmospheric pressure, calculated from equation 2.75

Fluid	Boiling point (K)	Vapour density (kg/m^3)	Latent heat (kJ/kg)	Surface tension (N/m)	ΔT(°C)
Ammonia (NH_3)	239.7	0.3	1350	0.028	2.0
Ethyl alcohol (C_2H_5OH)	338	2.0	840	0.021	0.51
Water	373	0.60	2258	0.059	1.9
Potassium	1047	0.486	1938	0.067	8.9
Sodium	1156	0.306	3913	0.113	26.4
Lithium	1613	0.057	19700	0.26	44.6

Correlation of data in nucleate boiling region. Nucleate boiling is very dependent on the heated surface and factors such as release of absorbed gas, surface roughness, surface oxidation and wettability greatly affect the surface to bulk liquid temperature difference. The nature of the surface may change over a period of time – a process known as conditioning. (The effect of pressure is also important.) For these reasons, reproducibility of results is often difficult. However, a number of authors have proposed correlations, some empirical and some based on physical models. Some of these are listed below. The straight line C–D on Fig. 2.23 and experimental results for nucleate boiling indicate that correlations are likely to be of a form:

$$\dot{q} = a\Delta T_{\text{sat}}^{m}$$

This may be rearranged in terms of a heat transfer coefficient, α,

$$\alpha = \frac{\dot{q}}{\Delta T_{\text{sat}}} = a\Delta T_{\text{sat}}^{m-1} \tag{2.76a}$$

the value of m is generally in the range 3–3.33, corresponding to n being in the range 0.67–0.7.

$$\alpha = b\dot{q}^{\frac{m-1}{m}} = b\dot{q}^{n} \tag{2.76b}$$

Rohsenow [29]. An early nucleate boiling correlation is that due to Rohsenow, following the example of turbulent-forced convective heat transfer correlations, Rohsenow who argued that

$$\text{Nu} = f(\text{Re},\text{Pr})$$

$$\text{Nu} = \frac{\alpha d}{k_{\text{l}}} \qquad \text{Re} = \frac{\rho_{\text{l}} v d}{\mu} \qquad \text{Pr}_{\text{l}} = \frac{\mu_{\text{l}} c_{\text{pl}}}{k_{\text{l}}}$$

If the fluid properties are all those for the liquid, this still left the problem of choosing a suitable velocity and representative length, d.

The velocity may be taken as the velocity with which the liquid flows towards the surface to replace that which has been vapourised

$$v = \frac{\dot{q}}{\rho_{\text{l}} L}$$

and the representative length is given by

$$d = \left[\frac{\sigma_{\text{l}}}{g(\rho_{\text{l}} - \rho_{\text{v}})}\right]^{0.5}$$

The correlation thus produced was:

$$\text{Nu} = \frac{1}{C_{\text{sf}}}\text{Re}^{1-x}\,\text{Pr}^{-y}$$

which is frequently presented in the form

$$\frac{c_{\text{pf}}\Delta T_{\text{sat}}}{L} = C_{\text{sf}}\left[\frac{\dot{q}}{\mu_{\text{l}}L}\left(\frac{\sigma_{\text{l}}}{g(\rho_{\text{l}}-\rho_{\text{v}})}\right)^{0.5}\right]^{x}\left[\frac{\mu_{\text{l}}c_{\text{pl}}}{k_{\text{l}}}\right]^{l+y} \tag{2.77}$$

For most fluids, the recommended values of the exponents were $x = 0.33$, $y = 0.7$.
This correlation then corresponds to

$$\dot{q} = [\text{constant depending upon fluid properties and surface}] \times \Delta T^3$$

The value of the constant C_{sf} depends upon the fluid and the surface and typical values range between 0.0025 and 0.015. Since, for a given value of ΔT_{sat} the heat flux is proportional to C_{sf}^3, the correlation is very sensitive to selection of the correct value.

It is arguable that the complexity of the correlation is not warranted because of the need for this factor.

Some values of C_{sf} for use in equation 2.77 are given in Table 2.4.

Mostinski [30]

$$\alpha = 0.106P_{\text{cr}}^{0.69}\left(1.8P_{\text{r}}^{0.17} + 4P_{\text{r}}^{1.2} + 10P_{\text{r}}^{10}\right)\dot{q}^{0.7} \tag{2.78}$$

Cooper [31]

$$\alpha = 55P_{\text{r}}^{(0.12-0.2\log\varepsilon)}(-\log P_{\text{r}})^{-0.55}M^{-0.5}\dot{q}^{0.67} \tag{2.79a}$$

where ε is the surface roughness in microns. Typically, a value of 1 may be used, thus simplifying the equation.

$$\alpha = 55P_{\text{r}}^{0.12}(-\log P_{\text{r}})^{-0.55}M^{-0.5}\dot{\text{q}}^{0.67} \tag{2.79b}$$

Table 2.4 Values of C_{sf} for use in equation 2.77 [29]

Fluid	Surface	C_{sf}
Water	Nickel	0.006
Water	Platinum	0.013
Water	Copper	0.013
Water	Brass	0.006
Carbon tetrachloride	Copper	0.013
Benzene	Chromium	0.010
n-Penthane	Chromium	0.015
Ethanol	Chromium	0.0027
Isopropanol	Copper	0.0025
n-Butanol	Copper	0.0030

P_r in equations 2.78 and 2.79 is the reduced pressure, defined as P/P_{cr} and P_{cr} is the critical pressure of the fluid.

Mostinski and Cooper are both dimensional equations, therefore the units must be consistent with the constants given. For the forms quoted here pressures are in bar and heat flux in W/m^2, giving heat transfer coefficients in W/m^2K.

Comprehensive references on liquid metal boiling are Subbotin [32] and Dwyer [33].

Burn out Correlations. As for nucleate boiling the critical peak flux or boiling crisis flux is also very dependent on surface conditions. For water at atmospheric pressure, the peak flux lies in the range 950–1300 kW/m^2 and is between three and eight times the value obtained for organic liquids. The corresponding temperature difference for both water and organics is between 20 °C and 50 °C. Liquid metals have the advantage of low viscosity and high thermal conductivity and the alkali metals in the pressure range 0.1–10 bar give peak flux values of 100–300 kW/m^2 with a corresponding temperature difference of around 5 °C.

A number of authors have provided relationships to enable the critical heat flux $\dot{q}_{cr}$ to be predicted. One of these was developed by Rohsenow and Griffith [34] who obtained the following expression:

$$\dot{q}_{cr} = 0.012 L\rho_v \left[\frac{\rho_l - \rho_v}{\rho_v}\right]^{0.6} \tag{2.80}$$

Another correlation due to Caswell and Balzhieser [35] applies to both metals and nonmetals.

$$\dot{q}_{cr} = 1.02 \times 10^{-6} \frac{L^2 \rho_v k_l}{c_p^{\gamma}} \left[\frac{\rho_l - \rho_v}{\rho_v}\right]^{0.65} \mathrm{Pr}^{0.71} \tag{2.81}$$

2.3.7.4 Boiling from wicked surfaces

There is a considerable literature on boiling from wicked surfaces. The work reported includes measurements on plane surfaces and on tubes, the heated surfaces can be horizontal or vertical and either totally immersed in the liquid or evaporating in the heat pipe mode. Water, organics and liquid metals have all been investigated. The effect of the wick is to further complicate the boiling process, since in addition to the factors referred to in the section on boiling from smooth surfaces, the wick provides sites for additional nucleation and significantly modifies the movement of the liquid and vapour phases to and from the heated surface.

At low values of heat flux, the heat transfer is primarily by conduction through the flooded wick. This is demonstrated in the results of Philips and Hinderman [36] who carried out experiments using a wick of 220.5 nickel foam 0.14 cm thick and one layer of stainless steel screen attached to a horizontal plate. Distilled water was used as the working fluid. Their results are shown in Fig. 2.25. The solid curve is the theoretical curve for conduction through a layer of water of the wick thickness.

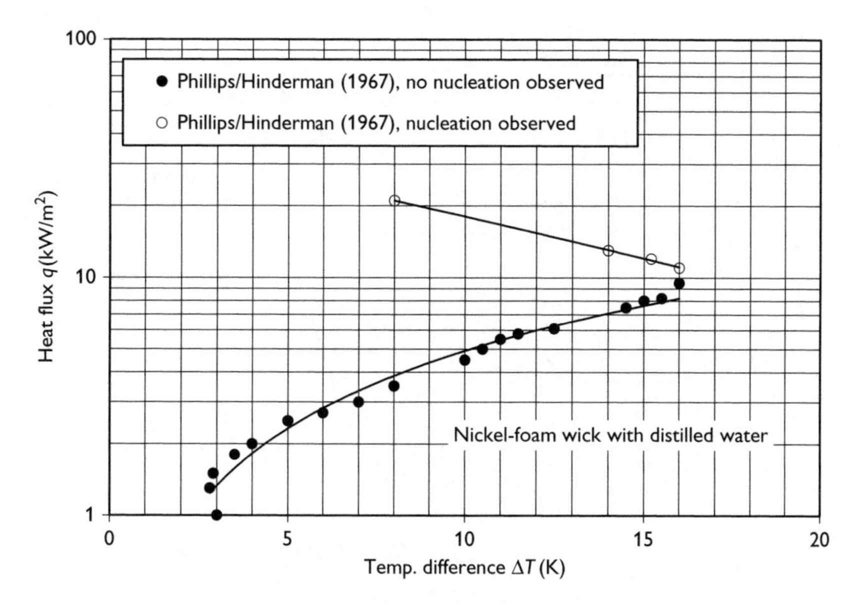

Fig. 2.25 Heat transfer from a submerged wick data from [36].

At higher values of heat flux nucleation occurs. Ferrell and Aileavitch [37] studied the heat transfer from a horizontal surface covered with beds of Monel beads. Results are reported on 30–40 and 40–50 mesh using water at atmospheric pressure as the working fluid and a total immersion to a depth of 7.5 cm. The bed depth ranged from 3 to 25 mm. They concluded that the heat transfer mechanism was conduction through a saturated wick liquid matrix to a vapour interface located in the first layer of beads. Agreement was good between the theoretical predictions of this model and the experimental results. No boiling was observed. Figure 2.26 shows these results together with the experimental values obtained for the smooth horizontal surface. It is seen that the latter agrees well with the Rohsenow correlation but that for low values of temperature difference the heat flux for the wicked surface is much greater than for the smooth surface. This effect has been observed by other workers, for example, Corman and Welmet [38], the curves cross over at higher values of heat flux, probably because of increased difficulty experienced by the vapour in leaving the surface.

This concept has been developed further by Brautsch and Kew [39] who examined boiling from a bare surface and the surface covered with mesh wicks. Samples of these results are shown in Fig. 2.27.

Brautsch [39–41] correlated the heat transfer from the mesh-covered surface by comparing it with a reference heat transfer coefficient obtained for boiling of water and R141b from a smooth plate. Having observed the boiling process using high-speed video, it was noted that the mesh acted to enhance nucleation due to

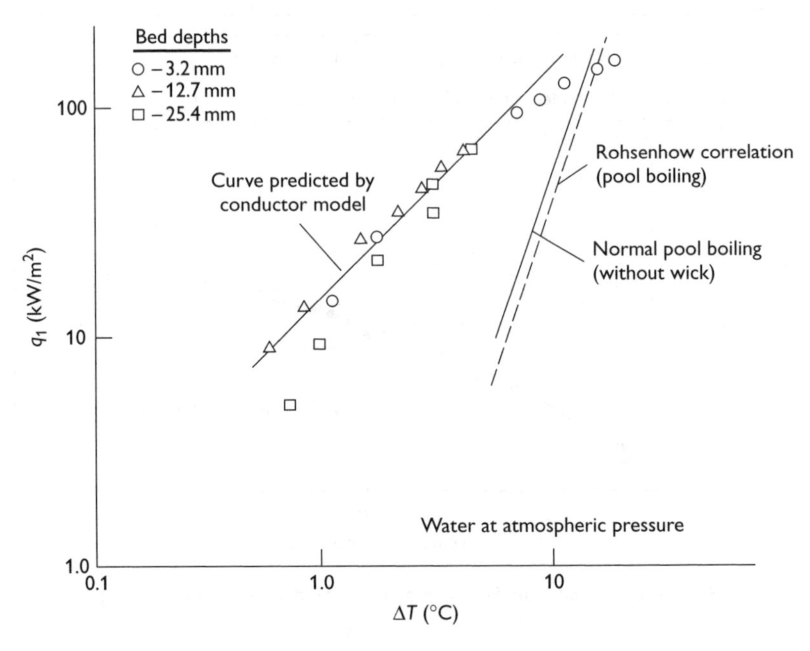

Fig. 2.26 Evaporation from a submerged wick compared with boiling from a smooth surface [37].

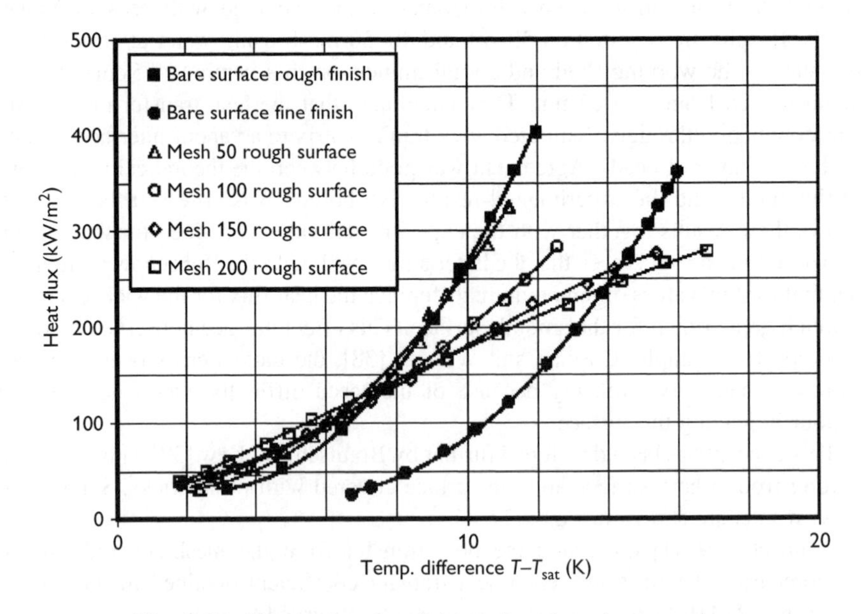

Fig. 2.27 Effects of surface finish and mesh on heat transfer.

the provision of additional nucleation sites, but the presence of the mesh impeded departure of bubbles from the surface. This yielded an equation of the form

$$\frac{\alpha}{\alpha_0} = E(1 - B) \tag{2.82}$$

where E and B were defined as enhancement and blocking factors, respectively, and correlated by

$$E = \left[1 + \left(\frac{d_b}{d_{m,i}}\right)^m \left(\frac{R_a}{R_{a_0}}\right)^n \frac{q_0}{q}\right]^r \tag{2.83}$$

and

$$B = \left(\frac{d_b}{d_{m,i}}\right)^u \left(\frac{R_a}{R_{a_0}}\right)^v K^{-1} \left(\left(\frac{q}{q_0}\right)^2 + \frac{q}{q_0}\right) \tag{2.84}$$

with

$$d_b = 0.851\beta_0 \sqrt{\frac{2\sigma}{g(\rho_l - \rho_v)}} \tag{2.85}$$

where β_o is the contact angle (rad), and the exponents m, n, r, u, v and the fitting factor K are summarised in Table 2.5.

The Cooper correlation (equation 2.79) was found to predict the reference heat transfer coefficient, α_0, well.

This correlation showed good agreement with experimental data from Asakavicius et al. [42] who tested multiple brass and stainless steel screen layers in Freon-113, ethyl alcohol and water; Liu et al. [43], who published experimental results for single layers of mesh 16 and mesh 50, both for the working fluids methanol and HFE-7100 and the results of Tse et al. [44] for single layers of mesh 24 and mesh 50 with distilled water as the working fluid.

Several researchers have worked on evaporation from wicks and selected results are summarised in Fig. 2.29. Abhat and Seban [45] reported on heat transfer

Table 2.5 Relevant material parameters, exponents and the fitting factor used in the equations

	All surface finishes
m	1
n	0.69
r	1.8
u	1.2
v	0.12
K	2.0E+07

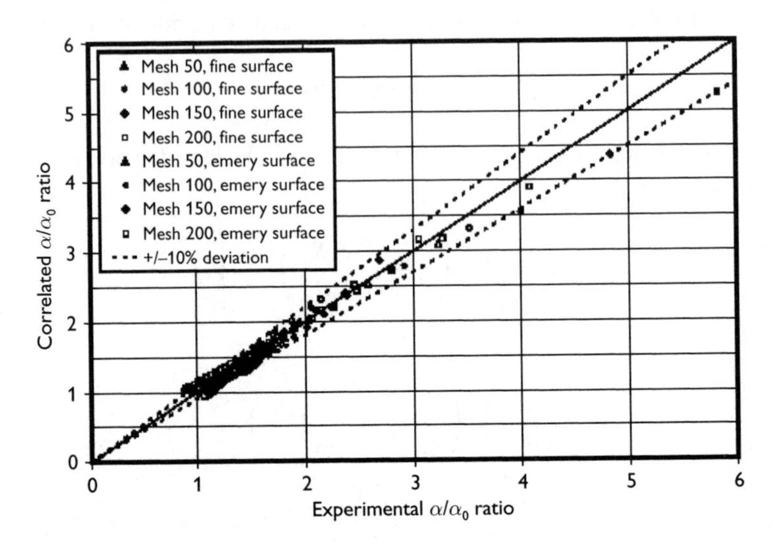

Fig. 2.28 Comparison between correlated and experimental heat transfer coefficient ratio for different mesh sizes on different heater surface roughnesses [39].

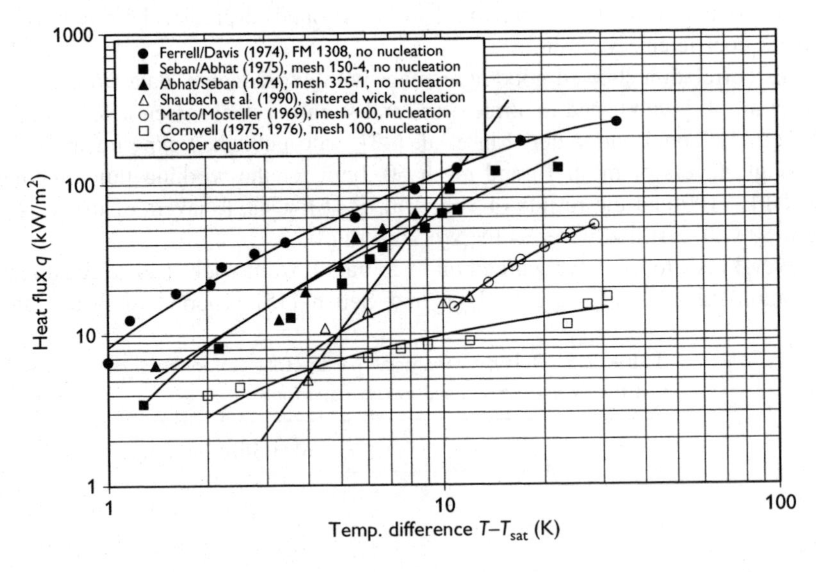

Fig. 2.29 Results from researchers measuring boiling and convection in the wick material, working fluid distilled water [40]. Data from Refs [45–50].

measurements on vertical tubes using water, ethanol and acetone as the working fluids. In this series of experiments, results were given for smooth surfaces, immersed wicks and evaporating wicks. The authors concluded that up to heat fluxes of 15 W/cm^2 the heat transfer coefficient for a screen or felt wicked tube was similar to that of the bare tube and also not very different from that for the evaporating surface.

Marto and Mosteller [46] used a horizontal tube surrounded by four layers of 100-mesh stainless steel screen and water as the working fluid. The outer container was of glass, so it was possible to observe the evaporating wick surface. They measured the superheat as a function of heat flux. They found that the critical radius was 0.013 mm compared to the effective capillary radius of 0.6 mm.

There is some evidence that the critical heat flux for wicked surfaces may be greater than that for smooth surfaces, for example, a report by Costello and Frea [51] suggests that the critical heat flux could be increased by about 20% over that for a smooth tube.

Reiss et al. [52] report very high values of critical heat flux for sodium in grooved heat pipes. They report values from 2–10 times the critical values reported by Ballzhieser et al. [53] in the temperature region around 550 °C. The authors observed the grooves as dark stripes on the outer side of the heat pipe and concluded that evaporation was from the grooves only. Their results are plotted in Fig. 2.30

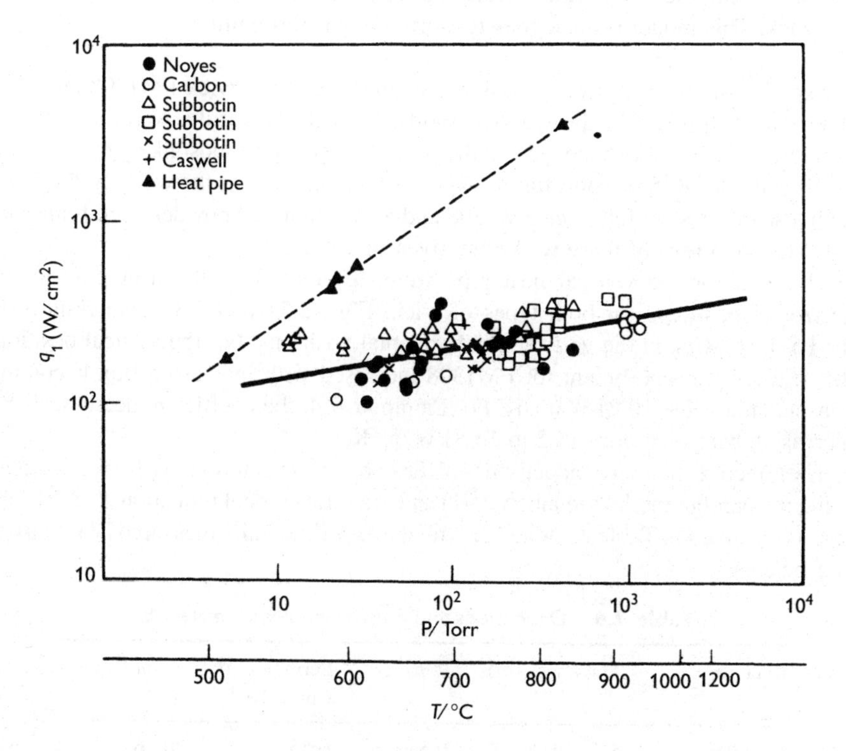

Fig. 2.30 Critical flux for sodium pool boiling compared to critical flux in grooved sodium heat pipes [52].

and the heat flux is calculated on the assumption that evaporation is from the groove only. Moss and Kelley [54] constructed a planar evaporator using a stainless steel wick 1/4 in. thick and water as the working fluid. The authors employed a neutron radiographic technique to measure the water thickness. In order to explain their results they proposed two models, in the first it was assumed that vapourisation occurs at the liquid–vapour interface, in the second model a vapour blanket was assumed to occur at the base of the wick.

Both the Moss and Kelley models can be used to explain most of the published data. These models are discussed in a paper by Ferrell et al. [47] who describe experiments aimed at differentiating between them. The two models are described by the authors as

(i) A layer of liquid filled wick adjacent to the heated surface with conduction across this layer and liquid vapourisation at the edge of the layer. In this model, liquid must be drawn into the surface adjacent to the capillary forces.
(ii) A thin layer of vapour-filled wick adjacent to the heated surface with conduction across this layer to the saturated liquid within the wick. In this model, the liquid is vapourised at the edge of this layer and the resulting vapour finds its way out of the wick along the surface and through large pore size passages in the wick. This model is analogous to conventional film boiling.

Experiments were carried out using a stainless steel felt wick FM1308 and both water and potassium as the working fluids. Though the results were not entirely conclusive, the authors believed that the second model is the most likely mechanism. Davis and Ferrell [55] report further work using potassium as the working fluid and both a stainless steel felt wick FM1308 and a steel-sintered powder wick Lamipore 7.4. The properties of these wicks are given in Table 2.6.

The data for the vertical heat pipe were different from the heat pipe in the horizontal positions for both types of wick. Figs. 2.31 and 2.32 give results for the FM1308 wick in the vertical and horizontal positions. For the vertical position, the heat transfer coefficient for FM1308 increased with increasing flux becoming constant at a value 10.2 kW/m^2K. For Lamipore 7.4, the coefficient decreases with increasing heat flux from 18.2 to 14.8 kW/m^2K.

The effective thermal conductivities of the two wick structures have been calculated using the parallel model (equation 2.91) and the series model (equation 2.92). These results are given in Table 2.7 together with the experimentally measured heat transfer coefficients.

Table 2.6 Dimensions and properties of wick materials

Material	Thickness (m)	Porosity	Permeability ($m^2 \times 10^{10}$)	Capillary rise (m)
FM1308	3.2×10^{-3}	0.58	0.55	0.26
Lamipore 7.4	1.5×10^{-3}	0.61	0.48	0.35

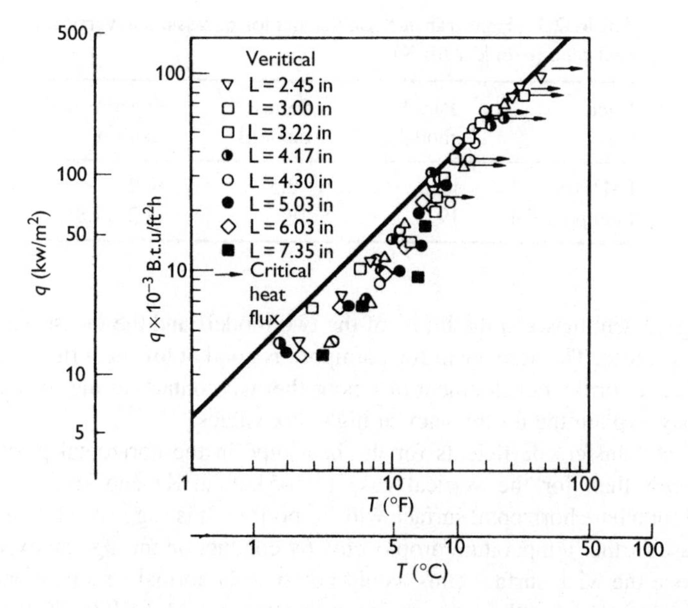

Fig. 2.31 Potassium boiling from vertical wick FM 1308 [55] (*L* is vertical height above pool).

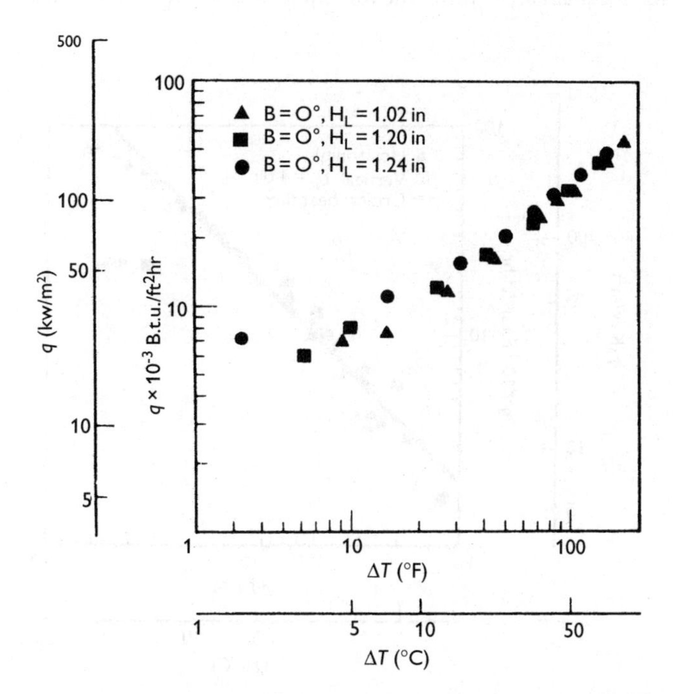

Fig. 2.32 Potassium boiling from horizontal wick FM 1308 [55].

Table 2.7 Heat transfer coefficients for potassium in vertical heat pipe (α in kW/m^2K)

Wick	Parallel model k_w	Series model k_w	Experimental result α
FM1308	9.6	8.5	10.2
Lamipore 7.4	19	18	18.2–14.8

The agreement between the limits of the two models and the measured value for FM1309 is close. The agreement for Lamipore is good at low heat fluxes, and there was evidence for the development of a poor thermal contact during the experiment which may explain the discrepancy at high flux values.

The heat transfer coefficients for the heat pipe in the horizontal position were much lower than for the vertical case (1.1–$5\,kW/m^2K$) and similar to results obtained for a bare horizontal surface with no boiling. It is suggested that in the horizontal case further temperature drop occurs by conduction through an excess liquid layer above the wick surface (this would not arise in normal heat pipe operation).

In the case of water the heat transfer coefficient is $11.3\,kW/m^2K$ for both the vertical and horizontal cases, as shown in Fig. 2.33. The authors conclude that the heat transfer mechanism is different for liquid metals from the mechanism which

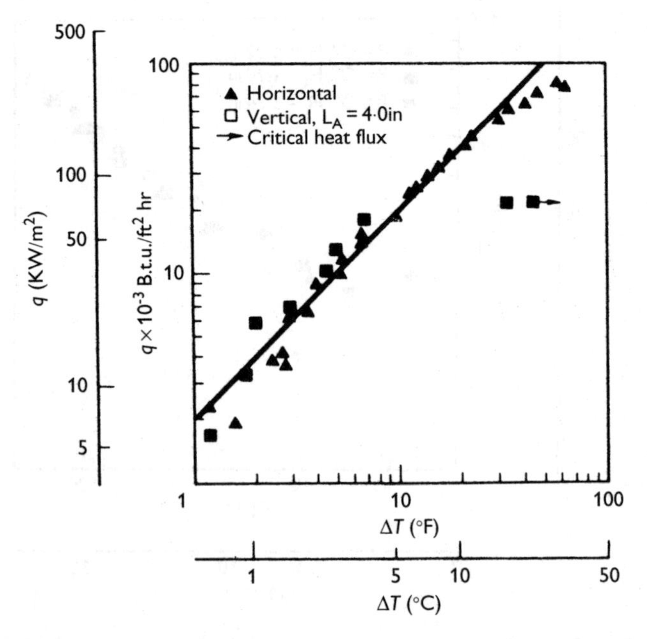

Fig. 2.33 Water boiling from FM1308 wick. Data for wick vertical and horizontal [55].

applies to water and other nonmetallic fluids. In the case of nonmetallic fluids, they suggest that the vapourisation process occurs within the porous media. This vapourisation is probably initiated by inert gas trapped in the porous media or nucleation on active sites on the heated surface. Once initiated the vapour phase spreads out to form a stable layer. The data for water show that high fluxes are observed for quite low values of temperature difference, whereas much larger values of temperature difference might be expected if the mechanism is one of conventional film boiling. The mechanism may be of nucleate boiling at activation sites on the heating surface and the wick adjacent to it. This was confirmed by the observations of Brautsch [40].

It is difficult to initiate bubbles in liquid metals and the experimental results of Ref. [40] strongly suggested that for these fluids the wick is saturated and that vapourisation occurs at the liquid surface on the outside of the wick. Hence, for liquid metals the heat transfer coefficient can be accurately predicted from equations 2.91 and 2.92.

One limit to the radial heat transfer to the evaporating fluid will be set by the 'wicking limit', that is, when the capillary forces are unable to feed sufficient liquid to the evaporator.

The limiting heat flux $\dot{q}_{\text{crit}}$ will be given by the expression

$$\dot{q}_{\text{crit}} = \frac{\dot{m}L}{\text{area of evaporator}}$$

where the mass flow in the wick is related to the pressure head, ΔP, by an expression such as equation 2.32.

Ferrell et al. [56] have derived a relationship for a planar surface with a homogeneous wick.

$$\dot{q}_{\text{crit}} = \frac{g\left(h_{\text{co}}\rho_{\text{l}}\dfrac{\sigma_{\text{l}}}{\sigma_{\text{lo}}} - \rho_{\text{l}} l \sin\phi\right)}{\dfrac{l_{\text{e}}\mu_{\text{l}}}{L\rho_{\text{l}}kd}\left[\dfrac{l_{\text{e}}}{2} + l_{\text{a}}\right]} \tag{2.86}$$

where h_{co} is the capillary height of the fluid in the wick measured at a reference temperature, σ_{lo}, ρ_{lo} are the fluid surface tension and density measured at the same reference temperature, σ_{l}, ρ_{l}, μ_{l} the fluid surface tension, density and viscosity measured at the operating temperature. All other symbols have their usual significance.

Ferrell and Davis [47] extended their equation by including a correction for thermal expansion of the wick.

$$\dot{q}_{\text{crit}} = \frac{g\left(h_{\text{co}}\rho_{\text{l}}\dfrac{\sigma_{\text{l}}}{\sigma_{\text{lo}}} - \rho_{\text{l}} l \sin\phi\,(1 + \alpha_{\text{t}}\Delta T)\right)}{\dfrac{l_{e}\mu_{l}}{L\rho_{\text{l}}kd\,(1 + \alpha_{\text{t}}\Delta T)}\left[\dfrac{l_{\text{e}}}{2} + l_{\text{a}}\right]} \tag{2.87}$$

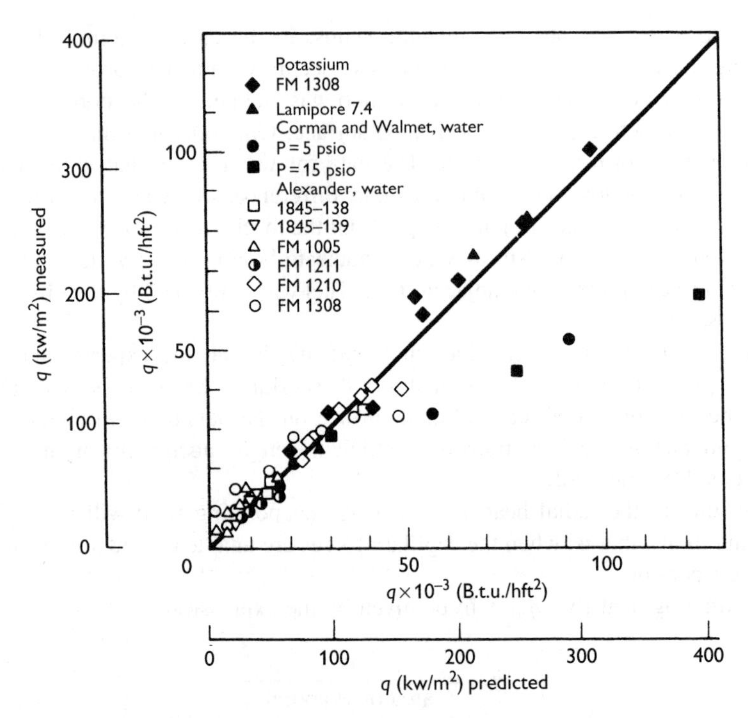

Fig. 2.34 Comparison of measured and predicted burnout for water and potassium [55].

where α_t is the coefficient of linear expansion of the wick and ΔT the difference between the operating and reference temperature.

Figure 2.34 shows a comparison of measured values of $\dot{q}_{crit}$ against predicted values from equation 2.87 for both water and potassium. The equation successfully predicts heat flux limits for potassium up to a value of $315\,kW/m^2$. It is also in good agreement for water upto $130\,kW/m^2$. Above this value, the experimental values fall below the values predicted by equation 2.87 showing that another mechanism is limiting the flux. The limiting factor is probably due to difficulty experienced by water vapour in leaving the heat surface through the wick. The reduction in heat flux below the predicted value for water further supports the view that for nonmetallic fluids vapourisation occurs within the wick. More recent work on radial heat flux measurements and observation of vapour/liquid proportions and nucleation within wicks has resulted in more data showing that nucleation is not detrimental to heat pipe performance.

2.3.7.5 Liquid–vapour interface temperature drop

Consider a liquid surface, there will be a continuous flux of molecules leaving the surface by evaporation. If the liquid is in equilibrium with the vapour above its surface, an equal flux of molecules will return to the liquid from the vapour and there will be no net loss or gain of mass. However, when a surface is losing mass by evaporation, clearly the vapour pressure and hence temperature of the vapour above the surface must be less than the equilibrium value. In the same way for net condensation to occur the vapour pressure and temperature must be higher than the equilibrium value.

The magnitude of the temperature drop can be estimated as follows:

First consider the vapour near the interface. The average velocity V_{av} in a vapour at temperature T_v and having molecular mass M is given by kinetic theory as

$$V_{av} = \sqrt{\frac{8k_B T_v}{\pi M}} \tag{2.88}$$

where k_B is Boltzmann's constant.

The average flow of molecules in any given direction is:

$$\frac{nV_{av}}{4} \text{ per unit area}$$

and the corresponding flow of heat per unit area is

$$\frac{MLnV_{av}}{4}$$

where n is the number of molecules per unit volume and L the latent heat.

For a perfect gas

$$P_v = nk_B T_v$$

hence the heat flux to the liquid surface from the vapour $= P_v L\sqrt{\frac{M}{2\pi k T_v}}$

The heat flux away from the liquid surface is given by $P_l L\sqrt{\frac{M}{2\pi k T_l}}$

Setting $T_v = T_l = T_s$ allows evaluation of the net heat flux across the surface

$$\begin{aligned} \dot{q} &= (P_l - P_v)L\sqrt{\frac{M}{2\pi k_B T_s}} \\ &= \frac{(P_l - P_v)L}{\sqrt{2\pi RT}} \end{aligned} \tag{2.89}$$

By using the Clapeyron equation,

$$\frac{dP}{dT} = \frac{L}{(v_v - v_l)T_{sat}}$$

and assuming that the vapour behaves as a perfect gas and the liquid volume is negligible

$$\frac{\Delta P}{\Delta T} = \frac{PL}{RT^2}$$

equation 2.90 may be written

$$\dot{q} = \frac{\Delta T L^2 P}{(2\pi R T_s)^{0.5}} \frac{1}{R T_s^2} \tag{2.90}$$

thus permitting calculation of the interfacial temperature change in the evaporator and condenser.

Values of $\dot{q}/P_l - P_v$ for liquids near their normal boiling points are given in Table 2.8.

2.3.7.6 Wick thermal conductivity

The effective conductivity of the wick saturated with the working fluid is required for calculating the thermal resistance of the wick at the condenser region and also under conditions of evaporation when the evaporation is from the surface for the evaporator region. Two models are used in the literature (see also Chapter 3).

(i) *Parallel case.* Here it is assumed that the wick and working fluid are effectively in parallel.

If K_l is the thermal conductivity of the working fluid and k_s is the thermal conductivity of the wick material and

$$\varepsilon = \text{Voidage fraction} = \frac{\text{Volume of working fluid in wick}}{\text{Total volume of wick}}$$

$$k_w = (1-\varepsilon)\,k_s + \varepsilon k_l \tag{2.91}$$

(ii) *Series case.* If the two materials are assumed to be in series

$$k_w = \frac{1}{(1-\varepsilon)/k_s + \dfrac{\varepsilon}{k_l}} \tag{2.92}$$

Additionally, convection currents in the wick will tend to increase the effective thermal conductivity.

Table 2.8 Interfacial heat flux as a function of pressure difference [57]

Fluid	T_b(K)	$q/P_l - P_v$(kW/cm^2 atm)
Lithium	1613	55
Zinc	1180	18
Sodium	1156	39
Water	373	21.5
Ethanol	351	13.5
Ammonia	238	15.2

2.3.7.7 Heat transfer in the condenser

Vapour will condense on the liquid surface in the condenser, the mechanism is similar to that discussed in Section 2.3.7.5 on the mechanism of surface evaporation and there will be a small temperature drop and hence thermal resistance. Further temperature drops will occur in the liquid film and in the saturated wick and in the heat pipe envelope.

Condensation can occur in two forms, either by the condensing vapour forming a continuous liquid surface or by forming a large number of drops. The former, film condensation occurs in most practical applications, including heat pipes and will be discussed here. Condensation is seriously affected by the presence of a noncondensable gas. However, in the heat pipe, vapour pumping will cause such gas to be concentrated at the end of the condenser. This part of the condenser will be effectively shut off and this effect is the basis of the gas-buffered heat pipe. The temperature drop through the saturated wick may be treated in the same manner as at the evaporator.

Film condensation may be analysed using Nusselt theory. The first analysis of film condensation was due to Nusselt and given in standard text books, e.g. [58]. The theory considers condensation onto a vertical surface and the resulting condensed liquid film flows down the surface under the action of gravity and is assumed to be laminar. Viscous shear between the vapour and liquid phases is neglected. The mass flow increases with distance from the top and the flow profile is shown in Fig. 2.35.

The average heat transfer coefficient, α, over a distance x from the top is given by:

$$\alpha = 0.943\left[\frac{L\rho_1^2 g k_1^3}{x\mu_1 (T_s - T_w)}\right]^{0.25} \tag{2.93}$$

where $(T_s - T_w)$ is the temperature drop across the liquid film.

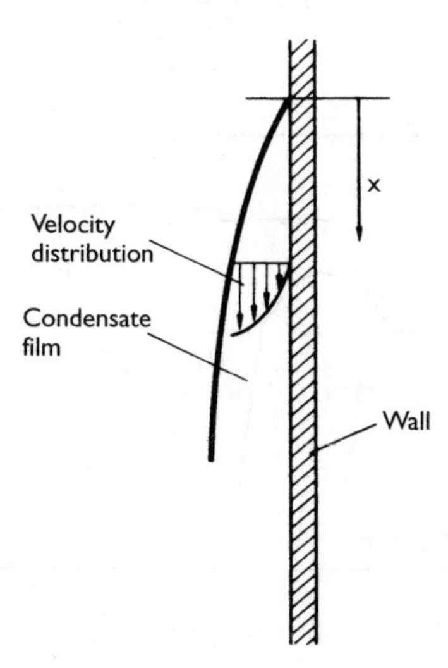

Fig. 2.35 Film condensation on a vertical surface.

2.4 APPLICATION OF THEORY TO HEAT PIPES AND THERMOSYPHONS

2.4.1 Wicked heat pipes

2.4.1.1 The merit number

It will be shown, with reference to the capillary limit, that if vapour pressure loss and gravitational head can be neglected then the properties of the working fluid which determine the maximum heat transport can be combined to form a figure of merit, M.

$$M = \frac{\rho_l \sigma_l L}{\mu_l} \tag{2.94}$$

The way in which M varies with temperature for a number of working fluids is shown in Fig. 2.36.

2.4.1.2 Operating limits

Upper limits to the heat transport capability of a heat pipe may be set by one or more factors. These limits were illustrated in Fig. 2.1.

Viscous, or vapour pressure, limit. At low temperature, viscous forces are dominant in the vapour flow down the pipe. Busse has shown that the axial heat flux increases as the pressure in the condenser is reduced, the maximum heat flux occurring when the pressure is reduced to zero. Busse carried out a two-dimensional analysis, finding that the radial velocity component had a significant effect, he derived the following equation.

$$\dot{q} = \frac{r_v L \rho_v P_v}{16 \mu_v l_{eff}} \tag{2.95}$$

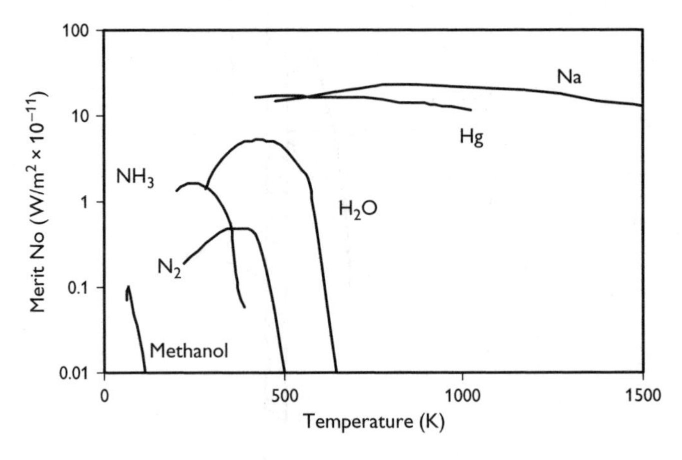

Fig. 2.36 Merit Number of selected working fluids.

where P_v and ρ_v refer to the evaporator end of the pipe. This equation agrees well with the published data [59]. The vapour pressure limit is described well in the ESDU Data Item on capillary-driven heat pipes [60]. It is stated that when the vapour pressure is very low, the minimum value must occur at the closed end of the condenser section, the vapour pressure drop can be constrained by this very low – effectively zero – pressure, and the low vapour pressure existing at the closed end of the evaporator section. The maximum possible pressure drop is therefore equal to the vapour pressure in the evaporator. Because the vapour pressure difference naturally increases as the heat transported by the heat pipe rises, the constraint on this pressure difference thus necessitates that $\dot{q}$ is limited. The limit is generally only of importance in some units during start-up. A criterion for avoidance of this limit is given in reference [60] as:

$$\frac{\Delta P_v}{P_v} < 0.1$$

Sonic limit. At a somewhat higher temperature choking at the evaporator exit may limit the total power handling capability of the pipe. This problem was discussed in Section 2.3.5. The sonic limit is given by

$$\dot{q} = 0.474L\,(\rho_v P_v)^{0.5} \tag{2.96}$$

There is good agreement between this formula and experimental results. Fig. 2.37 due to Busse [59] plots the temperature at which the sonic limit is equal to the viscous limit as a function of l_{eff}/d^2 for some alkali metals, where $d = 2r_v$, the vapour space diameter.

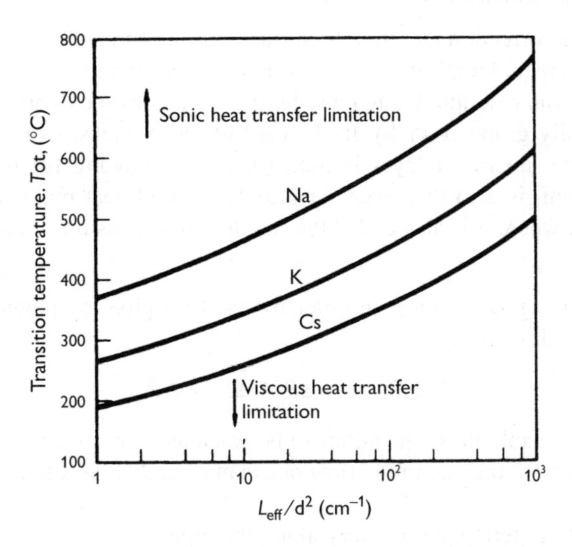

Fig. 2.37 Transition from viscous to sonic limit [59].

Entrainment limit. This was already discussed in Section 2.3.6. Equation 2.65 gave the entrainment limiting flux

$$\dot{q} = \sqrt{\frac{2\pi\rho_v L^2 \sigma_l}{z}} \tag{2.65}$$

Kemmes experiments suggest a rough correlation of this failure mode with the centre-to-centre wire spacing for Z, and that a very fine screen will suppress entrainment.

A number of workers have presented data on entrainment limits for specific cases. Tien and Chung [61] present equations which, it is claimed, enable the user to predict the maximum heat transfer rate due to the entrainment limitation. Some experimental data suggest that the equations give accurate results. Data are presented for 'gravity-assisted wickless heat pipes' (thermosyphons), gravity-assisted wicked heat pipes, horizontal heat pipes, grooved heat pipes and rotating heat pipes. Experiments with gravity-assisted heat pipes employing simple wick structures were carried out by Coyne Prenger and Kemme [62] in order to arrive at a correlation that could be used to predict the entrainment limits for all the data investigated.

This work led to the conclusions that the entrainment limits in the cases studied were vapour-dominated and were essentially independent of the fluid inventory. The entrainment limits were successfully correlated with a physical model based on a critical Weber number (see equation 2.62), the expression being

$$\dot{Q} = \sqrt{\frac{2\pi E_t}{\alpha_v} \frac{\delta}{\delta^*}} \tag{2.97}$$

where E_t is the dimensionless entrainment parameter, which equals $\frac{\sigma_l}{\rho_l L\delta}$, α_v the velocity profile correction factor, δ the surface depth, δ^* the reference surface depth.

The characteristic length in the Weber number formulation was the depth of the wick structure (an example being cited by the authors as the depth of grooves in a circumferentially grooved wick). In the case of mesh wicks, one half of the wire diameter of the innermost layer is used (this also allowing for mesh composite wicks). The limit is also discussed for gravity-assisted heat pipes by Nguyen-Chi and Groll [78] where it is also called the flooding limit – as the condenser becomes flooded.

Capillary limit (wicking limit). In order for the heat pipe to operate, equation 2.2 must be satisfied, namely

$$\Delta P_{c,max} \geq \Delta P_l + \Delta P_v + \Delta P_g \tag{2.2}$$

Expressions to enable these quantities to be calculated are given in Section 2.3.

An expression for the maximum flow rate m may readily be obtained if we assume

(i) the liquid properties do not vary along the pipe,
(ii) the wick is uniform along the pipe,

(iii) the pressure drop due to vapour flow can be neglected.

$$\dot{m}_{\max} = \left[\frac{\rho_l \sigma_l}{\mu_l}\right]\left[\frac{KA}{l}\right]\left[\frac{2}{r_e} - \frac{\rho_l g l}{\sigma_l}\sin\phi\right] \tag{2.98}$$

and the corresponding heat transport $\dot{Q}_{\max} = \dot{m}_{\max} L$ is given by

$$\dot{Q}_{\max} = \dot{m}_{\max} L = \dot{m}_{\max} = \left[\frac{\rho_l \sigma_l L}{\mu_l}\right]\left[\frac{KA}{l}\right]\left[\frac{2}{r_e} - \frac{\rho_l g l}{\sigma_l}\sin\phi\right] \tag{2.99}$$

The group, defined in equation 2.94 as

$$\frac{\rho_l \sigma_l L}{\mu_l}$$

depends only on the properties of the working fluid and is known as the figure of merit, or Merit No., M, plotted for several fluids on Fig. 2.36.

2.4.1.3 Burnout

Burnout will occur at the evaporator at high radial fluxes. A similar limit on peak radial flux will also occur at the condenser. These limits are discussed in Section 2.3.7. At the evaporator, equation 2.87 gives a limit that must be satisfied for a homogeneous wick. This equation, which represents a wicking limit, is shown to apply to potassium up to $315\,\text{kW/m}^2$ and probably is applicable at higher values for potassium and other liquid metals. For water and other nonmetallics, vapour production in the wick becomes important at lower flux densities ($130\,\text{kW/m}^2$ for water) and there are no simple correlations. For these fluids, experimental data in Table 3.3 should be used as an indication of flux densities that can be achieved. Tien [63] has summarised the problem of nucleation within the wick and the arguments for possible performance enhancement in the presence of nucleation. Because of phase equilibrium at the interface, liquid within the wick at the evaporator is always superheated to some degree, but it is difficult to specify the degree of superheat needed to initiate bubbles in the wick. While boiling is often taken as a limiting feature in wicks, possibly upsetting the capillary action, Cornwell and Nair [49] found that nucleation reduces the radial temperature difference and increases heat pipe conductance. One factor in support of this is the increased, thermal conductivity of liquid-saturated wicks in the evaporator section, compared with that in the condenser.

Tien believed that the boiling limit will become effective only when the bubbles generated within the wick become trapped there, forming a vapour blanket. This was confirmed by Brautsch [40]. Thus, some enhancement of radial heat transfer coefficient in the evaporator section can be obtained by nucleate boiling effects, however, excessive nucleate boiling disrupts the capillary action and reduces the effective area for liquid flow. It is useful to compare the measurements of radial heat flux as a function of the degree of superheat (in terms of the difference between wall temperature T_w and saturation temperature T_{sat}) for a number of surfaces and wick forms which have been the subject of recent studies. These are shown in Table 2.9. Abhat and Seban [45], in discussing their results as shown in Table 2.9, stated that for every pool depth, there was a departure from the performance recorded involving a slow increase in the evaporator temperature. In order to guarantee indefinite operation of the heat pipe without dryout, the authors therefore suggested that the operating flux should be considerably less than that shown in the table. They also found that their measurements of radial flux as a function of $T_w - T_{sat}$ were similar, regardless of whether the surface was flat, contained mesh or a felt. This is partly borne out by the results of Wiebe and Judd [65], also given in Table 2.9. Costello and Frea, however, disputed this [51].

The work of Cornwell and Nair [49] gives results similar to those of Abhat and Seban [45]. Some of Cornwell and Nair's results are presented in Fig. 2.39 and compared with theory. Cornwell and Nair found that some indication of the $\dot{q}/\Delta T$ curve for boiling within a wick could be obtained by assuming that boiling occurs only on the liquid-covered area of the heating surface (measured from observations) and that $\dot{q}$ based on this area may be expressed by a nucleate boiling type correlation

$$\dot{q} = C\Delta T^a$$

where C is a constant and a is in the range 2–6.

The results of Brautsch [40] followed the same trends as those of Cornwell and Nair. The enhancement due to a mesh wick at low heat fluxes, and the reduction

Table 2.9 Measured radial heat fluxes in wicks

Reference	Wick	Working fluid	Superheat (°C)	Flux (W/cm^2)
Wiebe and Judd [65]	Horizontal flat surface	Water	3	0.4
			6	1.2
			11	8
			17	20
Abhat and Seban [45]	Meshes and felts	Water	5	1.6
			6	5
			11	12
			17	20
Cornwell and Nair [49]	Foam	Water	5	2
			10	10
			20	18
	Mesh (100)	Water	5	7
			10	9
			20	13
Abhat and Nguyenchi [64]	Mesh*	Water	6	1
			10	8

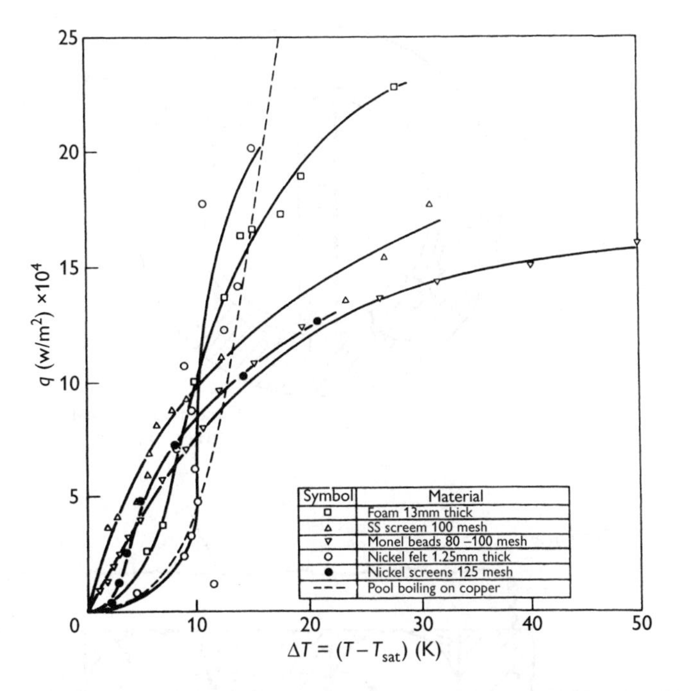

Fig. 2.38 Variation of heat flux with superheat for water in porous media [49].

in heat transfer due to the wick at higher heat fluxes was explained in terms of the competing effects of additional nucleation sites and vapour blocking the liquid in the wick, the resulting correlation is given as equation 2.82.

These analyses are restricted to situations where the vapour formed in the wick escapes through the wick surface and not out of the sides or through grooves in the heating surface.

A different approach was taken by Nishikawa and Fujita [66] who have investigated the effect of bubble population density on the heat flux limitations and the degree of superheat. While this work is restricted to flat surfaces, the authors' suggested nucleation factor may become relevant in studies where the wick contains channels or consists solely of grooves, possibly flooded, in the heat pipe wall.

The work of Saaski [67] on an inverted meniscus wick is of considerable interest. This concept, illustrated in Fig. 2.39, embodies in the high heat transfer coefficient of the circumferential groove while retaining the circumferential fluid transport capability of a thick sinter or wire mesh wick. With ammonia, heat transfer coefficients in the range 2–2.7 W/m^2K were measured at radial heat flux densities of 20 W/cm^2. These values were significantly higher than those obtained for other non-boiling evaporative surfaces. Saaski suggested that the heat transfer enhancement may be due to film turbulence generated by vapour shear or surface-tension-driven convection. He contemplated an increase in groove density as one way of increasing

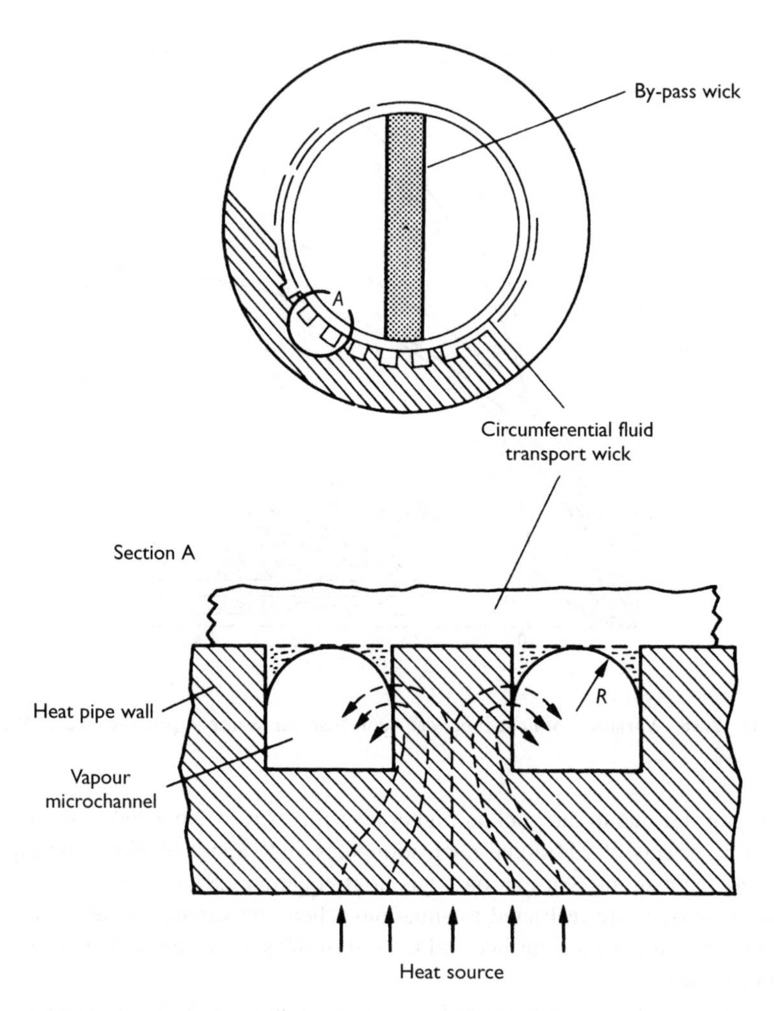

Fig. 2.39 High-performance inverted meniscus hybrid wick proposed by Saaski [67].

the heat transfer coefficient, as first results indicated a direct relationship between groove density and heat transfer coefficient. Vapour resulting from evaporation at the inverted meniscus interface flows along to open ends of the grooves, where it enters the central vapour core.

The main theoretical requirement for inverted meniscus operation, according to Saaski, is the maintenance of a sufficiently low vapour pressure drop in each channel of the wick. Recession of the inverted meniscus, which is the primary means of high evaporative heat transfer, reduces capability considerably. The equation given below describing the heat flux capability contains a term (j) which is a function of vapour microchannel pressure drop, being defined as the ratio of this pressure drop the maximum capillary priming potential.

Thus, $j = \dfrac{\Delta P_v}{2\sigma/r_c}$

The heat flux q_{max} is defined by

$$q_{max} = \frac{jN^2x^4}{8\pi}\left[\frac{\rho_v L\sigma}{\mu_v M}\right]\frac{1}{N^3 d_g{}^2 r_c}$$

$$x = \cos\frac{\Psi}{2}\Big/\left[1+\sin\frac{\Psi}{2}\right] \quad (2.100)$$

where Ψ is the groove angle (deg), M the molecular weight of working fluid, N the number of grooves per cm, d_g = heat pipe diameter at inner groove radius (q_{max} is given in W/cm^2 by equation 2.101).

Feldman and Berger [68] carried out a theoretical analysis to determine surfaces having potential in a high heat flux water heat pipe evaporator. Following a survey of the literature, circumferential grooves were chosen, these being of rectangular and triangular geometry. The model proposed that

(i) At heat fluxes below nucleate boiling, conduction is the main mode of heat transfer and vapourisation occurs at the liquid–vapour interface without affecting capillary action.
(ii) As nucleation progresses, the bubbles are readily expelled from the liquid in the grooves, with local turbulence, and convection becomes the main mode of heat transfer.
(iii) At a critical heat flux, nucleation sites will combine forming a vapour blanket, or the groove will dry out.

Both (ii) and (iii) result in a sharp increase in evaporator temperature.

It is claimed that film coefficients measured by other workers supported the computer model predictions, and Feldman stated that evaporator film coefficients as high as $8\,W/cm^2K$ could be obtained with water as the working fluid. Using triangular grooves, it was suggested that radial fluxes of up to $150\,W/cm^2$ could be tolerated. However, such a suggestion appears largely hypothetical, bearing in mind the measured values given in Table 2.9.

Winston et al. [69] reported further progress on the prediction of maximum evaporator heat fluxes, including a useful alternative to equation 2.86. Based on the modification of an equation first developed by Johnson and [70], the relationship given below takes into account vapour friction in the wick evaporator section

$$\dot{q}_{crit} = \frac{g\left[h_{co}\rho_{lo}\frac{\sigma_l}{\alpha_{lo}} - \rho_l(l_a + l_e)\sin\Phi\right]}{\dfrac{l_e\mu_l}{L\rho_l K\varepsilon(r_\varpi - r_v)}\left[\dfrac{l_e}{2}+l_a\right] + \dfrac{\mu_v}{K\rho_v}\cdot\dfrac{(r_\varpi - r_v)}{(\varepsilon-\varepsilon_l)L}} \quad (2.101)$$

The solution of the equation was accomplished by arbitrarily varying the porosity for liquid flow, ε_l, from zero to a maximum given by the wick porosity ε. The portion of the wick volume not occupied by liquid was assumed to be filled with vapour. Equation 2.101 has been used to predict critical heat fluxes, and a comparison with measured values for water has shown closer agreement than that obtained using equations 2.86 and 2.87, although the validity of these two latter equations is not disputed for liquid metal heat pipes.

2.4.1.4 Gravity-assisted heat pipes

The use of gravity-assisted heat pipes as opposed to reliance on simple thermosyphons has been given some attention. The main areas of study have centred on the need to optimise the fluid inventory and to develop wicks that will minimise entrainment. Deverall and Keddy [71] used helical arteries, in conjunction with meshes for evaporator liquid distribution, and obtained relatively high axial fluxes, albeit with sodium and potassium as the working fluids. This, together with the work of Kemme [22], led to the recommendation that more effort be devoted to the vapour flow limitations in gravity-assisted heat pipes. The work was restricted to heat pipes having liquid metal working fluids, and the significance in water heat pipes is uncertain.

However, for liquid metal heat pipes, Kemme discusses in some detail a number of vapour flow limitations in heat pipes operating with gravity assistance. As well as the pressure gradient limit discussed above, Kemme presents equations for the viscous limit, described by Busse as

$$\dot{q}_{\rm v} = \frac{A_{\rm v} D^2 L}{64 \mu_{\rm v} l_{\rm eff}} \rho_{\rm ve} P_{\rm ve} \tag{2.102}$$

where $A_{\rm v}$ is the vapour space cross-sectional area, D the vapour passage diameter and suffix e denotes conditions in the evaporator. Kemme also suggested a modified entrainment limitation to cater for additional buoyancy forces during vertical operation:

$$\dot{Q}_{\max} = A_{\rm v} L \left[\frac{\rho_{\rm v}}{A} \left(\frac{2\pi\gamma}{\lambda} + \rho g D \right) \right]^{1/2} \tag{2.103}$$

where λ is the characteristic dimension of the liquid–vapour interface and was calculated as $d_{\rm w}$ plus the distance between the wires of the mesh wick used. $\rho g d$ is the buoyancy force term.

Abhat and Nguyenchi [64] report work carried out at IKE, Stuttgart, on gravity-assisted copper/water heat pipes, retaining a simple mesh wick located against the heat pipe wall. Tests were carried out with heat pipes at a number of angles to the horizontal, retaining gravity assistance, with fluid inventories of up to five times that required to completely saturate the wick. Thus, a pool of liquid was generally present in at least a part of the evaporator section. The basis of the analysis carried out by Abhat and Nguyenchi to compare their experimental results with theory was the model proposed by Kaser [72] which assumed a liquid puddle in the heat pipe varying along the evaporator length from zero at the end of the heat pipes to a maximum at the evaporator exit. Results obtained by varying the operating temperature, tilt angle and working fluid inventory (see Fig. 2.40) were compared with Kaser's model, which was, however, found to be inadequate. Kaser believed that the limiting value of heat transport occurred when the puddle commenced receding from the end of the evaporator. However, the results of Abhat and Nguyenchi indicate that the performance is limited by nucleate boiling in the puddle. While these results are of interest, much more work is needed in this area. Strel'tsov [73] carried out a theoretical and experimental study to determine the optimum quantity of fluid to use in gravity-assisted units. Without quantifying the heat fluxes involved, he derived expressions to permit determination of the fluid inventory. Of particular interest was his observation that film evaporation, rather than nucleation, occurred under all conditions in the evaporator section, using water

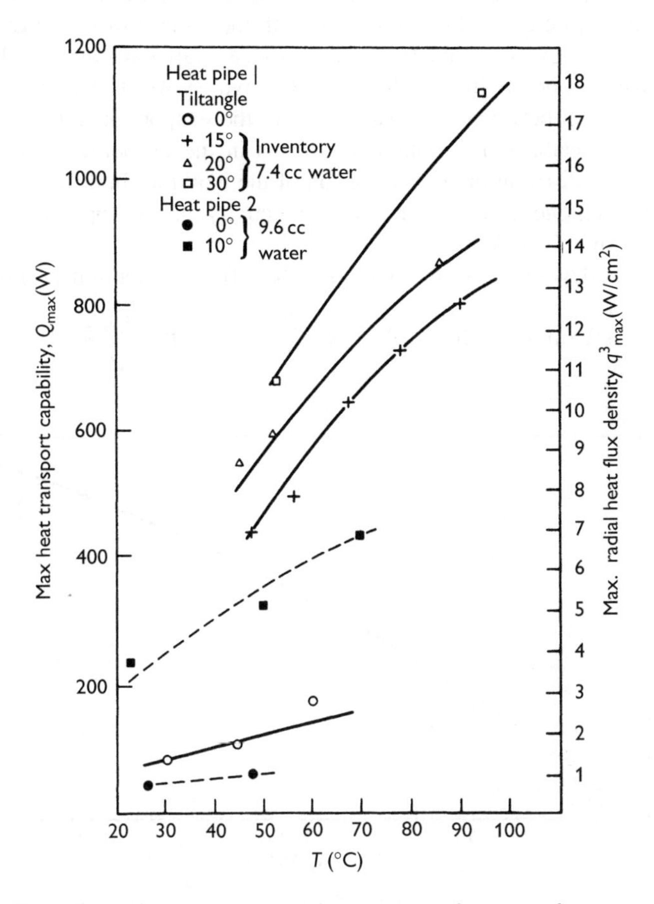

Fig. 2.40 The effect of temperature on heat pipe performance for various tilt angles (gravity-assisted angle measured from the horizontal) [64].

and several organic fluids, this seeming to contradict the findings of Abbat and Nguyenchi, assuming that Strel'tsov achieved limiting heat fluxes. Strel'tsov derived the following expression for the optimum fluid inventory for a vertical heat pipe:

$$G = (0.8l_c + l_a + 0.8l_e)\sqrt[3]{\frac{3\dot{Q}\mu_1\rho_1\pi^2D^2}{Lg}} \tag{2.104}$$

where $\dot{Q}$ is the heat transport (W).

For a given heat pipe design and assumed temperature level, the dependence of the optimum quantity of the working fluid is given by the expression

$$G = K\sqrt[3]{\dot{Q}}$$

where K is a function of the particular heat pipe under consideration.

The predicted performance of a heat pipe using methanol as the working fluid at a vapour temperature of 55 °C is compared with measured values in Fig. 2.41.

The use of arteries in conjunction with grooved evaporator and condenser surfaces has been proposed by Vasiliev and Kiselyov [74]. The artery is used to transfer condensate between the condenser and the evaporator, thus providing an entrainment-free path, while triangular grooves in the evaporator and condenser wall are used for distribution and collection of the working fluid. It is claimed that this design has a higher effective thermal conductivity than a simple thermalsyphon, particularly at high heat fluxes.

The equation developed for the maximum heat flux of such a heat pipe is

$$\dot{q}_{max}(\mathrm{W/m^2}) = 3.26 \times 10^{-2}\ a \cos\frac{\Psi}{2} \cos\theta \cdot \theta(\Psi) \cdot \frac{\sigma\rho_l L}{\mu_l} \tag{2.105}$$

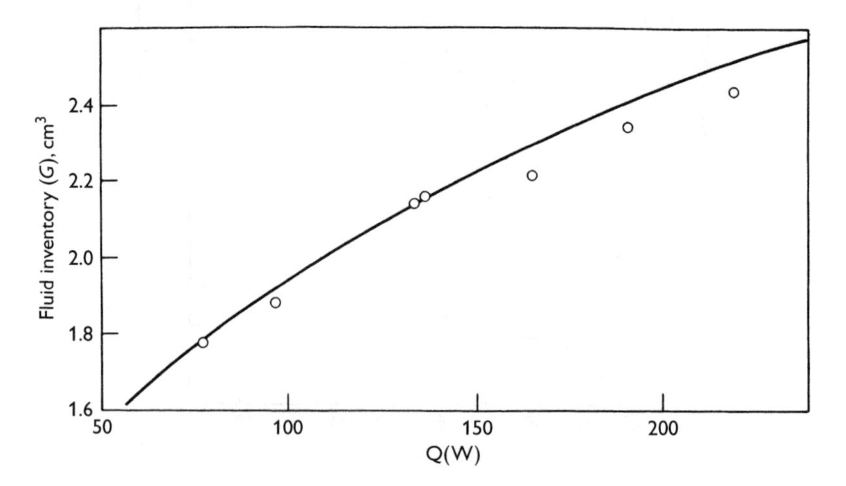

Fig. 2.41 The measured effect of fluid inventory on the performance of a methanol heat pipe and a comparison with results predicted by equation 2.104 [73].

where a is the groove width, Ψ the groove angle, θ the contact angle.

$$f(\Psi) = \left[tg\Psi + c \cdot tg\Psi - \frac{\left(\frac{\pi}{2} - \Psi\right)}{\cos^2 \Psi} \right]^3 \cos^2 \Psi$$

Other data on gravity-assisted heat pipes, in particular dryout limits, suggests that two contrary types of dryout can be shown to exist. Busse and Kemme [75] state that the 'axial dryout' arises from a lack of hydrostatic driving force and makes itself evident when the heat flux increases. The so-called aximuthal dryout is, however, caused by an excess of hydrostatic driving force and was found to be characterised by a concentration of the liquid flow on the lower part of the heat pipe cross section, without any lack of axial liquid return. It was recommended that graded capillary structures were the best technique for overcoming this limit.

Of particular interest is the use of wick-type structures in thermosyphons for enhancing internal heat transfer coefficients. As part of a project leading to the development of improved heat pipe exchangers, UKAEA Harwell inserted longitudinal grooves in the walls of thermosyphons. Using plain bore thermosyphons of near-rectangular cross section, of 1 m length, with 10 per cent of the vapour space filled with water, a series of measurements of evaporator and condenser heat transfer coefficients was made. Vapour temperatures varied from 40 °C to 120 °C, with power inputs of 2 and 3 kW [76].

Heat transfer coefficients varied from 15 to 21 $kW/m^2\,°C$, with a mean for the evaporator of 17 755 $W/m^2\,°C$ and for the condenser of 17 983 $W/m^2\,°C$.

However, with grooves on the inside wall along the complete l-m length, tests carried out with 3-kW power input revealed that condenser heat transfer coefficients of 50–100 $kW/m^2\,°C$ were achieved, a mean of 62 061 $W/m^2\,°C$ being recorded.

2.4.1.5 Total temperature drop

Referring back to 2.21, this shows the components of the total temperature drop along a heat pipe and the equivalent thermal resistances.

- R_1 and R_9 are the normal heat transfer resistances for heating a solid surface and are calculated in the usual way.
- R_2 and R_8 represent the thermal resistance of the heat pipe wall.
- R_3 and R_7 take account of the thermal resistance of the wick structure and include any temperature difference between the wall and the liquid together with conduction through the saturated wick. From the discussion in Section 2.3.7.4, it is seen that the calculation of R_3 is difficult if boiling occurs. R_7 is made up principally from the saturated wick, but if there is appreciable excess liquid then a correction must be added.
- R_4 and R_6 are the thermal resistances corresponding to the vapour liquid surfaces. They may be calculated from equation 2.8.7b but can usually be neglected.

- R_5 is due to the temperature drop ΔT_5 along the vapour column. ΔT_5 is related to the vapour pressure drop ΔP by the Clapeyron equation.

$$\frac{\mathrm{d}P}{\mathrm{d}T} = \frac{L}{TV_\mathrm{v}}$$

or combining with the gas equation

$$\frac{\mathrm{d}P}{\mathrm{d}T} = \frac{LP}{RT^2}$$

Hence

$$\Delta T_5 = \frac{RT^2}{LP}\Delta P_\mathrm{v} \qquad (2.106)$$

where ΔP_v is obtained from Section 2.5 and ΔT_5 can usually be neglected.
- R_s is the thermal resistance of the heat pipe structure, it can normally be neglected but may be important in the start-up of gas-buffered heat pipes.

It is useful to have an indication of the relative magnitude of the various thermal resistances, Table 2.10 (from Asselman and Green [57]) lists some approximate values per cm^2 for a water heat pipe.

Expressions for calculation of thermal resistance are given in Table 2.11.

2.4.2 Thermosyphons

2.4.2.1 Working fluid selection

Obviously, there is no equivalent of the wicking limit due to the absence of a wick in a thermosyphon. However, the temperature drop may be appreciable in which

Table 2.10 Representative values of thermal resistances

Resistance	cm^2/C °W
R_1	10^3–10
R_2	10^{-1}
R_3	10
R_4	10^{-5}
R_5	10^{-8}
R_6	10^{-5}
R_7	10
R_8	10^{-1}
R_9	10^3–10

Table 2.11 Summary of thermal resistances in a heat pipe

Term	Defining relation	Thermal resistance	Comment
1	$\dot{Q}_i = \alpha_e A_e \Delta T_1$	$R_1 = \dfrac{1}{\alpha_e A_e}$	
2	Plane geometry $\dot{q}_e = \dfrac{k\Delta T_2}{t}$	$R_2 = \dfrac{\Delta T_2}{A_e \dot{q}_e}$	For thin-walled cylinders $r_2 \ln \dfrac{r_2}{r_1} = t$
	Cylindrical geometry $\dot{q}_e = \dfrac{k\Delta T_2}{r_2 \ln \dfrac{r_2}{r_1}}$		
3	Equation 2.91 or 2.92	$R_3 = \dfrac{d}{k_w A_e}$	Correct for liquid metals. Gives upper limit for nonmetallics
4	$\dot{q}_e = \dfrac{L^2 P_v \Delta T_4}{(2\pi RT)^{0.5} RT^2}$	$R_4 = \dfrac{(2\pi RT)^{0.5} RT^2 L^2 P_v \Delta T_4}{L^2 P_v A_e}$	Can usually be neglected
5	$\Delta T_5 = \dfrac{RT^2 \Delta P_v}{LP_v}$	$R_5 = \dfrac{RT^2 \Delta P_v}{\dot{Q} LP_v}$	Can usually be neglected ΔP_v from Section 2.3.5
6	$\dot{q}_e = \dfrac{L^2 P_v \Delta T_6}{(2\pi RT)^{0.5} RT^2}$	$R_6 = \dfrac{(2\pi RT)^{0.5} RT^2 L^2 P_v \Delta T_4}{L^2 P_v A_c}$	Can usually be neglected
7	Equation 2.91 or 2.92	$R_7 = \dfrac{d}{k_w A_c}$	If excess working fluid is present, an allowance should be made for the additional resistance
8	Plane geometry $\dot{q}_e = \dfrac{k\Delta T_8}{t}$	$R_8 = \dfrac{\Delta T_8}{A_e \dot{q}_e}$	For thin-walled cylinders $r_2 \ln \dfrac{r_2}{r_1} = t$
	Cylindrical geometry $\dot{q}_e = \dfrac{k\Delta T_8}{r_2 \ln \dfrac{r_2}{r_1}}$	$R_8 = \dfrac{\Delta T_8}{A_e \dot{q}_e}$	

(*Continued*)

Table 2.11 *(Continued)*

Term	Defining relation	Thermal resistance	Comment
9	$\dot{Q}_i = \alpha_e A_e \Delta T_1$	$R_9 = \dfrac{1}{\alpha_c A_c}$	
	$\dot{Q}_s = \dfrac{\Delta T_s}{\left(\dfrac{l_e + l_a + l_c}{A_w k_w + A_{wall} k}\right)}$	$R_s = \dfrac{l_e + l_a + l_c}{A_w k_w + A_{wall} k}$	Axial heat flow by conduction

Notes:
R_X = Thermal Resistance = $\dfrac{\text{Temperature difference}}{\text{Heat flow}} = \left(\frac{\Delta T}{\dot{Q}}\right)_X$
Total heat flow $= \dot{Q}_i = Q + \dot{Q}_s$, $\dot{Q}, \dot{Q}_i$ and $\dot{Q}_s$ are defined in Fig. 2.21
Heat flux $= \dot{q} = \frac{\dot{Q}}{A}$
ΔT_X and R_X are defined in Fig. 2.21

Other symbols:

$\dot{q}_e, \dot{q}_c$	heat flux through the evaporator and condenser walls
k	thermal conductivity of the heat pipe wall
t	heat pipe wall thickness
r_1, r_2	inner and outer radii of cylindrical heat pipe
d	wick thickness
k_w	effective wick thermal conductivity
P_v	vapour pressure
L	latent heat
R^*	gas constant for vapour
T	absolute temperature
ΔP_v	total vapour pressure drop in the heat pipe
A_w	cross-sectional area of wick
A_{wall}	cross-sectional area of wall

case the fluid must be selected to minimise this. A figure of merit M′ may be defined for thermosyphon working fluids. This figure of merit that has the dimensions

$$\frac{\text{kg}}{\text{K}^{3/4}\text{S}^{5/2}}$$

is defined

$$M' = \left(\frac{L k_l^3 \sigma_l}{\mu_l}\right)^{\frac{1}{4}} \quad (2.107)$$

where k_l is the liquid thermal conductivity.

As with the heat pipe figure of merit, M′ should be maximised for optimum performance. Data on M′ for a variety of working fluids of use in the vapour temperature range −60 °C to 300 °C are presented in Ref. [77] in graphical and tabulated form.

The maximum value of M′ for several working fluids, together with the corresponding temperature is given in Table 2.12.

Table 2.12 Maximum thermosyphon Merit No. for selected fluids

Fluid	Temperature (°C)	M′ max $\left(\frac{kg}{K^{3/4}S^{5/2}}\right)$
Water	180	7542
Ammonia	−40	4790
Methanol	145	1948
Acetone	0	1460
Toluene	50	1055

The Thermosyphon Merit No. is relatively insensitive to temperature, for example, water remains above 4000 for the range 0.01–350 °C

2.4.2.2 Entrainment limit

This limit is described by Nguyen-Chi and Groll [78] where it is also called the flooding limit – as the condenser becomes flooded – and in the ESDU Data Item on thermosyphons [79] – where it is additionally described as the countercurrent flow limit. A correlation derived in Ref. [80] gives good agreement with the available experimental data. For thermosyphons, this gives the maximum axial vapour mass flux as

$$Q_{max}/AL = f_1 f_2 f_3 (\rho_v)^{0.5} [g(\rho_l - \rho_v)\sigma_l]^{0.25} \quad (2.108)$$

where f_1 is a function of the Bond number defined as

$$B_o = D\frac{g(\rho_l - \rho_v)^{0.5}}{\rho_l} \quad (2.109)$$

Values of f_1 can be obtained from Fig. 2.42, where it is plotted against B_o.

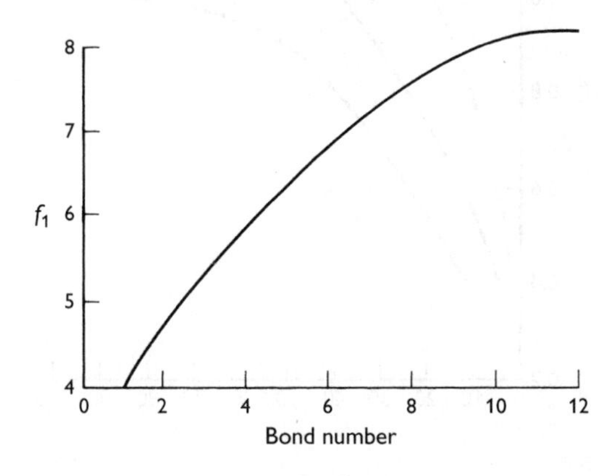

Fig. 2.42 Variation of factor f_1 with Bond number.

The factor f_2 is a function of a dimensionless pressure parameter, K_p, where K_p is defined in [79] as

$$K_p = \frac{P_v}{g(\rho_l - \rho_v)\sigma_1^{0.5}} \tag{2.110}$$

and

$$f_2 = K_p^{-0.17} \quad \text{when} \quad K_p \leq 4 \times 10^4$$
$$f_2 = 0.165 \quad \text{when} \quad K_p > 4 \times 10^4$$

The factor f_3 is a function of the inclination angle of the thermosyphon. When the thermosyphon is operating vertically, $f_3 = 1$; when the unit is inclined, the value of f_3 may be obtained by reading Fig. 2.43, where it is plotted against the inclination angle to the horizontal, ϕ, for various values of the Bond number. (Note that the product f_1, f_2, f_3 is sometimes known as the Kutateladze number.)

Note that much of the experimental evidence on the performance of thermosyphons indicates that the maximum heat transport capability occurs when the thermosyphon is at 60°–80° to the horizontal, i.e. not vertical. Data are given by Groll [81] and Terdtoon et al. [80] concerning this behaviour.

2.4.2.3 Thermal resistance and maximum heat flux

Payakaruk et al. [82] has investigated the influence of inclination angle on thermosyphon performance and have correlated the minimum resistance R_m to the

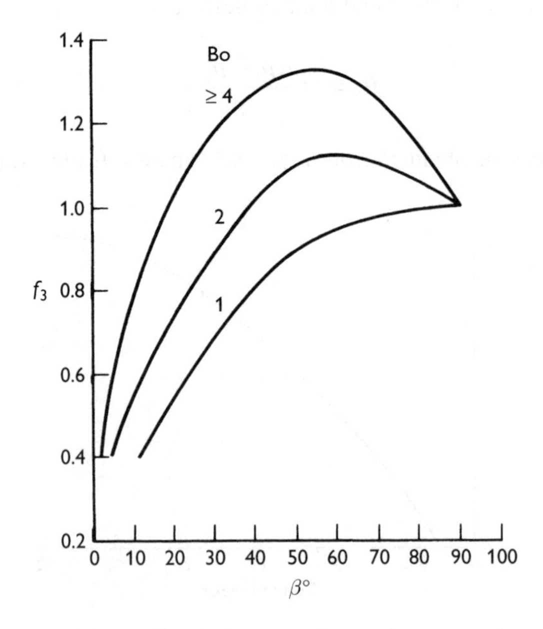

Fig. 2.43 Variation of factor f_3 *with* thermosyphon inclination angle and Bond number.

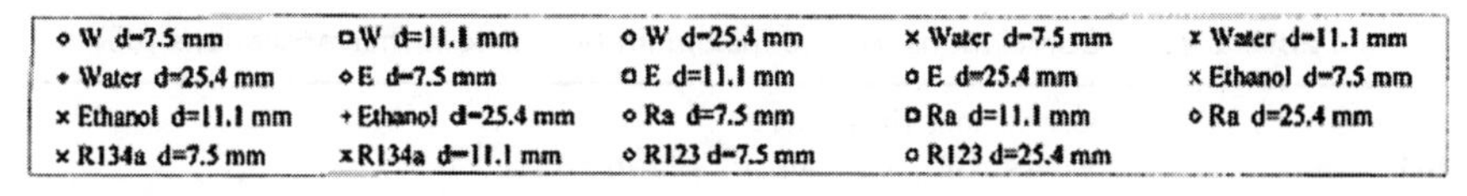

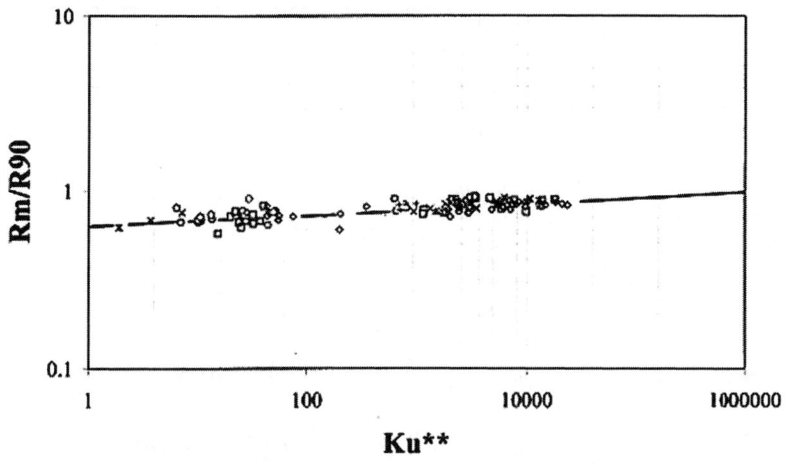

Fig. 2.44 R_m/R_{90} as a function of Ku** for a range of operating conditions compared with equation 2.111.

resistance of the thermosyphon, R_{90}, in the vertical position. This correlation is shown in equation 2.111.

$$R_{\mathrm{m}}/R_{90} = 0.647\ \mathrm{Ku}^{**0.0297} \tag{2.111}$$

where

$$\mathrm{Ku} = \frac{\dot{q}}{L\rho_{\mathrm{v}}\left[\sigma_{\mathrm{l}}g(\rho_{\mathrm{l}} - \rho_{\mathrm{v}})/\rho_{\mathrm{v}}^{2}\right]^{0.25}}$$

and

$$\mathrm{Ku}^{**} = \mathrm{Ku} \times \frac{l_{\mathrm{e}}}{2r_{\mathrm{v}}} \times \frac{\rho_{\mathrm{l}}}{\rho_{\mathrm{v}}}$$

Golobic and Gaspersic [96] have conducted a review of correlations for the prediction of thermosyphon maximum heat flux and developed a correlation based on the principle of corresponding states (Table 2.13). This formulation has the advantage little data regarding fluid properties – only molecular weight, critical temperature, critical pressure and Pitzer acentric factor are needed. The acentric factor may be calculated from

$$\omega = -1 - \log\left(p/p_{\mathrm{c(Tr=0.7)}}\right) \tag{2.112}$$

hence requiring knowledge of the vapour pressure at a reduced temperature of 0.7.

Table 2.13 Correlations for the maximum heat flux in thermosyphons [83]

Reference	Correlations Kutateladze number $\mathrm{Ku} = q_{co}/\{\Delta h_{lv}\rho_v^{0.5}[\sigma g(\rho_l - \rho_v)]^{0.25}\}$
Sakhuja [84]	$\mathrm{Ku} = \frac{0.725^2}{4} Bo^{1/2} \frac{(d_e/l_e)}{[1+(\rho_v/\rho_l)^{1/4}]^2}$
Nejat [85]	$\mathrm{Ku} = 0.09 Bo^{1/2} \frac{(d_e/l_e)^{0.9}}{[1+(\rho_v/\rho_l)^{1/4}]^2}$
Katto [86]	$\mathrm{Ku} = \frac{0.1}{1+0.491(l_e/d_e)Bo^{-0.3}}$
Tien and Chang [87]	$\mathrm{Ku} = \frac{3.2(d_e/l_e)}{4[1+(\rho_v/\rho_l)^{1/4}]^2}; \quad Bo \geq 30$
	$\mathrm{Ku} = \frac{3.2}{4}\left[\tanh\left(0.5Bo^{1/2}\right)\right]^2 \frac{(d_e/l_e)}{[1+(\rho_v/\rho_l)^{1/4}]^2}$
Harada et al. [88]	$\mathrm{Ku} = 9.64(d_e/l_e)\rho_v h_{lv} C\left(\frac{\sigma}{\rho_v}\right)^{1/2}; \quad C\left(\frac{\sigma}{\rho_v}\right)^{1/2} \geq 0.079$
	$\mathrm{Ku} = 14.1(d_e/l_e)\rho_v h_{lv}\left[C\left(\frac{\sigma}{\rho_v}\right)^{1/2}\right]^{1.15}; \quad C\left(\frac{\sigma}{\rho_v}\right)^{1/2} < 0.079$
	$C = 1.58\left(\frac{d_e}{\sigma}\right)^{0.4}; \quad \frac{d_e}{\sigma} < 0.318 \qquad C = 1; \quad \frac{d_e}{\sigma} \geq 0.318$
Gorbis and Savchenkov [89]	$\mathrm{Ku} = 0.0093(d_e/l_e)^{1.1}(d_e/l_e)^{-0.88}F_e^{-0.74} \times (1+0.03Bo)^2; \quad 2 < Bo < 60$
Bezrodnyi [90]	$\mathrm{Ku} = 2.55(d_e/l_e)\left\{\frac{\sigma}{p}\left[\frac{g(\rho_l-\rho_v)}{\sigma}\right]^{1/2}\right\}^{0.17}; \quad \frac{\sigma}{p}\left[\frac{g(\rho_l-\rho_v)}{\sigma}\right]^{1/2} \geq 2.5\cdot 10^{-5}$
	$\mathrm{Ku} = 0.425(d_e/l_e); \quad \frac{\sigma}{p}\left[\frac{g(\rho_l-\rho_v)}{\sigma}\right]^{1/2} < 2.5\cdot 10^{-5}$
Groll and Rösler [91]	$\mathrm{Ku} = f_{1(Bo)} f_2 f_{3(\varphi,Bo)}(d_e/l_e)$
	$f_2 = \left\{\frac{\sigma}{p}\left[\frac{g(\rho_l-\rho_v)}{\sigma}\right]^{1/2}\right\}^{0.17}; \quad \frac{\sigma}{p}\left[\frac{g(\rho_l-\rho_v)}{\sigma}\right]^{1/2} \geq 2.5\cdot 10^{-5}$
	$f_2 = 0.165; \quad \frac{\sigma}{p}\left[\frac{g(\rho_l-\rho_v)}{\sigma}\right]^{1/2} < 2.5\cdot 10^{-5}$
Prenger [92]	$\mathrm{Ku} = 0.747(d_e/l_e)[g\sigma(\rho_l-\rho_v)]^{0.295}(h_{lv}\rho_v)^{-0.045}$
Fukano et al. [93]	$\mathrm{Ku} = 2(d_e/l_e)^{0.83}F_e^{0.03}\left\{\frac{[\sigma g(\rho_l-\rho_v)]^{1/2}}{h_{lv}\rho_v}\right\}^{0.2}$
Imura et al. [94]	$\mathrm{Ku} = 0.16\{1-\exp[-(d_e/l_e)(\rho_l/\rho_v)^{0.13}]\}$

Table 2.13 (*Continued*)

Reference	Correlations Kutateladze number $Ku = q_{co}/\{\Delta h_{lv}\rho_v^{0.5}[\sigma g(\rho_l - \rho_v)]^{0.25}\}$
Pioro and Voroncova [95]	$Ku = 0.131\{1 - \exp[-(d_e/l_e)(\rho_l/\rho_v)^{0.13}\cos^{1.8}(\varphi - 55°)]\}^{0.8}$
Golobič and Gašperšič [96]	$q_{co} = \frac{0.16 d_e T_c^{1/3} p_c^{11/12} g^{1/4}}{l_e M^{1/4}} \tau \exp(2.530 - 8.233\tau + 1.387\omega + 17.096\tau^2$ $-4.944\tau\omega^2 + 15.542\tau^2\omega^2 - 23.989\tau^3 - 19.235\tau^3\omega - 18.071\tau^3\omega^2$
	$q_{co} = \frac{0.16 d_e T_c^{1/3} p_c^{11/12} g^{1/4}}{l_e M^{1/4}} \tau \exp(2.530 - 13.137\tau^2)$

Additional nomenclature and references for Table 2.13

Nomenclature	
Bo	Bond number $= d_e/[g(\rho_l - \rho_v)/\sigma]^{0.5}$
C	parameter in equation (T6)
d_e	evaporator diameter (m)
f_1, f_2, f_3	parameters in equation (T9)
F_e	fill ratio, liquid fill volume/evaporator volume
g	gravitational acceleration (m/s^2)
Ku	Kutateladze number $= q_{co}/\{\Delta h_{lv}\rho_v^{0.5}[\sigma g(\rho_l - \rho_v)]^{0.25}\}$
l_e	evaporator length (m)
M	molecular weight (kg/kmol)
p_c	critical pressure (Pa)
q_{co}	maximum heat flux (W/m^2)
T	temperature (K)
T_r	reduced temperature $= T/T_c$
T_c	critical temperature (K)
x_1	fluid properties parameter, equation (2)
x_2	density ratio parameter, equation (3)
Δh_{lv}	heat of evaporation (J/Kg)
φ	inclination angle (°)
μ	dipole moment (debye)
ρ_l	liquid density (kg/m^3)
ρ_v	vapour density (kg/m^3)
σ	surface tension (N/m)
τ	temperature function $= 1 - T_r$
ω	Pitzer acentric factor

2.5 SUMMARY

The basic principles of fluid mechanics and heat transfer relevant to the design and performance evaluation of heat pipes have been outlined and these have been related to studies of heat pipes and thermosyphons. Particular attention has been paid to identifying the operating limits of heat pipes and thermosyphons.

REFERENCES

[1] Cotter, T.P. Theory of heat pipes, LA-3246-MS, 26 March 1965.
[2] Shaw, D.J. Introduction to Colloid and Surface Chemistry. 2nd Ed. Butterworth, 1970.
[3] Semenchenke, V.K. Surface Phenomena in Metals and Alloys. Pergamon, 1961.
[4] Bohdansky, J., Schins, H.E. The surface tension of the alkali metals, 3. J. Inorg. Nucl. Chem. Vol. 29, pp 2173–2179, 1967.
[5] Jasper, J.J. 'The surface tension of pure liquid compounds,' J. Phys. Chem. Ref. Data., Vol. 1, 841, 1972.
[6] IAPWS Release on Surface Tension of Ordinary Water Substance, International Association for the Properties of Water and Steam, September 1994.
[7] Bohdansky, J., Schins, H.E.J. The temperature dependence of liquid metals. J. Chem. Phys., Vol. 49, p 2982, 1968.
[8] Fink, J.K. and Leibowitz, L. Thermodynamic and transport properties of sodium liquid and vapor, ANL/RE-95/2, Argonne National Laboratory, January 1995.
[9] Iida, T. and Guthrie, R.I.L. The Physical Properties of Liquid Metals. Clarendon Press, Oxford, 1988.
[10] Walden, P. Uber den zusammenhang der kapillaritaskonstanten mit der latenten verdampfungswarme der losungsmittel, Z. Phys. Chem., Vol. 65., pp 267–288, 1909.
[11] Sugden, S. A relation between surface tension, density and chemical composition, J. Chem. Soc., Vol. 125, pp 1177–1189, 1924.
[12] Quale, O.R. The parachors of organic compounds. Chem. Rev., Vol. 53, pp 439–585, 1953.
[13] Hewitt, G.F. (Ed.), Handbook of Heat Exchanger Design, pp 5.1.5–2, Begell House, 1992.
[14] Busse, C.A. Pressure drop in the vapour phase of long heat pipes. Thermionic Conversion Specialists Conference, Palo Alto, CA, October, 1967.
[15] Kay J.M. and Nedderman R.M. An introduction to fluid mechanics and heat transfer 3rd Edn CUP, Cambridge, 1974.
[16] Grover, G.M., Kemme, J.E., Keddy, E.S. Advances in heat pipe technology. 2nd International Symposium on Thermionic Electrical Power Generation. Stresa, Italy, May 1968.
[17] Ernst, D.M. Evaluation of theoretical heat pipe performance. Thermionic Specialist Conference, Palo Alto, CA, October, 1967.
[18] Bankston, C.A., Smith, J.H. Incompressible laminar vapour flow in cylindrical heat pipes. ASME Paper No. 7 1-WAJHT-1 5, 1971.
[19] Rohani, A.R., Tien, C.L. Analysis of the effects of vapour pressure drop on heat pipe performance. Int. J. Heat Mass Trans., Vol. 17, pp 61–67, 1974.
[20] Deverall, J.E., Kemme, J.E., Flarschuetz, L.W. Some limitations and startup problems of heat pipes. Los Alames Scientific Laborating Report LA-4518, November 1970.

[21] Kemme, J.E. Ultimate heat pipe performance. I.E.E.E. Thermionic Specialist Conference. Framlingham MA, 1968.

[22] Kemme, J.E. High performance heat pipes. I.E.E.E. Thermionic Specialist Conference. Palo Alto, CA, October 1967.

[23] Tower, L.K. and Hainley, D.C. An improved algorithm for the modelling of vapour flow in heat pipes. Paper 13-8, Proceedings of 7th International Heat Pipe Conference, Minsk, Byelorussia, May 1990.

[24] Kim, B.H. and Peterson, G.P. Analysis of the critical Weber number at the onset of liquid entrainment in capillary-driven heat pipes, International J. Heat Mass Transfer. Vol. 38, No. 8, pp 1427–1442, 1995.

[25] Cheung, H. A critical review of heat pipe theory and applications. Report No. UCRL50453, July 1968.

[26] Drazin P.G. and Reid, W.H. Hydrodynamic Stability, pp 14–22. Cambridge University Press, Cambridge, U.K. 1981.

[27] Nukiyama, S. Maximum and minimum values of heat transmitted from metal to boiling water under atmospheric pressure. J. Soc. Mech. Eng. Japan, Vol. 37, p 367, 1934.

[28] Hsu, Y.Y. On the size range of active nucleation cavities on a heating surface. J. Heat Transf., Trans A.S.M.E., August 1962.

[29] Rohsenow, W.M. A method of correlating heat transfer data for surface boiling of liquids. Trans. A.S.M.E., Vol. 74, 1955.

[30] Mostinski, I.L. Application of the rule of corresponding states for the calculation of heat transfer and critical heat flux., Teploenergetika Vol. 4, p 66 English Abst. Br. Chem. Eng., Vol. 8, p 580, 1963.

[31] Cooper, M.G. Saturated nucleate pool boiling – a simple correlation, 1st UK National Heat Transfer Conference IChemE Symposium Series No. 86, 2, pp 785–793, 1984.

[32] Subbotin, V.I. Heat transfer in boiling metals by natural convection. USAEC-Tr-72 10, 1972.

[33] Dwyer, O.E. On incipient boiling wall superheats in liquid metals. Int. J. Heat Mass Transf, Vol. 12, pp 1403–1419, 1969.

[34] Rosenhow, W.M., Griffith, P. Correlation of maximum heat flux data for boiling of saturated liquids. A.S.M.E. – A.LC.E. Heat Transfer Symposium, Louisville, KY, 1955.

[35] Caswell, B.F., Balzhieser, R.E. The critical heat flux for boiling metal systems. Chemical Engineering Progress Symposium Series of Heat Transfer, Los Angeles, Vol. 62, No. 64, 1966.

[36] Philips, E.C., Hinderman, J.D. Determination of capillary properties useful in heat pipe design. A.S.M.E. – A.I.Ch.E. Heat Transfer Conference Minneapolis, MN, August 1967.

[37] Ferrell, J.K., Aileavitch, J. Vaporisation heat transfer in capillary wick structures. Chem. Eng. Symp. Series No. 66, Vol. 02, 1970.

[38] Corman, J.C., Welmet, C.E. Vaporisation from capillary wick structures. A.S.M.E. – A.I.Ch.E. Heat Transfer Conference, Tulsa, Oklahoma, Paper 71-HT-35, August 1971.

[39] Brautsch, A. and Kew P.A. The effect of surface conditions on boiling heat transfer from mesh wicks, Proceedings of 12th International Heat Transfer Conference, Elsevier SAS, Grenoble, 2002.

[40] Brautsch A. Heat transfer mechanisms during the evaporation process from mesh screen porous structures, Ph.D. Thesis, Heriot-Watt University, 2002.

[41] Brautsch A. and Kew P.A. Examination and visualization of heat transfer processes during evaporation in capillary porous structures, J. Appl. Therm. Eng., Vol. 22, pp 815–824, 2002.

[42] Asakavicius J.P., Zukauskas V.A., Gaigalis V.A., Eva V.K. Heat transfer from freon-113, ethyl alcohol and water with screen wicks, Heat Transf – Soviet Res., Vol. 11, No. 1. 1979.
[43] Liu J.W., Lee D.J., Su A. Boiling of methanol and HFE-7100 on heated surface covered with a layer of mesh, Int. J. Heat Mass Transf., Vol. 44, pp 241–246, 2001.
[44] Tse J.Y., Yan Y.Y., Lin T.F. Enhancement of pool boiling heat transfer in a horizontal water layer through surface roughness and screen coverage, J. Heat Transf., Vol. 32, pp 17–26. 1996.
[45] Abhat, A., Seban, R.A. Boiling and evaporation from heat pipe wicks with water and acetone J. Heat Transf., August 1974.
[46] Marto, P.S., Mosteller, W.L. Effect of nucleate boiling and the operation of low temperature heat pipes. A.S.M.E. – A.I.Ch.E. Heat Transfer Conference, Minneapolis, MN, August 1969.
[47] Ferrell J.K., Davis R., Winston H. Vaporisation heat transfer in heat pipe wick materials. Proceedings of the 1st International Heat Pipe Conference, Stuttgart, Germany, 1973.
[48] Shaubach R.M., Dussinger P.M., Bogart J.E., Boiling in heat pipe evaporator wick structures. Proceedings of the 7th International Heat Pipe Conference, Minsk. 1990.
[49] Cornwell K., Nair B.G. Boiling in wicks Proceedings of Heat Pipe Forum Meeting, National Engineering Laboratory, Department of Industry. Report No. 607, East Kilbride, Glasgow, 1975.
[50] Cornwell K., Nair B.G. Observation of boiling in porous media. Int. J. Heat and Mass Transf., Vol. 19, pp 236–238. 1976.
[51] Costello, C.P., Frea, WJ. The roles of capillary wicking and surface deposits in attainment of high pool boiling burnout heat fluxes. A.I.Ch.E. J., Vol. 10, No. 3, p 393, 1964.
[52] Reiss, F. et al. Pressure balance and maximum power density at the evaporator gained from heat pipe experiments. 2nd International symposium on Thermionic Electrical Power Generation. Stresa, Italy, May 1968.
[53] Ballzhieser, R.E. et al. Investigation of liquid metal boiling heat transfer. Air Force Propulsion Lab. Wright Patterson, AFB/Ohio, AFAPL-TR 66–85, 1966.
[54] Moss, R.A. and Kelley, A.J. Neutron radiographic study of limiting heat pipe performance. mt. J. Heat Mass Transf., Vol. 13, No. 3, pp 491–502, 1970.
[55] Davis, W.R., Ferrell, J.K. Evaporative heat transfer of liquid potassium in porous media. A.I.A.A./A.S.M.E. 1974. Thermophysics and Heat Transfer Conference, Boston, MA, July 1974.
[56] Ferrell, J.K. et al. Vaporisation heat transfer in heat pipe wick materials. A.I.A.A. Thermophysics Conference, San Antonio, Texas, April 1972.
[57] Asselman, G.A.A., Green, D.B. Heat Pipes. Phillips Tech. Rev., Vol. 33, No. 4, pp 104–113, 1973.
[58] Rogers, G.F.C. and Mayhew, Y.W., Engineering thermodynamics—work and heat transfer, 4th Edn., Longman, 1992.
[59] Busse, C.A. Theory of ultimate heat transfer limit of cylindrical heat pipes. Int. J. Heat Mass Transf., Vol. 16, pp 169–186, 1973.
[60] Anon. Heat pipes—performance of capillary-driven designs. Data Item No.70012, Engineering Sciences Data Unit, London, September 1979.
[61] Tien, C.L. and Chung, K.S. Entrainment limits in heat pipes. A.I.A.A. Paper 78–382, Proceedings of III mt. Heat Pipe Conference, Palo Alto, A.I.A.A.. Report CP784, New York, 1978.

[62] Coyne Prenger, F. and Kemme, J.E. Performance limits of gravity-assist heat pipes with simple wick structures. Proceedings of IV International Heat Pipe Conference, London. Oxford, Pergamon Press, 1981.
[63] Tien, C.L. Fluid mechanics of heat pipes. Annu. Rev. Fluid Mech. Am. Inst. Phy., Vol. 7, pp 167–185, 1975.
[64] Abhat, A. and Nguyenchi, H. Investigation of performance of gravity assisted copper-water heat pipes. Proceedings of 2nd International Heat Pipe Conference, Bologna ESA Report SP112, Vol. 1, 1976.
[65] Wiebe, J.R. and Judd, R.L. Superheat layer thickness measurements in saturated and subcooled nucleate boiling. Trans. ASME, J. Heat Transf., Paper 71-HT-43, 1971.
[66] Nishikawa, K. and Fujita, Y. Correlation of nucleate boiling heat transfer based on bubble population density. Int. J. Heat Mass Transf., Vol. 20, pp 233–245, 1977.
[67] Saaski, E.W. Investigation of an inverted meniscus heat pipe wick concept, NASA Report CR-137724, 1975.
[68] Feldman, K.T. and Berger, M.E. Analysis of a high heat flux water heat pipe evaporator. Tech. Report ME-6 273, ONR-012-2, U.S. Office of Naval Research, 1973.
[69] Winston, H.M., Ferrell, J.K. and Davis, R. The mechanism of heat transfer in the evaporator zone of the heat pipe. Proceedings of 2nd International Heat Pipe Conference, Bologna. ESA Report SP112, Vol. 1, 1976.
[70] Johnson, J.R. and Ferrell, J.K. The mechanism of heat transfer in the evaporator zone of the heat pipe. ASME Space Technical and Heat Transfer Conference, Los Angeles, Paper 70-HT/SpT-12, 1970.
[71] Deverall, J.E. and Keddy, E.S. Helical wick structures for gravity-assisted heat pipes. Proceedings of 2nd International Heat Pipe Conference, Bologna ESA Report SP112, Vol. 1, 1976.
[72] Kaser, R.V. Heat pipe operation in a gravity field with liquid pool pumping. Unpublished paper, McDonnell Douglas Corporation, July 1972.
[73] Strel'tsov, A.I. Theoretical and experimental investigation of optimum filling for heat pipes. Heat Transf-Soviet Res, Vol. 7, No. 1, Jan/Feb 1975.
[74] Vasiliev, L.L. and Kiselyov, V.G. Simplified analytical model of vertical arterial heat pipes. Proceedings of 5th International Heat Transfer Conference, Japan Vol. 5, Paper HE 2.3, 1974.
[75] Busse, C.A. and Kemme, J.E. Dry-out phenomena in gravity-assist heat pipes with capillary flow. Int. J. Heat Mass Transf, Vol. 23, pp 643–654, 1980.
[76] Davies, M.J. and Chaffey, G.H. Development and demonstration of improved gas to gas heat pipe heat exchangers for the recovery of residual heat. Report EUR 7127 EN, Commission of the European Communities, 1981.
[77] Anon. Thermophysical properties of heat pipe working fluids: operating range between −60 °C and 300 °C. Data Item No. 80017, Engineering Sciences Data Unit, London, August 1980.
[78] Nguyen-Chi, H. and Groll, M. Entrainment or flooding limit in a closed two-phase thermosyphon, Proceedings of IV International Heat Pipe Conference, London. Pergamon Press, Oxford, 1981.
[79] Anon. Heat pipes-performance of two-phase closed thermosyphons. Data Item No. 81038, Engineering Sciences Data Unit, London, October 1981.
[80] Terdtoon, P. et al. Investigation of effect of inclination angle on heat transfer characteristics of closed two-phase thermosyphons. Proceedings of 7th International Heat Pipe Conference, Paper B9P, Minsk, May 1990.

[81] Groll, M. Heat pipe research and development in Western Europe. Heat Recov Syst and CHP, Vol. 9, No. 1, pp 19–66, 1989.
[82] Payakaruk, T., Terdtoon, P. and Ritthidech S. Correlations to predict heat transfer characteristics of an inclined closed two-phase thermosyphon at normal operating conditions. Appl. Therm. Eng. Vol. 20, pp 781–790, 2000.
[83] Golobic, I. and Gaspersic, B. Corresponding states correlation for maximum heat flux in two-phase closed thermosyphon. Int. J. Refrig., Vol. 20, No. 6, pp 402–410, 1997.
[84] Sakhuja, R. K. Flooding constraint in wickless heat pipes ASME Publ. paper No. 73-WA/ HT-7, 1973.
[85] Nejat, Z. Effects of density ratio on critical heat flux in closed and vertical tubes. Int. J. Multiphase Flow, Vol. 7, pp 321–327, 1981.
[86] Katto, Y. Generalized correlation for critical heat flux of the natural convection boiling in confined channels. Trans. Japan. Soc. Mech. Engrs, Vol. 44, pp 3908–3911, 1978.
[87] Tien, C.L. and Chang, K.S. Entrainment limits in heat pipes. AIAA J., Vol. 17, pp 643–646, 1979.
[88] Harada, K., Inoue, S., Fujita, J., Suematsu, H. and Wakiyama, Y. Heat transfer characteristics of large heat pipes. Hitachi Zosen Tech. Rev. Vol. 41, pp 167–174, 1980.
[89] Gorbis, Z.R. and Savchenkov, G.A. Low temperature two-phase closed thermosyphon investigation 2nd International Heat Pipe Conference, Bologna, Italy, pp 37–45, 1976.
[90] Bezrodnyi, M.K. Isledovanie krizisa teplomassoperenosa v nizkotemperaturnih besfiteljinyih teplovih trubah, Teplofizika Visokih Temperature, Vol. 15, pp 371–376, 1977.
[91] Groll, M. and Rosler, S. Development of advanced heat transfer components for heat recovery from hot waste gases. Final Report CEC Contract No.EN3E-0027-D(B), 1989.
[92] Prenger, F.C. Performance limits of gravity-assisted heat pipes. 5th International Heat Pipe Conference Tsukuba, Japan, pp 1–5, 1984.
[93] Fukano, T., Kadoguchi, K. and Tien, C.L. Experimental study on the critical heat flux at the operating limit of a closed two-phase thermosyphon. Heat Transf – Jan. Res, Vol. 17 pp 43–60, 1988.
[94] Imura, H., Sasaguchi, K. and Kozai, H. Critical heat flux in a closed two-phase thermosyphon. Int. J. Heat Mass Transf, Vol. 26, pp 1181–1188, 1983.
[95] Pioro, I.L. and Voronenva, M.V. Rascetnoe opredelenie predelnogo teplovogo potoka pri kipenii zidkostej v dvuhfaznih termosifonah, Inz. fiz. zurnal. Vol. 53, pp 376–383, 1987.
[96] Golobic:, I. and Gaspersic, B. Generalized method for maximum heat transfer performance in two-phase closed thermosyphon, International Conference CFCs, The Day After Padova, Italy, pp 607–616, 1994.

3

HEAT PIPE COMPONENTS AND MATERIALS

In this Chapter, we will discuss the main components of the heat pipe and the materials used. Since the 4th Edition of Heat Pipes was written, the materials and components of heat pipes have remained essentially the same. Nevertheless, life tests have had an opportunity to extend over a further 10-year period, and some working fluids have lost their attractiveness. This may be dictated by health and safety considerations (see also Chapter 5) or by environmental pressures – for example the use of chlorofluorocarbons is now banned, and in some countries in Europe the application of hydrofluorocarbons (HFCs) is being phased out in favour of fluids that contribute less to global warming.

The temperature range affected by these trends is principally between −50 °C and +100 °C. This affects products in the electronics thermal control, domestic and heat recovery areas, as well as (although less important in the context of global warming) spacecraft – see Chapters 7 and 8 on applications.

The issue of compatibility and the results of life tests on heat pipes and thermosyphons remain critical aspects of heat pipe design and manufacture. In particular, the generation of noncondensable gases that adversely affects the performance of heat pipes in either short-term or long-term must be taken particularly seriously in the emerging technology of micro-heat pipes and arrays of such units (see also Chapter 6, where micro-heat pipes are discussed).

An aspect of heat pipes that has always been of interest to researchers is the compatibility of water with steel (mild or stainless variants). The superior properties of water and the low cost of some steels, together with their strength, make the combination of water–steel attractive, if sometimes elusive. Later in this Chapter, substantial discussion of the combination is given.

A considerable quantity of data on heat pipe life tests was accumulated in the 1960s–1980s. To many, this was the most active period of heat pipe research and development, and in many applications heat pipes are now routine components in many terrestrial and space applications. It is therefore important to retain much of the early life test data (often now discarded in paper copies of reports from company libraries etc.). Thus, apart from examples where the fluids have been discarded due to environmental or other considerations (for example refrigerants such as R11 and R113), the historical data are retained in this new edition of *Heat Pipes*. The majority of the fluids, materials and operating conditions remains the same, so the life test data remain valid and should be of not just archival value to those entering the heat pipe field. These data, and some additional results, are given in Section 3.5.1.

However, the importance of life tests at each laboratory, once new designs are made, remains great, as explained in Chapter 5.

The three basic components of a heat pipe are as follows:

(i) The working fluid
(ii) The wick or capillary structure
(iii) The container.

In the selection of a suitable combination of the above, inevitably a number of conflicting factors may arise, and the principal bases for selection are discussed below.

3.1 THE WORKING FLUID

A first consideration in the identification of a suitable working fluid is the operating vapour temperature range and a selection of fluids is shown in Table 3.1. Within the approximate temperature band several possible working fluids may exist, and a variety of characteristics must be examined in order to determine the most acceptable of these fluids for the application being considered. The prime requirements are as follows:

(i) Compatibility with wick and wall materials
(ii) Good thermal stability
(iii) Wettability of wick and wall materials
(iv) Vapour pressures not too high or low over the operating temperature range
(v) High latent heat
(vi) High thermal conductivity
(vii) Low liquid and vapour viscosities
(viii) High surface tension
(ix) Acceptable freezing or pour point.

The selection of the working fluid must also be based on thermodynamic considerations which are concerned with the various limitations to heat flow occurring within the heat pipe. These were also discussed in Chapter 2 and are the viscous, sonic, capillary, entrainment and nucleate boiling limitations.

Table 3.1 Heat pipe working fluids

Medium	Melting point (°C)	Boiling point at atmos. press. (°C)	Useful range (°C)
Helium	−271	−261	−271 to −269
Nitrogen	−210	−196	−203 to −160
Ammonia	−78	−33	−60 to 100
Pentane	−130	28	−20 to 120
Acetone	−95	57	0 to 120
Methanol	−98	64	10 to 130
Flutec PP2[1]	−50	76	10 to 160
Ethanol	−112	78	0 to 130
Heptane	−90	98	0 to 150
Water	0	100	30 to 200
Toluene	−95	110	50 to 200
Flutec PP9[1]	−70	160	0 to 225
Thermex[2]	12	257	150 to 350
Mercury	−39	361	250 to 650
Caesium	29	670	450 to 900
Potassium	62	774	500 to 1000
Sodium	98	892	600 to 1200
Lithium	179	1340	1000 to 1800
Silver	960	2212	1800 to 2300

Note: (The useful operating temperature range is indicative only.) Full properties of most of the above are given in Appendix 1.
[1] Included for cases where electrical insulation is a requirement.
[2] Also known as Dowtherm A, an eutectic mixture of diphenyl ether and diphenyl.

Many of the problems associated with long-life heat pipe operation are a direct consequence of material incompatibility. This involves all three components of the heat pipe and is discussed fully later. One aspect peculiar to the working fluid, however, is the possibility of thermal degradation. With certain organic fluids, it is necessary to keep the film temperature below a specific value to prevent the fluid breaking down into different compounds. A good thermal stability is therefore a necessary feature of the working fluid over its likely operating temperature range.

The surface of a liquid behaves like a stretched skin except that the tension in the liquid surface is independent of surface area. All over the surface area of a liquid there is a pull due to the attraction of the molecules tending to prevent their escape. This surface tension varies with temperature and pressure, but the variation with pressure is frequently small.

The effective value of surface tension may be considerably altered by the accumulation of foreign matter at the liquid/vapour liquid/liquid or solid surfaces. Prediction of surface tension is discussed in Chapter 2.

In heat pipe design, a high value of surface tension is desirable in order to enable the heat pipe to operate against gravity and to generate a high capillary driving force.

In addition to high surface tension, it is necessary for the working fluid to wet the wick and container material. That is the contact angle must be zero, or at least very small. In spite of suggestions that additives can improve the performance of heat pipes – for example by the addition of small amounts of a long-chain alcohol to water heat pipes [47] – such a practice is not generally recommended. Those designing and assembling heat pipes for testing should not be tempted to consider additives which may be claimed to improve the 'wettability' of surfaces. In two-phase systems, additives tend to get left behind when phase change occurs!

The vapour pressure over the operating temperature range must be sufficiently great to avoid high vapour velocities which tend to set up a large temperature gradient, entrain the refluxing condensate in the countercurrent flow, or cause flow instabilities associated with compressibility. However, the pressure must not be too high because this will necessitate a thick-walled container.

A high latent heat of vapourisation is desirable in order to transfer large amounts of heat with a minimum fluid flow, and hence to maintain low pressure drops within the heat pipe. The thermal conductivity of the working fluid should also preferably be high in order to minimise the radial temperature gradient and to reduce the possibility of nucleate boiling at the wick/wall interface.

The resistance to fluid flow will be minimised by choosing fluids with low values of vapour and liquid viscosity.

A convenient means for quickly comparing working fluids is provided by the Merit number, introduced in Chapter 2, defined as $\sigma_l L \rho_l / \mu_l$ where σ_l is the surface tension, L the latent heat of vapourisation, ρ_l the liquid density and μ_l the liquid viscosity. Figure 3.1 gives the Merit number at the boiling point for working fluids, covering temperature ranges between 200 and 1750 K. One obvious feature is the marked superiority of water with its high latent heat and surface tension, compared with all organic fluids, such as acetone and the alcohols. The final fluid selected is, of course, also based on cost, availability, compatibility and the other factors listed above. Graphical and tabulated data on merit numbers of a range of common working fluids are present in [1]. (Note that this reference also presents thermosyphon figures of merit.)

A high Merit number is not the only criterion for the selection of the working fluid, and other factors may, in a particular situation, be of greater importance. For example on grounds of cost, potassium might be chosen rather than caesium or rubidium which are one hundred times more expensive. Also, over the temperature range 1200–1800 K, lithium has a higher Merit number than most metals, including sodium. However, its use requires a container made from an expensive lithium-resistant alloy, whereas sodium can be contained in stainless steel. It may therefore, be cheaper and more convenient to accept a lower performance heat pipe made from sodium/stainless steel.

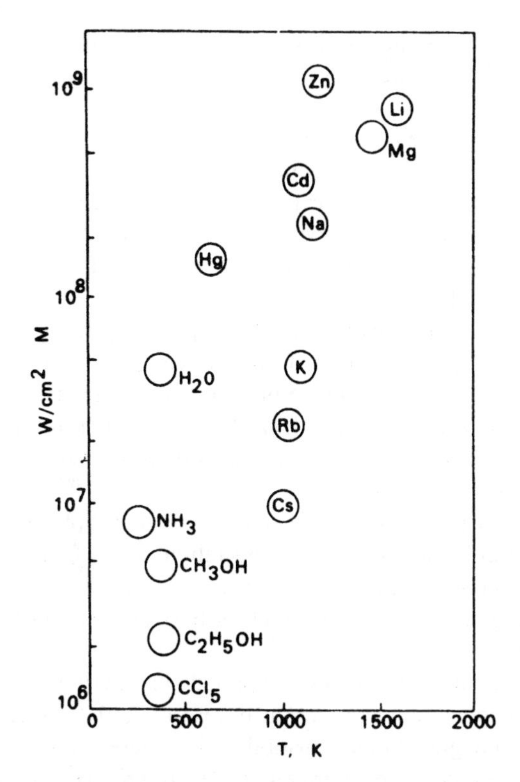

Fig. 3.1 Merit number for selected working fluids for their boiling point (Courtesy Philips Technical Review).

Working fluids used in heat pipes range from helium at 4 K up to lithium at 2300 K. Figure 3.1 shows the superiority of water over the range 350–500 K, where the alternative organic fluids tend to have considerably lower Merit numbers. At slightly lower temperatures, 270–350 K, ammonia is a desirable fluid, although it requires careful handling to retain high purity, and acetone and the alcohols are alternatives having lower vapour pressures. These fluids are commonly used in heat pipes for space applications. Water and methanol, both being compatible with copper, are often used for cooling electronic equipment.

Where HFCs are acceptable, R134A and R407C have been investigated as heat pipe/thermosyphon working fluids in the context of solar collectors [48]. In a comparison with a third fluid, R22 (now being phased out), it was found that R407C gave a superior performance to the other two fluids, but it is unclear why water was not selected. Some of these 'new' HFCs have boiling ranges, and care should be exercised in their use in heat pipes. (See also Chapter 7 where renewable energy uses of heat pipes are discussed).

For temperatures over 500 K and up to 650 K, the high temperature organic heat transfer fluids, normally offered for single-phase duties, may be used. These are available from companies such as The Dow Chemical Company and are eutectics of diphenyl and diphenyl oxide. The boiling point is around 260 °C at atmospheric pressure.

Unfortunately, they have a low surface tension and poor latent heat of vapourisation. As with many other organic compounds, diphenyls are readily broken down when film temperatures exceed the critical value. However, unlike many other fluids having similar operating temperature ranges, these eutectic mixtures have a specific boiling point, rather than a boiling range. Other fluids such as silicons are being studied for use at above 600 K.

One of the most comprehensive sets of compatibility tests was that carried out at IKE, Stuttgart [3, 4], on a number of organic working fluids. These concentrated on thermosyphons made using a range of boiler and austenitic steels, and the data obtained are summarised in Table 3.2. It can be seen from the test results that organic fluids operating at temperatures much in excess of 300 °C tend to be unsuitable for long-term use in heat pipes.

It was pointed out by the research workers [15] that diphenyl and naphthalene, the two working fluids most suitable for operation in this temperature regime, can both suffer from decomposition created by overheating at the evaporator section. This leads to the generation of non-condensable gas that can be vented via, for example, a valve.

The use of naphthalene is also reported in thermosyphons by Chinese researchers [2]. Vapour temperatures in excess of 250 °C were achieved in the experiments but no degradation/compatibility data were given.

More recently [68] the temperature range 400–700 K has received attention from workers at NASA Glenn Research Centre. The suggestion is that metallic halides might be used as working fluids within this temperature range. The halides are typically compounds of lithium, sodium, potassium, rubidium and copper, with fluorine, iodine, bromine and iodine [69]. The suggestion in much of the literature is that these will be reactive, and it should be pointed out that mixtures of iodine and sulphur were investigated as heat pipe fluids by the UK Atomic Energy Authority, some 25–30 years ago. Appropriate data to allow an accurate assessment of their potential are not yet available.

Moving further up the temperature scale, one now enters the regime of liquid metals. Mercury has a useful operating temperature range of about 500–950 K and has attractive thermodynamic properties. It is also liquid at room temperature, which facilitates handling, filling and start-up of the heat pipe.

Apart from its toxicity, the main drawback to the use of mercury as a working fluid in heat pipes, as opposed to thermal syphons, is the difficulty encountered in wetting the wick and wall of the container. There are few papers specifically devoted to this topic, but Deverall [5] at Los Alamos and Reay [6] have both reported work on mercury wetting.

Japanese work on mercury heat pipes using type 316L stainless steel as the container material showed that good thermal performance could be achieved once full start-up had been overcome, but the materials compatibility proved a problem,

Table 3.2 Compatibility tests with organic working fluids [4]

Test duration (Year)	Working fluids	Structural materials (operating temperatures)		
		Compatible	Fairly compatible	Incompatible
4.5–5	*N*-Octane	St35 (230 °C)	X10CrNiT189 (200 °C, 250 °C)	
	Diphenyl		St35 (270 °C) X10CrNiTi189 (300 °C)	St35 (300 °C) X10CrNiTi189 (350 °C)
	Diphenyl Oxide	St35 (220 °C)		
3	Toluene	St35 (250 °C) 13CrMo44 (250 °C) X2CrNiMo1812 (280 °C)		
	Naphthalene	St35 (270 °C)	13CrMo44 (270 °C)	
1	Diphenyl	Ti99.4 (270 °C)		CuNi10Fe (250 °C)
	Diphenyl	13CrMo44 (250 °C) X2CrNiMo1812 (250 °C)		13CrMo44 (400 °C) X2CrNiMo181 2 (400 °C)
	QM			13CrMo44 (320 °C, 400 °C) X2CrNiMo181 2 (350 °C, 400 °C)
	OMD			13CrMo44 (350 °C, 400 °C) X2CrNiMo181 2 (350 °C, 400 °C)
	Toluene	Ti99.4 (250 °C) CuNi10Fe (280 °C)		
	Naphthalene	X2CrNiMo1812 (320 °C) Ti99.4 (300 °C)	CuNi10Fe (320 °C)	

due to corrosion [71]. Such problems were not reported in the series of experiments, including a 1-month 'life test' by Macarino and colleagues at IMGC in Turin, Italy, using a unit supplied by the Joint Research Centre, Ispra. One of a series of gas-controlled heat pipes for accurate temperature measurements, the unit was also made in stainless steel and operated at a vapour temperature of around 350 °C, [72].

Bienert [7], in proposing mercury/stainless steel heat pipes for solar energy concentrators, used Deverall's technique for wetting the wick in the evaporator section of the heat pipe and achieved sufficient wetting for gravity-assisted operation. He argued that nonwetting in the condenser region of the heat pipe should enhance dropwise condensation which would result in higher film coefficients than those obtainable with film condensation. Work at Los Alamos has suggested that magnesium can be used to promote mercury wetting [8].

Moving yet higher up the vapour temperature range, caesium, potassium and sodium are acceptable working fluids and their properties relevant to heat pipes are well documented (see Appendix 1). Above 1400 K, lithium is generally a first choice as a working fluid but silver has also been used [9]. Working on applications of liquid metal heat pipes for nuclear and space-related uses, Tournier and others at the University of New Mexico [73] suggest that lithium is the best choice of working fluid at temperatures above 1200 K. For those interested in the start-up of liquid metal, in particular lithium heat pipes, the research at the University of New Mexico is well worth studying.

3.2 THE WICK OR CAPILLARY STRUCTURE

The selection of the wick for a heat pipe depends on many factors, several of which are closely linked to the properties of the working fluid. Obviously the prime purpose of the wick is to generate capillary pressure to transport the working fluid from the condenser to the evaporator. It must also be able to distribute the liquid around the evaporator section to any areas where heat is likely to be received by the heat pipe. Often these two functions require wicks of different form particularly where the condensate has to be returned over a distance of, say, 1 m, in 0 gravity. Where a wick is retained in a 'gravity-assisted' heat pipe, the role may be to enhance heat transfer and to circumferentially distribute liquid.

It can be seen from Chapter 2 that the maximum capillary head generated by a wick increases with decreasing pore size. The wick permeability, another desirable feature, increases with increasing pore size, however. For homogeneous wicks, there is an optimum pore size, which is a compromise. There are three main types in this context. Low-performance wicks in horizontal and gravity-assisted heat pipes should permit maximum liquid flow rate by having a comparatively large pore size, as with 100 or 150 mesh. Where pumping capability is required against gravity, small pores are needed. In space the constraints on size and the general high-power capability needed necessitates the use of non-homogeneous or arterial wicks aided by small pore structures for axial liquid flow.

Another feature of the wick, which must be optimised is its thickness. The heat transport capability of the heat pipe is raised by increasing the wick thickness. However, the increased radial thermal resistance of the wick created by this would work against increased capability and would lower the allowable maximum evaporator heat flux. The overall thermal resistance at the evaporator also depends on the conductivity of the working fluid in the wick. (Table 3.3 gives measured values of evaporator heat fluxes for various wick/working fluid combinations.) Other necessary properties of the wick are compatibility with the working fluid, and wettability. It should be easily formed to mould into the wall shape of the heat pipe and should preferably be of a form that enables repeatable performance to be obtained. It should be cheap.

A guide to the relative cost of wicks may be given by noting that the mass-produced heat pipes for electronics applications use sintered copper powder or woven wire mesh, while where aluminium gravity-assisted units can be used (as in some solar collectors), extruded grooves are employed. Micro-heat pipes may not need a separate 'wick'. The corners of the pipe (where non-circular) may generate capillary action (see Chapter 6).

3.2.1 Homogeneous structures

Of the wick forms available, meshes and twills are the most common. These are manufactured in a range of pore sizes and materials, the latter including stainless steel, nickel, copper and aluminium. Table 3.4 shows measured values of pore size and permeabilities for a variety of meshes and twills. Homogeneous wicks fabricated using metal foams, and more particularly felts, are becoming increasingly useful, and by varying the pressure on the felt during assembly, varying pore sizes can be produced. By incorporating removable metal mandrels, an arterial structure can also be moulded in the felt.

Foams (see www.Porvair.com for data on a variety of foams) and felts are growing in importance for a variety of heat transfer duties. As well as incorporating arteries in felts, like foams their structure can be 'graded' to allow the structure to be designed for 'local' conditions that may differ from those in other parts of a heat pipe wick (e.g. radially or axially). The research at Auburn University in the United States [74] was directed at examining whether the heat transfer limits of heat pipes could be increased by gradation of the pore structure. In particular, variables could include the porosity, fibre diameter, pore diameter and pore diameter distribution. The particular layouts selected by the Auburn research team – which resemble composite mesh wicks in concept – were unsuccessful due to vapour being unable to escape into the vapour space, but the concept, with different variants, has merit.

Fibrous materials have been widely used in heat pipes, and generally have small pore sizes. The main disadvantage is that ceramic fibres have little stiffness and usually require a continuous support for example by a metal mesh. Thus, while the fibre itself may be chemically compatible with the working fluids, the supporting materials may cause problems. Semena and Nishchik discuss metallic fibre (and sinter) wick properties in reference [42].

Table 3.3 Measured radial evaporator heat fluxes in heat pipes (These are not necessarily limiting values)

Working fluid	Wick	Vapour temp. (°C)	Rad. flux (W/cm^2)
Helium [20]	s/s mesh	−269	0.09
Nitrogen [20]	s/s mesh	−163	1.0
Ammonia [21]	various	20–40	5–15
Ethanol [22]	4 × 100 mesh s/s	90	1.1
Methanol [23]	nickel foam	25–30	0.03–0.4
Methanol [23]	nickel foam	30	0.24–2.6
Methanol [23]	1 × 200 mesh (horiz.)	25	0.09
Methanol [23]	1 × 200 mesh (−2.5 cm head)	25	0.03
Water [21]	various	140–180	25–100
Water [22]	mesh	90	6.3
Water [22]	100 mesh s/s	90	4.5
Water [23]	nickel felt	90	6.5
Water [24]	sintered copper	60	8.2
Mercury [20]	s/s mesh	360	180
Potassium [20]	s/s mesh	750	180
Potassium [21]	various	700–750	150–250
Sodium [20]	s/s mesh	760	230
Sodium [21]	various	850–950	200–400
Sodium [25]	3 × 65 mesh s/s	925	214
Sodium [26]	508 × 3600 mesh s/s twill	775	1250
Lithium [20]	niobium 1% zirconium	1250	205
Lithium [27]	niobium 1% zirconium	1500	115
Lithium [27]	SGS-tantalum	1600	120
Lithium [28]	W-26 Re grooves	1600	120
Lithium [28]	W-26 Re grooves	1700	120
Silver [20]	tantalum 5% tungsten	–	410
Silver [28]	W-26 Re grooves	2000	155

Interest has also been shown in carbon fibre as a wick material. Carbon fibre filaments have many fine longitudinal grooves on their surface, have high capillary pressures and are, of course, chemically stable. A number of heat pipes have been successfully constructed using carbon fibre wicks, including one having a length of 100 m. This demonstrated a heat transport capability three times that of one having a metal mesh wick [12]. The use of carbon fibre reinforced structures as wall material cannot be ruled out in future aerospace applications. Other designs incorporating this wick material have been reported. [13, 14]

Table 3.4 Wick pore size and permeability data

Material and mesh size	Capillary height[1] (cm)	Pore radius (cm)	Permeability (m^2)	Porosity (%)
Glass fibre [29]	25.4	–	0.061×10^{-11}	–
Refrasil sleeving [29]	22.0	–	0.104×10^{-10}	–
Refrasil (bulk) [30]	–	–	0.18×10^{-10}	–
Refrasil (batt) [30]	–	–	1.00×10^{-10}	–
Monel beads [31]				
30–40	14.6	0.052[2]	4.15×10^{-10}	40
70–80	39.5	0.019[2]	0.78×10^{-10}	40
100–140	64.6	0.013[2]	0.33×10^{-10}	40
140–200	75.0	0.009	0.11×10^{-10}	40
Felt metal [32]				
FM1006	10.0	0.004	1.55×10^{-10}	–
FM1205	–	0.008	2.54×10^{-10}	–
Nickel powder [29]				
200 μ	24.6	0.038	0.027×10^{-10}	–
500 μ	>40.0	0.004	0.081×10^{-11}	–
Nickel fibre [29]				
0.01 mm dia.	>40.0	0.001	0.015×10^{-11}	68.9
Nickel felt [33]	–	0.017	6.0×10^{-10}	89
Nickel foam [33]				
Ampornik 220.5	–	0.023	3.8×10^{-9}	96
Copper foam [33]				
Amporcop 220.5	–	0.021	1.9×10^{-9}	91
Copper powder (sintered) [32]	156.8	0.0009	1.74×10^{-12}	52
Copper powder (sintered) [34]				
$45-56\,\mu$	–	0.0009	–	28.7
$100-145\,\mu$	–	0.0021	–	30.5
$150-200\,\mu$	–	0.0037	–	35
Nickel 50 [29]	4.8	–	–	62.5
50 [35]	–	0.0305	6.635×10^{-10}	–
Copper 60 [32]	3.0	–	8.4×10^{-10}	–
Nickel 60 [34]	–	0.009	–	–
100 [35]	–	0.0131	1.523×10^{-10}	–
100 [36]	–	–	2.48×10^{-10}	–
120 [32]	5.4	–	6.00×10^{-10}	–
120[3] [32]	7.9	0.019	3.50×10^{-10}	–
$2^5 \times 120$ [37]	–	–	1.35×10^{-10}	–
120 [38]	–	–	1.35×10^{-10}	–

(Continued)

Table 3.4 (*Continued*)

Material and mesh size	Capillary height[1] (cm)	Pore radius (cm)	Permeability (m^2)	Porosity (%)
S/s 180 (22 °C) [39]	8.0	–	0.5×10^{-10}	–
2 × 180 (22 °C) [39]	9.0	–	0.65×10^{-10}	–
200 [34]	–	0.0061	0.771×10^{-10}	–
200 [32]	–	–	0.520×10^{-0}	–
Nickel 200 [29]	23.4	0.004	0.62×10^{-10}	68.9
2 × 200 [37]	–	–	0.81×10^{-10}	–
Phosp./bronze 200 [40]	–	0.003	0.46×10^{-10}	67
Titanium 2 × 200 [34]	–	0.0015	–	67
4 × 200 [34]	–	0.0015	–	68.4
250 [36]	–	–	0.302×10^{-10}	–
Nickel[3] 2 × 250 [34]	–	0.002	–	66.4
4 × 250 [34]	–	0.002	–	66.5
325 [34]	–	0.0032	–	–
Phosp/bronze [38]	–	0.0021	0.296×10^{-10}	67
S/s (twill) 80[4] [41]	–	0.013	2.57×10^{-10}	–
90[4] [41]	–	0.011	1.28×10^{-10}	–
120[4] [41]	–	0.008	0.79×10^{-10}	–
250 [37]	–	0.0051	–	–
270 [37]	–	0.0041	–	–
400 [37]	–	0.0029	–	–
450 [41]	–	0.0029		

[1] Obtained with water unless stated otherwise.
[2] Particle diameter.
[3] Oxidised.
[4] Permeability measured in direction of warp.
[5] Denotes number of layers.

Sintered powers are available in spherical form in a number of materials, and fine pore structures may be made, possibly incorporating larger arteries for added liquid flow capability. Leaching has been used to produce fine longitudinal channels, and grooved walls in copper and aluminium heat pipes have been applied for heat pipes in zero-gravity environments. (In general grooves are unable to support significant capillary heads alone in earth gravity applications. Also entrainment may limit the axial heat flow.) (Covering grooves with a mesh prevents this.) (Fig. 3.2)

Polymers have been proposed for use as heat pipe wall and wick materials in the past, and for example early flexible heat pipes illustrated in earlier editions of *Heat Pipes* included a polymer section to give this flexibility. The specific porosity/pore size requirements used in the wicks of some modern loop heat pipes (LHPs) has

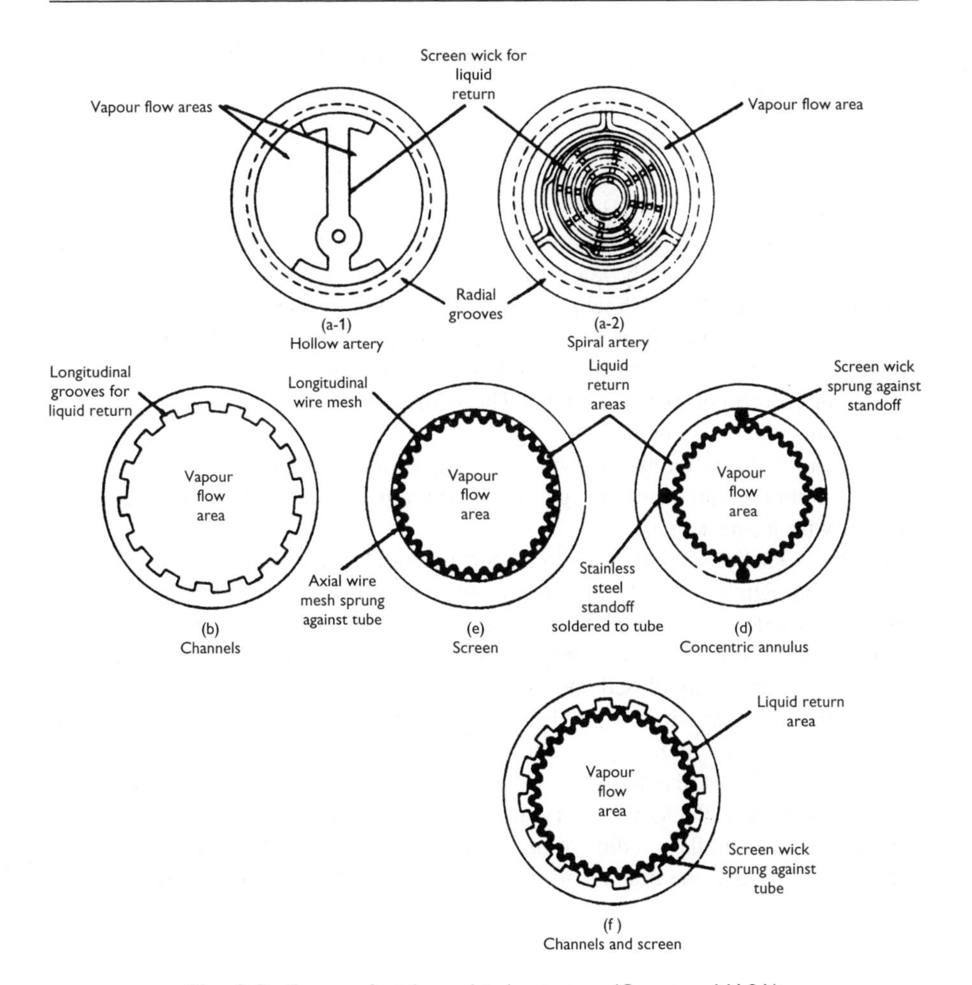

Fig. 3.2 Forms of wick used in heat pipes (Courtesy NASA).

encouraged users to examine a variety of porous structures, including ceramics and polyethylene, as well as porous nickel. As stated by Figus and colleagues at Astrium SAS in France [75], high-performance porous media are needed to achieve high evaporator heat fluxes (greater than $10\,000\,\text{W/m}^2\text{K}$). Early polymer wicks had very small pore sizes but low permeabilities ($10^{-4}\,\text{m}^2$), orders of magnitude lower than those of the wicks listed in Table 3.4, for example. PTFE has also been used by Astrium in LHP evaporators. (Note the wicks in Table 3.4 are principally for 'conventional' heat pipes, not LHPs).

In December 2001 [76, 77], the capillary pumped loop (CPL) successfully demonstrated, during a flight on the Shuttle spacecraft (STS-108) that a CPL with two parallel evaporators using polyethylene wicks, could start-up and operate for substantial periods in high-power (1.5 kW) and low-power (100 W) modes. Over the preceding 2-year fully charged storage period, no gas generation had been observed (see Fig. 1.10 in Chapter 1).

3.2.2 Arterial wicks

Arterial wicks are necessary in high-performance heat pipes for spacecraft, where temperature gradients in the heat pipe have to be minimised to counter the adverse effect of what are generally low thermal conductivity working fluids. An arterial wick developed at IRD is shown in Fig. 3.3. The bore of the heat pipe in this case was only 5.25 mm. This heat pipe, developed for the European Space Organisation (ESA) was designed to transport 15 W over a distance of 1 m with an overall temperature drop not exceeding 6 °C. The wall material was aluminium alloy and the working fluid acetone.

The aim of this wick system was to obtain liquid transport along the pipe with the minimum pressure drop. A high driving force was achieved by covering the six arteries with a fine screen.

In order to achieve the full heat transport potential of the arterial wick, the artery must be completely shut off from the vapour space. The maximum capillary driving force is thus determined by the pore size of the screen. Thus a high degree of quality control was required during the manufacturing process to ensure that the artery was successfully closed and the screen undamaged.

A further consideration in the design of arterial heat pipes is that of vapour or gas blockage of the arteries. If a vapour or gas bubble forms within or is vented into the artery, then the transport capability is seriously reduced. Indeed, if the bubble completely blocks the artery then the heat transport capability is dependent on the effective capillary radius of the artery, i.e. there is an effective state of open artery. In order that the artery will reprime following this condition, the heat load

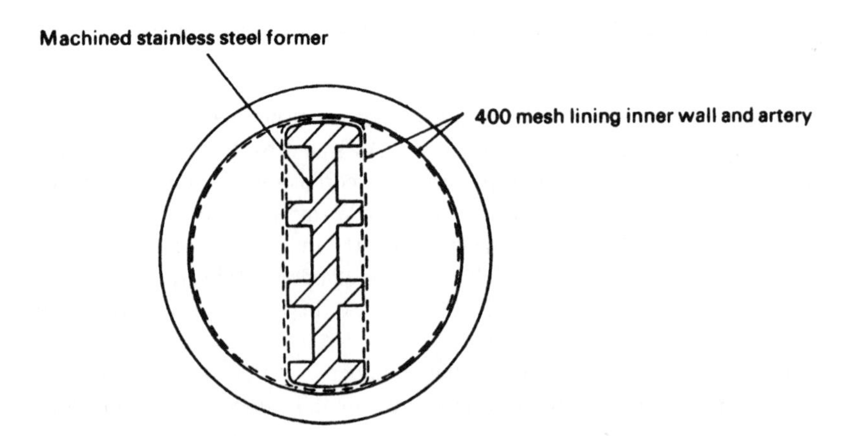

Fig. 3.3 Arterial wick developed at NEI-IRD.

must be reduced to a value below the maximum capability associated with the open artery.

The implications of wick design and working fluid properties in arterial heat pipes are as follows:

(i) The working fluid must be thoroughly degassed prior to filling to minimise the risk of non-condensable gases blocking the artery.
(ii) The artery must not be in contact with the wall to prevent nucleation within it.
(iii) A number of redundant arteries should be provided to allow for some degree of failure.
(iv) Successful priming (i.e. refilling) of the artery, if applied to spacecraft, must be demonstrated in a one 'g' environment, it being expected that priming in a zero 'g' environment will be easier.

As mentioned above, arterial heat pipes have been developed principally to meet the demands of spacecraft thermal control. These demands have increased rapidly over the last decade, and while mechanical pumps can be used in two-phase 'pumped loops', the passive attraction of the heat pipe has spurred the development of several derivatives of 'conventional' arterial heat pipes.

One such derivative is the monogroove heat pipe, shown in Fig. 3.4. This heat pipe comprises two axial channels, one for the vapour and one for the liquid flow. A narrow axial slot between these two channels creates a high capillary pressure difference, while in this particular design circumferential grooves are used for liquid distribution and to maximise the evaporation and condensation heat transfer coefficient, [10, 11]. The principal advantage of this design is the separation of the liquid and vapour flows, eliminating the entrainment, or countercurrent flow, limitation.

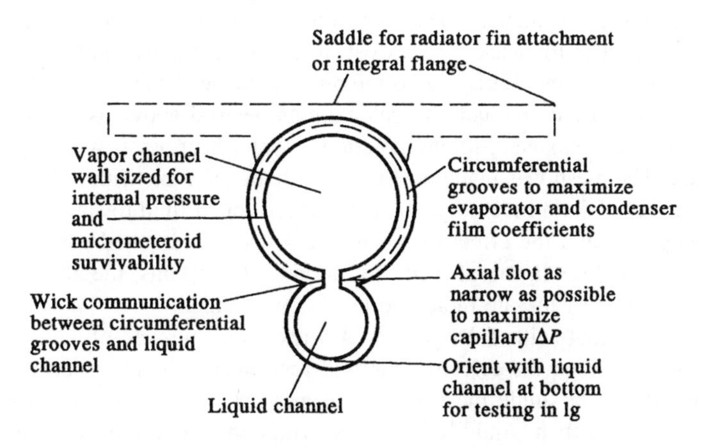

Fig. 3.4 Monogroove heat pipe configuration.

3.3 THERMAL RESISTANCE OF SATURATED WICKS

One feature mentioned in discussions of the desirable properties of both the wick and the working fluid is the thermal conductivity. Expressions are available for predicting the thermal conductivities of saturated wicks of several types, and these are discussed below. The conductivity is an important factor in determining the allowable wick thickness.

3.3.1 Meshes

Gorring and Churchill [43] present solutions for the determination of the thermal conductivity of heterogeneous materials that are divided into three categories; dispersions, packed beds and continuous pairs. No satisfactory solution for a mesh is given because a mesh is a limiting case of dispersion, i.e. the particles are in contact but not tightly packed. However, since the conductivity of dispersions is less than that of packed beds, an estimate of mesh conductivity can be made using Rayleigh's expression for the effective conductivity of a dispersion consisting of a square array of uniform cylinders:

$$k_{\mathrm{w}} = \left(\frac{\beta - \varepsilon}{\beta + \varepsilon}\right) k_{\mathrm{l}} \tag{3.1}$$

where $\beta = \left(1 + \frac{k_{\mathrm{s}}}{k_{\mathrm{l}}}\right) / \left(1 - \frac{k_{\mathrm{s}}}{k_{\mathrm{l}}}\right)$,

k_{s} is the thermal conductivity of solid phase,
k_{l} thermal conductivity of liquid phase and
ε volume fraction of solid phase.

A recent study of the effect of the number of layers of mesh in the wick on the overall heat pipe performance has thrown some doubt on the validity of conduction heat transfer models for wicks [78]. Most of the comparisons were made with the model of Kar and Dybbs [79], and no reference is made to the work of Goring and Churchill [43] and the conductivity calculations derived above as a basis of their work. The experiments were limited to copper–water heat pipes, with copper mesh as the wick. The mesh was made of 0.109 mm diameter wire, with 3.94 strands per mm. The effect of fluid inventory (too little or too much fluid to saturate the wick) was investigated, and the effect of mesh layers was investigated using heat pipes with one, two and six layers of wick. In the latter three units, the amount of fluid to fully saturate the wick, no more and no less was added.

It was found that with fully saturated wicks, the number of layers of wick had little influence on the effective thermal resistance. The difference in thermal resistance between a single mesh layer unit and one with six layers was only 40 per cent, much less than would be predicted with conduction models. However, no alternatives to conduction models were proposed.

3.3.2 Sintered wicks

The exact geometric configuration of a sintered wick is unknown because of the random dispersion of the particles and the varying degree of deformation and fusion, which occurs during the sintering process (see Chapter 4). For this reason, it is suggested that the sintered wick be represented by a continuous solid phase containing a random dispersion of randomly sized spheres of liquid.

Maxwell [44] has derived an expression that gives the thermal conductivity of such a heterogeneous material:

$$k_{\mathrm{w}} = k_{\mathrm{s}}\left[\frac{2+k_{\mathrm{l}}/k_{\mathrm{s}}-2\varepsilon(1-k_{\mathrm{l}}/k_{\mathrm{s}})}{2+k_{\mathrm{l}}/k_{\mathrm{s}}+\varepsilon(1-k_{\mathrm{l}}/k_{\mathrm{s}})}\right] \tag{3.2}$$

Gorring and Churchill show that this expression agrees reasonably well with experimental results.

Work in Japan [18] has confirmed the applicability of Maxwell's equation above for screen wicks also. However, it is stressed that when the effective thermal conductivity is predicted from equation 3.2, it is necessary to accurately estimate the porosity. In order to do so, the intermeshing between the screen layers should be taken into account. This can be done by accurately measuring the overall thickness of the wick.

3.3.3 Grooved wicks

The radial thermal resistance of grooves will be radically different in the evaporator, and condenser sections. This occurs because of the differences in the mechanisms of heat transfer. In the evaporator the land or fin tip plays no active part in the heat transfer process. The probable heat flow path is conduction via the fin, conduction, across a liquid film at the meniscus and evaporation at the liquid–vapour interface.

In the condenser section, grooves will be flooded and the fin tip plays an active role in the heat transfer process. The build up of a liquid film at the fin tip will provide the major resistance to heat flow.

The thickness of the liquid film is a function of the condensation rate and the wetting characteristics of the working fluid.

Since the mechanism in the evaporator section is less complex and should provide the greatest resistance, we will concentrate on the analysis of that region.

Joy [45] and Eggers and Serkiz [46] propose identical models that assume one-dimensional heat conduction along the fin and one-dimensional conduction near the fin tip across the liquid to the liquid/vapour interface where evaporation occurs. In the liquid, the average heat flow length is taken to be a quarter of the channel width and the heat flow area, the channel half width times the input length.

Thus,

$$\frac{\Delta T}{Q} = \frac{\mathrm{a}}{k_{\mathrm{s}}Nfl_{\mathrm{e}}} + \frac{1}{4k_{\mathrm{l}}l_{\mathrm{e}}N} + \frac{1}{h_{\mathrm{e}}\pi bNl_{\mathrm{e}}} \tag{3.3}$$

where N is the number of channels, a the channel depth, b the channel half width and f the fin thickness.

Kosowski and Kosson [49] have made measurements of the maximum heat transport capability and radial thermal resistance of an aluminium grooved heat pipe using Freon 21 and Freon 113[1] and ammonia as the working fluids. The relevant dimensions of their heat pipes were as follows:

$$
\begin{aligned}
N &= 30 \\
a &= 0.89\,\text{mm} \\
2b &= 0.76\,\text{mm} \\
\text{Pipe Outside dia.} &= 12.7\,\text{mm} \\
\text{Heat Pipe No.1 } l_e &= 304.8\,\text{mm},\ l_c = 477.6\,\text{mm} \\
\text{Heat Pipe No.2 } l_e &= 317.5\,\text{mm},\ l_c = 503\,\text{mm}
\end{aligned}
$$

The following heat transfer coefficients (based upon the outside area) were measured:

Fluid	h_c W/m^2 °C	h_c W/m^2 °C
Freon 21 (heat pipe No.1)	1134	1700
Freon 113 (heat pipe No.2)	652	1134
Ammonia (heat pipe No.3)	2268	2840

Converting the heat transfer coefficients into a thermal resistance:

Fluid	R °C/W(evap)	R °C/W(cond)
Freon 21	0.0735	0.031
Freon 113	0.122	0.044
Ammonia	0.035	0.0175

The contribution due to fin conduction is 0.0018 °C/W, ($f = 0.25$ mm, $k_w = 200$ W/m °C) and is negligible. This bears out Kosowski and Kosson's observation that the percentage fill has little effect upon the thermal resistance.

The evaporation term is also small, and the most significant contribution to the resistance is the liquid conduction term. Comparing the theory and experiment

[1] Note that R21 is not recommended for use now because of its toxicity. Other refrigerants, in particular those classed as chlorofluorocarbons (CFCs) and including R113 are banned or limited in their use because of their ozone layer depletion effects. 'New' refrigerants may be substituted as appropriate. R141b has been used in heat pipes, for example.

results suggest that the theory overpredicts the conduction resistance by a considerable amount (50–300 per cent). It would therefore be more accurate to use the integration mean heat flow length

$$\left(1-\frac{\pi}{4}\right)b,$$

rather than $\frac{b}{2}$ such that

$$\frac{\Delta T}{Q}=\frac{\left(1-\frac{\pi}{4}\right)}{2k_1 l_e N} \tag{3.4}$$

knowing the duty and allowable temperature drop into the vapour space, the number of grooves can be calculated for various geometries and working fluids.

In the heat pipe condenser section, or when the channels are mesh-covered, the fin tip plays an active part in the heat transfer process, and the channels are completely filled. In this case, the parallel conduction equation is used:

$$k_w = k_s\left\{1-\varepsilon\left(1-\frac{k_1}{k_s}\right)\right\} \tag{3.5}$$

where ε, the liquid void fraction, is given by:

$$\varepsilon=\frac{2b}{2b+f},$$

where f is the fin thickness.

3.3.4 Concentric annulus

In this case, the capillary action is derived from a thin annulus containing the working fluid. Thus:

$$k_w = k_1 \tag{3.6}$$

This case may also be used to analyse the effects of loose fitting mesh and sintered wicks.

3.3.5 Sintered metal fibres

Data tabulated by the Engineering Sciences Data Unit [50] on a variety of the more common wicks used in heat pipes include expressions for thermal conductivity. (Also included are expressions for porosity, minimum capillary radius, permeability and equivalent diameter.)

The following equation is given for the thermal conductivity of sintered metal fibre wicks:

$$k_w = \varepsilon^2 k_l + (1-\varepsilon)^2 k_s + \frac{4\varepsilon(1-\varepsilon)k_l\, k_s}{k_l + k_s} \tag{3.7}$$

Values obtained using this equation may be compared with data obtained experimentally by Semena and his co-workers [42, 51]. A more detailed discussion of metal felt wicks, with a presentation of data on other wick properties, is given by Acton [52].

3.4 THE CONTAINER

The function of the container is to isolate the working fluid from the outside environment. It has, therefore, to be leak-proof, to maintain the pressure differential across its walls and to enable the transfer of heat to take place into and from the working fluid.

Selection of the container material depends on several factors. These are as follows:

(i) Compatibility (both with working fluid and the external environment).
(ii) Strength-to-weight ratio.
(iii) Thermal conductivity.
(iv) Ease of fabrication, including weldability, machineability and ductility.
(v) Porosity.
(vi) Wettability.

Most of these are self-explanatory. A high strength-to-weight ratio is more important in spacecraft applications, and the material should be non-porous to prevent the diffusion of gas into the heat pipe. A high thermal conductivity ensures minimum temperature drop between the heat source and the wick.

The thermal conductivity of some wall materials is given in Appendix 2.

3.5 COMPATIBILITY

Compatibility has already been discussed in relation to the selection of the working fluid, wick and containment vessel of the heat pipe. However, this feature is of prime importance and warrants particular attention here.

The two major results of incompatibility are corrosion and the generation of non-condensable gas. If the wall or wick material is soluble in the working fluid, mass transfer is likely to occur between the condenser and the evaporator, with solid material being deposited in the latter. This will result either in local hot spots or blocking of the pores of the wick. Non-condensable gas generation is probably the most common indication of heat pipe failure and, as the non-condensables tend to accumulate in the heat pipe condenser section, which gradually becomes blocked, it is easy to identify because of the sharp temperature drop that exists at the gas/vapour interface.

Some compatibility data is of course available in the general scientific publications and from trade literature on chemicals and materials. However, it has become common practice to carry out life tests on representative heat pipes, the main aim of these tests being to estimate long-term materials compatibility under heat pipe operating conditions. At the termination of life tests, gas analyses and metallurgical examinations as well as chemical analysis of the working fluid may be carried out. (See also Section 4.2, Chapter 4.)

Many laboratories have carried out life tests, and a vast quantity of data have been published. However it is important to remember that while life test data obtained by one laboratory may indicate satisfactory compatibility, different assembly procedures at another laboratory, involving, for example a non-standard materials treatment process, may result in a different corrosion or gas generation characteristic. Thus, it is important to obtain compatibility data whenever procedural changes in cleaning or pipe assembly are made.

Stainless steel is a suitable container and wick material for use with working fluids such as acetone, ammonia and liquid metals from the point of view of compatibility. Its low thermal conductivity is a disadvantage, and copper and aluminium are used where this feature is important. The former is particularly attractive for mass-produced units using water as the working fluid. Plastic has been used as the container material, and at very high temperatures ceramics and refractory metals such as tantalum have been given serious consideration. In order to introduce a degree of flexibility in the heat pipe wall, stainless steel bellows have been used, and in cases where electrical insulation is important, a ceramic or glass-to-metal seal has been incorporated. This must of course be used in conjunction with electrically non-conducting wicks and working fluids.

3.5.1 Historical compatibility data

Long-term life tests on cryogenic heat pipes started a little later than those for higher temperature units. However, there are comprehensive data from European Space Agency sources [80] on stainless steel (container was type 304L and wick type 316) heat pipes using as working fluids methane, ethane, nitrogen or oxygen, arising out of tests extending over a period of up to 13 years.

The test units were 1 m in length and either 3.2 or 6.35 mm outside diameter. Heat transport capability was up to 5 Wm (meaning that the pipe transport 5 W over 1 m, or, for example, 10 W over 0.5 m), and vapour temperatures 70–270 K. Tests were completed in the mid-1990s.

The main outcomes were as follows:

- All pipes retained maximum heat transport capability.
- All pipes maintained maximum tilt capability (capillary pumping demonstration).

- The evaporator heat transfer coefficient remained constant.
- No incompatibility or corrosion was evident in the oxygen and nitrogen pipes.
- Slight incompatibility, resulting in non-condensable gas extending over 1 per cent of the heat pipe length and therefore affecting condenser efficiency, was noted in the ethane and methane units.

Unlike the oxygen and nitrogen TIG-welded pipes, the ethane and methane units were hard-brazed, and the implication is that the gas generation was attributed to this.

A comprehensive review of material combinations in the intermediate temperature range has been carried out by Basiulis and Filler [53] and is summarised below. Results are given in the paper over a wider range of organic fluids, most produced by Dow Chemicals, than given in Table 3.5.

Tests in excess of 8000 h with ammonia/aluminium were reported, but only 1008 h had been achieved at the same time of data compilation with aluminium/acetone. No temperatures were specified by Basiulis for these tests: other workers have exceeded 16 000 h with the latter combination.

Later, the life test work at IKE, Stuttgart, was published [54], involving tests on about 40 heat pipes. The tests indicated that copper/water heat pipes could be operated without degradation over long periods (over 20 000 h), but severe gas generation was observed with stainless steel/water heat pipes. IKE had some reservations concerning acetone with copper and stainless steel. While compatible, it was stressed that proper care had to be given to the purity of both the acetone and the metal. The same reservation applied to the use of methanol.

Exhaustive tests on stainless steel/water heat pipes were also carried out at Ispra [55], where experiments were conducted with vapour temperatures as high as 250 °C.

It was found that neither the variation of the fabrication parameters nor the addition of a large percentage of oxygen to the gas plug resulted in a drastic reduction of the hydrogen generation at 250 °C. Hydrogen was generated within 2 h of start-up in some cases. The stainless steel used was type 316 and such procedures

Table 3.5 Compatibility data (Low-temperature working fluids)

Wick material	Working fluids					
	Water	Acetone	Ammonia	Methanol	Dow-A	Dow-E
Copper	RU	RU	NU	RU	RU	RU
Aluminium	GNC	RL	RU	NR	UK	NR
Stainless steel	GNT	PC	RU	GNT	RU	RU
Nickel	PC	PC	RU	RL	RU	RL
Refrasil fibre	RU	RU	RU	RU	RU	RU

RU, recommended by past successful usage; RL, recommended by literature; PC, probably compatible; NR, not recommended; UK, unknown; GNC, generation of gas at all temperatures; GNT, generation of gas at elevated temperatures, when oxide present.

as passivation and out-gassing were ineffective in arresting generation. However, it was found that the complimentary formation of an oxide layer on the steel did inhibit further hydrogen generation.

Gerrels and Larson [56], as part of a study of heat pipes for satellites, also carried out comprehensive life tests to determine the compatibility of a wide range of fluids with aluminium (6062 alloy) and stainless steel (type 321). The fluids used included ammonia, which was found to be acceptable. It is important, however, to ensure that the water content of the ammonia is very low, only a few parts per million concentration being acceptable with aluminium and stainless steel.

The main conclusions concerning compatibility made by Gerrals and Larson are given below, data being obtained for the following fluids.

n-pentane	CP-32 (Monsanto experimental fluid)
n-heptane	CP-34 (Monsanto experimental fluid)
benzene	Ethyl alcohol
toluene	Methyl alcohol
water (with stainless steel)	ammonia
	n-butane

The stainless steel used with the water was type 321.

All life tests were carried out in gravity refluxing containers, with heat being removed by forced air convection, and being put in by immersion of the evaporator sections in a temperature-controlled oil bath.

Preparation of the aluminium alloy was as follows: initial soak in a hot alkaline cleaner followed by deoxidation in a solution of 112 g sodium sulphate and 150 ml concentrated nitric acid in 850 ml water for 20 min at 60 °C. In addition, the aluminium was either machined or abraded in the area of the welds. A mesh wick of commercially pure aluminium was inserted in the heat pipes. The capsules were TIG welded under helium in a vacuum purged inert gas welding chamber. Leak detection followed welding, and the capsules were also pressure tested to 70 bar. A leak check also followed the pressure test.

The type 321 stainless steel container was cleaned before fabrication by soaking in hot alkaline cleaner and pickling for 15 min at 58 °C in a solution of 15 per cent by volume of concentrated nitric acid, 5 per cent by volume of concentrated hydrochloric acid and 80 per cent water. In addition the stainless steel was passivated by soaking for 15 min at 65 °C in a 15 per cent solution of nitric acid. Type 316 stainless steel was used as the wick. The capsule was TIG welded in air with argon purging[2].

The boiling-off technique was used to purge the test capsules of air.

[2] In both series of U.S.A. tests mixtures of materials were used. This is not good practice in life tests as any degradation may not be identified as being caused by the particular single material.

In the case of methyl alcohol, reaction was noted during the filling procedure, and a full life test was obviously not worth proceeding with.

Sealing of the capsules was by a pinch-off, followed by immersion of the pinched end in an epoxy resin for final protection.

The following results were obtained:

n-Pentane: Tested for 750 h at 150 °C. Short-term instabilities noted with random fluctuation in temperature of 0.2 °C. On examination of the capsule, very light brownish areas of discolouration were noted on the interior wall but the screen wick appeared clean. No corrosion evidence was found. The liquid removed from the capsule was found to be slightly brown in colour.

n-Heptane: Tested for 600 h at 160 °C. Slight interior resistances noted after 465 h, but on opening the capsule at the end of the tests, the interior, including the screen, was clear, and the working fluid clean.

Benzene: Tested for 750 h at 150 °C plus vapour pressure 6.7 bar. Very slight local areas of discolouration found on the wall, the wick was clean, there was no evidence of corrosion and the liquid was clear. It was concluded that benzene was very stable with the chosen aluminium alloy.

Toluene: Test run for 600 h at 160 °C. A gradual decrease in condenser section temperature was noted over the first 200 h of testing but no change was noted following this. On opening the capsule, slight discolouration was noted locally on the container wall. This seemed to be a surface deposit, with no signs of attack on the aluminium. The screen material was clean and the working fluid clear at the end of the test.

Water: (stainless steel) Tested at 150 °C plus for 750 h. Vapour pressure 6.7 bar. Large concentration of hydrogen was found when analysis of the test pipe was performed. this was attributed in part to poor purging procedure as there was discolouration in the area of the welds, and the authors suggest that oxidation of the surfaces had occurred. A brown precipitate was also found in the test heat pipe.

CP-32: Tested for 550 h at 158 °C. A brownish deposit was found locally on interior surfaces. Screen clean, but the working fluid was darkened.

CP-34: Tested for 550 h at 158 °C. Gas generation was noted. Also extensive local discolouration on the capsule wall near the liquid surface. No discolouration on the screen. The fluid was considerably darkened.

Ammonia: Tested at 70 °C for 500 h. Some discolouration of the wall and mesh was found following the tests. This was attributed to some non-volatile impurity in the ammonia which could have been introduced when the capsule was filled. In particular, the lubricant on the valve at the filling position could have entered with the working fluid. (This was the only test pipe on which the filling was performed through a valve.)

n-Butane: Tested for 500 h at 68 °C. It was considered that there could have been non-condensable gas generation, but the fall off in performance was attributed to some initial impurity in the n-butane prior to filling. The authors felt that this impurity could be isobutane. Further tests on purer n-butane gave better results, but the impurity was not completely removed.

Gerrals and Larson argued thus concerning the viability of their life tests: 'It should be emphasised that the present tests were planned to investigate the compatibility of a particular working fluid–material combination for long-term (5 years) use in vapour chamber radiator under specified conditions. The reference conditions call for a steady state temperature of 143 °C for the primary radiator fluid at the inlet to the radiator and a 160 °C short-term peak temperature. The actual temperature to which the vapour chamber working fluid is exposed must be somewhat less than the primary radiator fluid temperature, since some temperature drop occurs from the primary radiator fluid to the evaporative surface within the vapour chamber. It is estimated that in these capsule tests the high temperature fluids were exposed to temperatures at least 10 °C higher than the peak temperature and at least 20 °C greater than long-term steady state temperatures that the fluids would experience in the actual radiator. Although the time of operation of these capsule tests is only about 1% of the planned radiator lifetime, the conditions of exposure were much more severe. It seems reasonable then, to assume that if the fluid–material combination completed the capsule tests with no adverse effects, it is a likely candidate for a radiator with a 5 year life.'

On the basis of the above tests, Gerrels and Larson selected the following working fluids:

6061 Aluminium at temperatures not exceeding 150 °C:

Benzene
n-Heptane
n-Pentane

6061 Aluminium at temperatures not exceeding 65 °C:

Ammonia
n-Butane

The following fluids were felt not to be suitable:

Water (in type 321 stainless steel)
CP-32, CP-34, Methyl alcohol, Toluene – in 6061 aluminium

Gerrels and Larson point out that Los Alamos Laboratory obtained heat pipe lives in excess of 3000 h without degradation using a combination of water and type 347 stainless steel.

Other data [59] suggest that alcohols in general are not suitable with aluminium.

Summarising the data of Gerrels and Larson, ammonia was recommended as the best working fluid for vapour chambers operating below 65 °C and *n*-pentane the best for operation above this temperature, assuming that aluminium is the container material.

At the other end of the temperature scale, long lives have been reported [8] for heat pipes with lithium or silver as the working fluid. With a tungsten rhenium (W-26 Re) container, a life of many years was forecast with lithium as the working fluid, operating at 1600 °C. At 1700 °C significant corrosion was observed after 1 year, while at 1800 °C the life was as short as 1 month. W-26 Re/silver heat pipes were considered capable of operating at 2000 °C for 10 900 h. Some other results are presented in Table 3.6 updated using information summarised in [58].

Confirmation on the importance of the purity of the working fluid in contributing to a satisfactory life with lithium heat pipes is given by the work at the Commissariat *à* l'Energie Atomique in Grenoble [16]. In pointing out that failures in liquid metal heat pipes most commonly arise due to impurity-driven corrosion mechanisms, a rigorous purifying procedure for the lithium is carried out. This involves forced circulation on a cold and on a hot trap filled with Ti-Zr alloy as a getter to lower the oxygen and nitrogen content, respectively. The lithium is then distilled under vacuum before filling the heat pipe.

Although at the time of writing, no internal examination of the heat pipes was reported, data on extended tests on units using lead at more modest temperatures (around 1340 °C) have been obtained over 4800 h at Los Alamos Laboratory [19]. Using one heat pipe fabricated from a molybdenum tube with a Ta-10% W wall and a W wick, no degradation in performance was noted over the test period.

One subject of much argument is the method of conducting life tests and their validity when extrapolating likely performance over a period of several years. For example, on satellites, where remedial action in the event of failure is difficult, if not impossible, to implement, a life of 7 years is a standard minimum requirement.[3] It is therefore necessary to accelerate the life tests so that reliability over a longer period can be predicted with a high degree of accuracy.

Life tests on heat pipes are commonly regarded as being primarily concerned with the identification of any incompatibilities that may occur between the working fluid and wick and wall materials. However, the ultimate life test would be in the form of a long-term performance test under likely operating conditions. However, if this is carried out, it is difficult to accelerate the life test by increasing, say, the evaporator heat flux, as any significant increase is likely to cause dry-out as the pipe will be operating well in excess of its probable design capabilities. Therefore, any accelerated life test that involves heat flux increases of the order of, say, four

[3] European Space Agency requirement.

Table 3.6 Compatibility data (life tests on high-temperature heat pipes)

Working	Material		Vapour temp. (°C)	Duration (h)
	Wall	Wick		
Caesium	Ti		400	>2000
	Nb + 1% Zr		1000	8700
Potassium	Ni		600	>6000
	Ni		600	16 000
	Ni		600	24 500
	304, 347 ss		510, 650	6100
Sodium	Hastelloy X		715	>8000
	Hastelloy X		715	>33 000
	316 ss		771	>4000
	Nb + 1% Zr		850	>10 000
	Nb + 1% Zr		1100	1000
	304, 347 ss		650–800	7100
Bismuth	Ta		1600	39
	W		1600	118
Lithium	Nb + 1% Zr		1100	4300
	Nb + 1% Zr		1500	>1000
	Nb + 1% Zr		1600	132
	Ta		1600	17
	W		1600	1000
	SGS-Ta		1600	1000
	TMZ		1500	9000
	W + 26% Re		830–1000	7700
Lead	Nb + 1% Zr		1600	19
	SGS-Ta		1600	1000
	W		1600	1000
	Ta		1600	>280
Silver	Ta		1900	100
	W		1900	335
	W		1900	1000
	Re	W	2000	300

over that required under normal operating conditions must be carried out in the reflux mode, with regular performance tests to ensure that the design capability is still being obtained.

An alternative possibility as a way of accelerating any degradation processes, and one that may be carried out with the evaporator up if the design permits, is to raise the operating temperature of the heat pipe. One drawback of this method is the effect that increased temperature may have on the stability of the working fluid itself. Acetone cracking, for example, might be a factor where oxides of metals are present, resulting in the formation of diacetone alcohol that has a much higher boiling point than pure acetone.

Obviously there are many factors to be taken into account when preparing a life test programme, including such questions as the desirability of heat pipes with valves, or completely sealed units as used in practice. This topic is of major importance, and life test procedures are discussed more fully in Chapter 4.

One of the most comprehensive life test programmes is that being carried out by Hughes Aircraft Co. [65]. Additional experience has been gained with some of the material combinations discussed above and a summary of recommendations based on these tests is given in Table 3.7.

The lack of support given to a nickel/water combination is based on an Arrhenius type accelerated life test carried out by Anderson [70]. Work carried out on nickel wicks in water heat pipes, generally with a copper wall, has, in the authors' experience, not created compatibility problems, and this area warrants further investigation. With regard to water/stainless steel combinations, for some years the subject of considerable study and controversy, the Hughes work suggests that type 347 stainless steel is acceptable as a container with water. Tests had been progressing since December 1973 on a 347 stainless steel container, copper wicked water heat pipe operating at 165 °C, with no trace of gas generation. (Type 347 stainless steel contains no titanium but does contain niobium.) Surprisingly, a type 347 wick caused rapid gas generation. The use of Dowtherm A (equivalent to Thermex, manufactured by ICI) is recommended for moderate temperatures only, breakdown of the fluid progressively occurring above about 160 °C. With careful materials preparation Thermex appears compatible with mild steel and the Hughes data is limited somewhat by the low operating temperature conditions.

Hughes emphasised the need to carry out rigorous and correct cleaning procedures, and also stressed that the removal of cleaning agents and solvents prior to filling with working fluid is equally important.

3.5.2 Compatibility of water and steel – A discussion

Water is an ideal working fluid for heat pipes, because of its high latent heat, thus requiring a relatively low inventory, its low cost and its high 'figure of merit'. As shown already in this Chapter, water is compatible with a number of container materials, the most popular being copper. However, ever since heat pipes were first conceived, experimenters have experienced difficulty in operating a water/steel (be it mild, boiler or stainless steel) heat pipe without obtaining the generation of hydrogen in the container. This gas generation always manifests itself as a cold plug of gas at the condenser section of the heat pipe – blocking off surface for heat rejection. There is a sharply defined interface between the water vapour and the non-condensable gas, making the presence readily identifiable. The effect is similar to that when air is present in a conventional domestic radiator system but of course the reasons may be different.

The earlier description in Chapter 1 of the Perkins Tube and its derivatives, which used iron or steel with water as the working fluid, shows that the fluid/wall combination has demonstrated a significant life, although gas generation did occur.

Table 3.7 Hughes Aircraft Compatibility Recommendations

	Recommended	Not recommended
Ammonia	Aluminium	Copper
	Carbon steel	
	Nickel	
	Stainless steel	
Acetone	Copper	
	Silica	
	Aluminium[1]	
	Stainless steel[1]	
Methanol	Copper	Aluminium
	Stainless steel	
	Silica	
Water	Copper	
	Monel	
	347 Stainless steel[2]	Stainless steel
		Aluminium
		Silica
		Inconel
		Nickel
		Carbon steel
Dowtherm A	Copper	
	Silica	
	Stainless steel[3]	
Potassium	Stainless steel	Titanium
	Inconel	
Sodium	Stainless steel	Titanium
	Inconel	

Note: Type 347 stainless steel as specified in AISI codes does not contain tantalum. AISI type 348, which is otherwise identical except for a small tantalum content, should be used in the United Kingdom (Authors). Work on water in steel heat pipes is also discussed later in this Chapter.

[1] The use of acetone with aluminium and/or stainless steel presented problems to the authors, but others have had good results with these materials. The problem may be temperature-related use with caution.

[2] Recommended with reservations.

[3] This combination should be used only where some non-condensible gas in the heat pipe is tolerable, particularly at higher temperatures.

3.5.2.1 The mechanism of hydrogen generation and protective layer formation

The reaction responsible for the generation of hydrogen is

$$Fe + 2H_2O \rightarrow Fe(OH)_2 + H_2$$

Corrosion of steel is negligible in the absence of oxygen. Therefore, in a closed system such as a steel heat pipe with water as the working fluid, the corrosion only takes place until all the free oxygen is consumed. When oxygen is deficient, low alloy steels develop a hydrated magnetite layer in neutral solutions, by decomposition of the ferrous hydroxide. Further dehydration, or a chemical reaction of iron with water, occurs.

$$3Fe(OH)_2 \rightarrow Fe_3O_4 + 2H_2O + H_2$$

The conversion of $Fe(OH)_2$ leads to a protective layer of Fe_3O_4 especially with mild/carbon steels. In high-temperature water, this Fe_4O_3 layer is responsible for the good corrosion resistance of boiler steels in power generation plant boilers.

A number of different methods have been developed to obtain non-porous and adhesive magnetite layers, and these are discussed below.

In addition, several other observations have been made concerning factors which influence the long-term compatibility of steel with water:

- The pH value of the water should be greater than 9 (some researchers quote a range of 6–11 as being satisfactory – see below for a discussion). This value can be adjusted by conditioning of the water. The high pH value has also been reported to favour the growth of effective passive layers.
- The water must be fully degassed and desalted (or demineralised) before conditioning.
- The protective layer can be destroyed if the metal is locally overheated above 570 °C (thus the passivation reaction is best carried out once the container has been welded or subjected to any other high-temperature procedures during assembly).

3.5.2.2 Work specifically related to passivation of mild steel

There are a number of potential/active heat pipe applications where the advantages of using water with mild steels is attractive. For example, in domestic radiators attempts have been made to construct two-phase units with more rapid responses than those of a single-phase heating system. Japanese and Korean companies have marketed such products, and research in the United Kingdom has also investigated such combinations.

Work some years ago in the then Czechoslovakia [57] discusses three approaches to the prevention of hydrogen generation:

- Preparation of heat pipes with an inhibitor added to the working fluid.
- Preparation of pipes provided with a protective oxide layer.
- Preparation of heat pipes involving a protective layer in conjunction with an inhibitor.

All the tests were carried out with the condenser above the evaporator, and with the evaporator in a sand bath heated to 200 °C. The test period was 6000 h.

Table 3.8 Heat pipe specification and analytical results after 6000 h

Code	Surface treatment	Working fluid	pH	Inhibitor concentration	Temperature difference (°C)	Amount of H_2
X	None	Water	8.32	None	33.2	Maximum
A	None	+ Inhibitor	10.52	0.04	6	Trace
b	Oxidation + vapour at 550 °C	Water	8.01	None	10	Trace
c_1	Oxidation + vapour at 550 °C	+ Inhibitor	9.05	0.03	3	None
c_2	Oxidation + vapour + catalyst at 550 °C	+ Inhibitor	7.01	0.45	13	Trace

Measurements were made of the temperature difference along the pipes (higher values are poor), and the gases inside were analysed for hydrogen. All results are compared in Table 3.8. A steel heat pipe without any treatment was tested to form the basis for comparison – pipe (X). As can be seen from the Table, this exhibited a high-temperature drop along its length, indicating gas generation.

3.5.2.3 Use of an inhibitor

The inhibitor selected was an anodic type based on a chromate, to minimise oxidation. K_2CrO_4 was selected and used in a concentration of 5 g for 1 l of water, with a pH of 7.87. The pipe internal surface had been previously degassed. The inhibitor acts to decelerate the anodic partial reaction ($Fe \rightarrow Fe^{2+} + 2e$). The chromate anion is first adsorbed on the active sites of the metal surface and the film formed is chromium III oxide, Cr_2O_3, or chromium III hydroxide–chromate, $Cr(OH)CrO_4$. The passivating concentration varies from 0.1 per cent to 1.0 per cent.

The pipe (A) with an inhibitor and no protective oxide layer exhibited a temperature difference along its length of 6 °C and traces of hydrogen were found on analysis of the gases in the heat pipe.

3.5.2.4 Production of a protective layer

As discussed above, the natural ageing process of mild steel leads to a protective layer of magnetite being formed, which reduces hydrogen generation rates. It was found in tests on naturally passivated steel in the laboratory that while the performance stabilised after 2000 h of operation, there was in the intervening period sufficient generation of hydrogen to reduce significantly the performance, again evident in an increased temperature drop along the length.

Therefore, it was decided to accelerate the passivation process prior to filling and sealing the heat pipe by oxidising the pipe surface with superheated steam vapour. The efficiency of this process is determined by the thickness and porosity of the magnetite layer formed. Optimum thickness of the layer was found to be 3–5 μm. Both of these properties are a function of temperature and the time of oxidation, and can also be positively influenced by adding an oxidation catalyst, ammonium molybdate $(NH_4)_2MoO_4$ to the water vapour. This leads to molybdenum trioxide being deposited on the surfaces being treated. Note that the stability of magnetite reduces above 570 °C and this temperature should not be exceeded during the process (as mentioned above).

The heat pipe tested (*b*) with vapour oxidation carried out at 550 °C had, at the end of the test period, working fluid with a pH of 8.01 and had a temperature difference along its length of 10 °C, larger than that with the inhibitor alone. Only a trace of hydrogen was found on analysis.

3.5.2.5 Pipes with both inhibitor and oxide layer

A combination of magnetite layer and inhibitor introduced into the working fluid was then tested. It was pointed out that the pH of the inhibitor solution should be maintained in the range $6 < pH < 11$ so that the inhibitor did not break down the magnetite layer.

The optimum performance was achieved with pipe c_1, which was passivated using superheated steam without addition of the oxidation catalyst, and also contained an inhibitor. The catalyst led to inferior performance in the pipe c_2, which again had an inhibitor present. The poor performance of this latter heat pipe was attributed to faults in the protective layer due to the exceeding of a limiting layer thickness and by dissolution of MoO_3 in the aqueous solution of potassium chromate.

The temperature drop in pipe c_1 was 3 °C, and there was no indication of hydrogen formation. An analysis of the working fluid showed that the pH is strongly shifted towards the alkaline region after the 6000 h test.

(Note that the temperature difference is that measured along the condenser section of the heat pipe.)

In a further communication [66], the Czech research team reported on the extension of the tests to 18 000 hours. They concluded, on the basis of the very small (1 °C) increase in temperature difference for pipes (*a*) and (c_1) measured over the additional 12 000 h, that the inhibitor stabilises the performance of water/steel heat pipes, but a superior performance is achieved with the combination of inhibitor and oxidation. However, this latter process was substantially more costly.

The use of inhibitors tested for up to 35 000 h has been reported from the Ukraine [82]. The steel used was Steel 10 GOST 8733–87. Best results with only a minor increase in thermal resistance were achieved with a chromate-based inhibitor. Although the vapour temperature was only 90 °C, the authors recommended that accelerated life tests should be done for at least 35 000 h at 200–250 °C.

Work in China [83] on long-term compatibility of steel/water units, with a carbon content slightly higher than the steel used in the RRR (0.123%), gave similar results to those of Novotna et al. The Chinese tests were carried out over a period of 29 160 h, with temperature drops along the heat pipes with passivation, with or without an inhibitor, never exceeding 6 °C, and frequently being substantially less. Since this, carbon steel/water heat pipes have become a regular feature of heat recovery plant in China, with one company producing over 1000 heat exchangers [64].

Reported in 2003, the extensive activities in China on steel/water heat pipes have continued [84]. Tests in heat recovery duties where exhaust gases are up to 700 °C have been carried out safely. Typified by the data in [63], the National Technological Supervision Administration of China has, with others, formulated Standards for the technical specification of carbon steel–water heat pipe heat exchangers and boiler heat pipe economisers.

Further work in France, following an almost identical path to that in Czechoslovakia, confirms the above results. The authors [17] concluded that: 'In accordance with other studies, the addition of chromates in a neutral solution of distilled water and the formation of a magnetite layer are sufficient conditions to prevent corrosion and hydrogen release in water/mild steel heat pipes at any operating temperature.'

The importance of deoxygenation of the water even in systems that are not sealed to such a high integrity as heat pipes is demonstrated by work on district heating pipeline corrosion [81]. Effective passivation was achieved in conjunction with deaerated water with an oxygen content of 40 μg/kg, whereas if normal water with an oxygen content of 4 mg/kg was used, pitting on the steel surfaces appeared.

3.5.2.6 Comments on the water-steel data

The above data suggest that, with correct treatment of the water and the internal wall of the heat pipe/thermosyphon, steel and water are compatible in terms of the rate of non-condensable gas generation.

There is a cost associated with the prevention of gas generation, both in treating the water and in achieving a suitable magnetite (or other material/form) layer on the container wall. If any subsequent manufacturing process (e.g. welding, heat treatment, enamel coating) involves a high-temperature operation downstream of passivation, this may need to be rescheduled – it is always important when considering a heat pipe procedure that involves changes to surface structures internally to remember that subsequent manufacturing processes may degrade such procedures. Similarly, the heat pipe manufacturer may have little or no control over the system installer.

3.6 HOW ABOUT WATER AND ALUMINIUM?

It is interesting to observe that there is, on the other hand, little or no data on the compatibility of aluminium and water as heat pipe combinations, and few attempts have been made by the heat pipe fraternity to overcome the perceived incompatibility using this combination.

The Ukrainian research mentioned in Ref. [82] did look at aluminium-water thermosyphons, using alloy 6060 and it was found that some corrosion could be slowed by selection of optimum pH conditions (5–6.5) and passivation method. However, long-term compatibility was not demonstrated.

Geiger and Quataert [67] have carried out corrosion tests on heat pipes using tungsten as the wall material and silver (Ag), gold (Au), copper (Cu), gallium (Ga), germanium (Ge), indium (In) and tin (Sn) as working fluids. The results from tests carried out at temperatures of up to 1650 °C enable Table 3.6 to be extended above 2000 °C, albeit for heat pipes having comparatively short lives. Of the above combinations, tungsten and silver proved the most satisfactory, giving a life of 25 h at 2400 °C, with a possible extension if improved quality tungsten could be used.

3.7 HEAT PIPE START-UP PROCEDURE

Heat pipe start-up behaviour is difficult to predict and may vary considerably depending upon many factors. The effects of working fluid and wick behaviour and configuration on start-up performance have been studied qualitatively, and a general description of start-up procedure has been obtained [60].

During start-up, vapour must flow at a relatively high velocity to transfer heat from the evaporator to the condenser, and the pressure drop through the centre channel will be large. Since the axial temperature gradient in a heat pipe is determined by the vapour pressure drop, the temperature of the evaporator will be initially much higher than that of the condenser. The temperature level reached by the evaporator will, of course, depend on the working fluid used. If the heat input is large enough, a temperature front will gradually move towards the condenser section. During normal heat pipe start-up, the temperature of the evaporator will increase by a few degrees until the front reached the end of the condenser. At this point, the condenser temperature will increase until the pipe structure becomes almost isothermal (when lithium or sodium are used as working fluids, this process occurs at temperature levels where the heat pipe becomes red hot, and the near isothermal behaviour is visible).

Heat pipes with screen-covered channels behave normally during start-up as long as heat is not added too quickly. Kemme found that heat pipes with open channels did not exhibit straightforward start-up behaviour. Very large temperature gradients were measured, and the isothermal state was reached in a peculiar manner. When heat was first added, the evaporator temperature levelled out at 525 °C (sodium being the working fluid) and the front, with a temperature of 490 °C, extended only a short distance into the condenser section. In order to achieve a near isothermal condition more heat was added, but the temperature of the evaporator did not increase uniformly, a temperature of 800 °C being reached at the end of the evaporator farthest from the condenser. Most of the evaporator remained at 525 °C and a sharp gradient existed between these two temperature regions.

Enough heat was added so that the 490 °C front eventually reached the end of the condenser. Before this occurred, however, temperatures in excess of 800 °C were observed over a considerable portion of the evaporator. Once the condenser became almost isothermal, its temperature rapidly increased and the very hot evaporator region quickly cooled in a pattern, which suggested that liquid return flow was in fact taking place. From this point, the heat pipe behaved normally.

In some instances during start-up, when the vapour density is low and its velocity high, the liquid can be prevented from returning to the evaporator. This is more likely to occur when open return channels are used for liquid transfer than when porous media are used.

Further work by van Andel [61] on heat pipe start-up has enabled some quantitative relationships to be obtained which assist in ensuring that satisfactory start-up can occur. This is based on the criterion that burn-out does not occur, i.e. the saturation pressure in the heated zones should not exceed the maximum capillary force. If burn-out is allowed to occur, drying of the wick results, inhibiting the return flow of liquid.

A relationship that gives the maximum allowable heat input rate during the start-up condition is

$$Q_{\mathrm{max}} = 0.4\pi r_{\mathrm{c}}^{2} \times 0.73\,L(P_{\mathrm{E}}\rho_{\mathrm{E}})^{\frac{1}{2}} \tag{3.8}$$

where r_{c} is the vapour channel radius, L the latent heat of vapourisation, and P_{E}, ρ_{E} are the vapour pressure and vapour density in the evaporator section.

It is important to meet the start-up criteria when a heat pipe is used in an application that may involve numerous starting and stopping actions, for example in cooling a piece of electronic equipment or cooling brakes. One way in which the problem can be overcome is to use an extra heat source connected to a small branch heat pipe when the primary role of cooling is required, thus reducing the number of start-up operations. The start-up time of gas buffered heat pipes is quicker.

Busse [62] has made a significant contribution to the analysis of the performance of heat pipes, showing that before sonic choking occurs, a viscous limitation that can lie well below the sonic limit can be met. This is described in detail in Chapter 2.

Where it is required to calculate transient behaviour of heat pipes during start-up and in later transient operation, time constants and other data may be calculated using equations presented in Ref. [58].

REFERENCES

[1] Anon. Thermophysical properties of heat pipe working fluids: operating range between −60 °C and 300 °C. Data Item No. 80017, Engineering Sciences Data Unit, London, 1980.

[2] Zhang, J., Li, J., Yang, J. and Xu, T. Analysis of heat transfer in the condenser of naphthalene thermosyphon at small inclination. Paper H3-1 Proceedings of 10th International Heat Pipe Conference, Stuttgart, 21–25 September 1997.

[3] Groll, M. et al. Heat recovery units employing reflux heat pipes as components. Final Report, Contract EE-81-133D(B). Commission of the European Communities Report EUR9166EN, 1984.

[4] Groll, M. Heat pipe research and development in Western Europe. Heat Recov. Syst. & CHP., Vol. 9, No. 1, pp 19–66, 1989.

[5] Deverall, J.E. Mercury as a heat pipe fluid. ASME Paper 70 HT/Spt-8, American Society of Mechanical Engineers, 1970.

[6] Reay, D.A. Mercury wetting of wicks. Proceedings of the 4th C.H.I.S.A. Conference, Prague, September 1972.

[7] Bienert, W. Heat pipes for solar energy collectors. 1st International Heat Pipe Conference, Stuttgart, Paper 12–1, October 1973.
[8] Kemme, J.E. et al. Performance investigations of liquid metal heat pipes for space and terrestrial applications. AIAA Paper 78–431, Proceedings of the III International Heat Pipe Conference, Palo Alto, May 1978.
[9] Quataert, D., Busse, C.A. and Geiger, F. Long time behaviour of high temperature tungsten-rhenium heat pipes with lithium or silver as the working fluid. Paper 4–4. Proceedings of the III International Heat Pipe Conference, Palo Alto, May 1978.
[10] Dobran, F. Heat pipe research and development in the Americas. Heat Recov. Syst. CHP, Vol. 9, No. 1, pp 67–100, 1989.
[11] Reay, D.A. Heat transfer enhancement – a review of techniques. Heat Recov. Syst. CHP, Vol. 11, No. 1, pp 1–40, 1991.
[12] Takaoka, T. et al. Development of long heat pipes and heat pipe applied products. Fujikura Technical Review, pp 77–93, 1985.
[13] Schimizu, A. et al. Characteristics of a heat pipe with carbon fibre wick. Proceedings of the 7th International Heat Pipe Conference, Minsk, 1990. Hemisphere, New York, 1991.
[14] Kaudinga, J.V. et al. Experimental investigation of a heat pipe with carbon fibre wick. Proceedings of the 7th International Heat Pipe Conference, Minsk, 1990. Hemisphere, New York, 1991.
[15] Roesler, A. et al. Performance of closed two-phase thermosyphons with high-temperature organic working fluids. Proceedings of the 7th International Heat Pipe Conference, Minsk, 1990. Hemisphere, New York, 1991.
[16] Bricard, A. et al. High temperature liquid metal heat pipes. Proceedings of the 7th International Heat Pipe Conference, Minsk, 1990. Hemisphere, New York, 1991.
[17] Bricard, A. et al. Recent advances in heat pipes for hybrid heat pipe heat exchangers. Proceedings of the 7th International Heat Pipe Conference, Minsk, 1990. Begel Corporation, 1995.
[18] Kozai, H. et al. The effective thermal conductivity of screen wick. Proceedings of the 3rd International Heat Pipe Symposium, Tsukuba. Japan Association for Heat Pipes, Sagamihara, Japan, 1988.
[19] Merrigan, M.A. An investigation of lead heat pipes. Proceedings of the 7th International Heat Pipe Conference, Minsk, 1990. Hemisphere, New York, 1991.
[20] Lidbury, J.A. A helium heat pipe. Nimrod Design Group Report NDG-72-11, Rutherford Laboratory, England, 1972.
[21] Groll, M. Wärmerohre als Baudemente in der Wärme-und Kältetechnik. Brennst-Waerme-kraft., Vol. 25, No. 1, 1973 (German).
[22] Marto, P.J. and Mosteller, W.L. Effect of nucleate boiling on the operation of low temperature heat pipes. ASME Paper 69-HT-24.
[23] Phillips, E.C. Low temperature heat pipe research program. NASA CR-66792, 1970.
[24] Keser, D. Experimental determination of properties of saturated sintered wicks. 1st International Heat Pipe Conference, Stuttgart, 1973.
[25] Moritz, K. and Pruschek, R. Limits of energy transport in heat pipes. Chem. Ing. Technik., Vol. 41, No. 1, 2, 1969 (German).
[26] Vinz, P. and Busse, C.A. Axial heat transfer limits of cylindrical sodium heat pipes between $25\,W/cm^2$ and $15.5\,kW/cm^2$. 1st International Heat Pipe Conference, Paper 2–1, Stuttgart, 1973.
[27] Busse, V.A. Heat pipe research in Europe. Euratom Report. EUR 4210 f, 1969.

[28] Quataert, D., Busse, C.A. and Geiger, F. Long term behaviour of high temperature tungsten-rhenium heat pipes with lithium or silver as working fluid. 1st International Heat Pipe Conference, Paper 4–4, Stuttgart, 1973.

[29] Schroff, A.M. and Armand, M. Le Caloduc. Rev. Tech. Thomson-CSF., Vol. 1, No. 4, 1969 (French).

[30] Farran, R.A. and Starner, K.E. Determining wicking properties of compressible materials for heat pipe applications. Proceedings of Aviation and Space Conference, Beverley Hills, California, June 1968.

[31] Ferrell, J.K. and Alleavitch, J. Vaporisation heat transfer in capillary wick structures, Department of Chemical Engineering Report, North Carolina University, Raleigh, USA, 1969.

[32] Freggens, R.A. Experimental determination of wick properties for heat pipe applications. 4th Intersociety Energy Conference Engineering Conference, Washington DC, 22–26 September 1969, pp 888–897.

[33] Phillips, E.C. and Hinderman, J.D. Determination of properties of capillary media useful in heat pipe design. ASME Paper 69-HT-18, 1969.

[34] Birnbreier, H. and Gammel, G. Measurement of the effective capillary radius and the permeability of different capillary structures. 1st International Heat Pipe Conference, Paper 5–4, Stuttgart, October 1973.

[35] Langston, L.S. and Kunz, H.R. Liquid transport properties of some heat pipe wicking materials. ASME Paper 69-HT-17, 1969.

[36] McKinney, B.G. An experimental and analytical study of water heat pipes for moderate temperature ranges. NASA-TM-X53849. Marshall Space Flight Center, Alabama: June 1969.

[37] Calimbas, A.T. and Hulett, R.H. An avionic heat pipe. ASME Paper 69-HT-16, New York, 1969.

[38] Katzoff, S. Heat pipes and vapour chambers for thermal control of spacecraft. AIAA Paper 67–310, 1967.

[39] Hoogendoorn, C.J. and Nio, S.G. Permeability studies on wire screens and grooves. 1st International Heat Pipe Conference, Paper 5–3, Stuttgart, October 1973.

[40] Chun, K.R. Some experiments on screen wick dry-out limits. ASME Paper 71-WA/HT-6, 1971.

[41] Ivanovskii, M.N. et al. Investigation of heat and mass transfer in a heat pipe with a sodium coolant. High Temp., Vol. 8, No. 2, pp 299–304, 1970.

[42] Semena, M.G. and Nishchik, A.P. Structure parameters of metal fibre heat pipe wicks, J. Eng. Phy., Vol. 35, No. 5, pp 1268–1272, 1957.

[43] Gorring, R.L. and Churchill, S.W. Thermal conductivity of heterogeneous materials. Chem. Eng. Prog., Vol. 57, No. 7, July 1961.

[44] Maxwell, J.C. A treatise on electricity and magnetism, Vol. 1, 3rd Edn. OUP, 1954 (1981), reprinted by Dover, New York.

[45] Joy, P. Optimum cryogenic heat pipe design. ASME Paper 70-HT/SpT-7, 1970.

[46] Eggers, P.E. and Serkiz, A.W. Development of cryogenic heat pipes. ASME Paper 70-WA/Ener-1, New York, 1970.

[47] Zhang, N. Innovative heat pipe systems using a new working fluid. Int. Commun. Heat Mass Transf. Vol. 28, No. 8, pp 1025–1033, 2001.

[48] Esen, M. Thermal performance of a solar cooker integrated vacuum-tube collector with heat pipes containing different refrigerants. Sol. Energy, Vol. 76, pp 751–757, 2004.

[49] Kosowski, N. and Kosson, R. Experimental performance of grooved heat pipes at moderate temperatures. AIAA Paper 71–409, 1971.
[50] Anon. Heat pipes — properties of common small-pore wicks. Data Items No. 79013, Engineering Sciences Data Unit, London, 1979.
[51] Semena, M.G. and Zaripov, V.K. Influence of the diameter and length of fibres on material heat transfer of metal fibre wicks of heat pipes. Therm. Eng., Vol. 24, No. 4, pp 69–72, 1977.
[52] Acton, A. Correlating equations for the properties of metal felt wicks. Advances in Heat Pipe Technology, Proceedings of the IV International Heat Pipe Conference. Pergamon Press, Oxford, 1981.
[53] Basiulis, A. and Filler, M. Operating characteristics and long life capabilities of organic fluid heat pipes. AIAA Paper 71–408. 6th AIAA Thermophys. Conference, Tullahoma, Tennessee, April 1971.
[54] Kreeb, H., Groll, M. and Zimmermann, P. Life test investigations with low temperature heat pipes. 1st International Heat Pipe Conference, Stuttgart, Paper 4–1, October 1973.
[55] Busse, C.A., Campanile, A. and Loens, J. Hydrogen generation in water heat pipes at 250 °C. 1st International Heat Pipe Conference, Stuttgart, Paper 4–2, October 1973.
[56] Gerrels, E.E. and Larson, J.W. Brayton cycle vapour chamber (heat pipe) radiator study. NASA CR-1677, General Electric Company, Philadelphia, NASA, February 1971.
[57] Novotna, I. et al. Compatibility of steel-water heat pipes. Proceedings of the 3rd International Heat Pipe Symposium. Japan Association for Heat Pipes, Tsukuba, Japan, 1988.
[58] Anon. Heat pipes — general information on their use, operation and design. Data Item No. 80013, Engineering Sciences Data Unit, London, 1980.
[59] Freon, T.F. and Solvent, E.I. Dupont de Nemours and Company Inc. Technical Bulletin FSR-1, 1965.
[60] Kemme, J.E. Heat pipe capability experiments. Los Alamos Scientific Laboratory, Report LA-3585, August 1966.
[61] Van Andel, E. Heat pipe design theory. Euratom Center for Information and Documentation. Report EUR No. 4210 e, f, 1969.
[62] Busse, C.A. Theory of the ultimate heat transfer limit of cylindrical heat pipes. Int. J. Heat Mass Transf., Vol. 16, pp 169–186, 1973.
[63] Anon. Technology Specification for Carbon Steel-Water Gravity Heat Pipe, Su Q/B-25-86. Approved by Jiangsu Province Standard Bureau, 1986.
[64] Ya Ming Liang et al. Applications of heat pipe heat exchangers in energy saving and environmental protection. Proceedings of the International Conference on Energy and Environment, Shanghai, 8–10 May, 1995.
[65] Basiulis, A., Prager, R.C. and Lamp. T.R. Compatability and reliability of heat pipe materials. Proceedings of the 2nd International Heat Pipe Conference, Bologna. ESA Report SP 112, 1976.
[66] Novotna, I. et al. A contribution to the service life of heat pipes. Proceedings of the 7th International Heat Pipe Conference, Minsk, 1990. Begel Corporation, 1995.
[67] Geiger, F. and Quataert, D. Corrosion studies of tungsten heat pipes at temperatures up to 2650 °C. Proceedings of the 2nd International Heat Pipe Conference, Bologna. ESA Report SP 112, 1976.
[68] Devarakonda, A. and Olminsky, J.K. An evaluation of halides and other substances as potential heat pipe fluids. Proceedings 2nd International Energy Conversion Engineering Conference, Vol. 1, pp. 471–477, 2004.

[69] Libby, L.M. and Libby, W.F. One parameter equation of state for metals and certain other solids. Proc. Nat. Acad. Sci. USA., Vol. 69, No. 11, pp 3305–3306, November 1972.
[70] Anderson, W.T. Hydrogen evolution in nickel-water heat pipes. AIAA Paper 73-726, 1973.
[71] Yamamoto, T., Nagata, K., Katsuta, M. and Ikeda, Y. Experimental study of mercury heat pipes. Exp. Therm. Fluid Sci., Vol. 9, pp 39–46, 1994.
[72] Marcarino, P. and Merlone, A. Gas-controlled heat pipes for accurate temperature measurements. Appl. Therm. Eng., Vol. 23, pp 1145–1152, 2003.
[73] Tournier, J.-M. and El-Genk, M.S. Startup of a horizontal lithium molybdenum heat pipe from a frozen state. Int. J. Heat Mass Transf., Vol. 46, pp 671–685, 2003.
[74] Williams, R.W. and Harris, D.K. The heat transfer limit of step-graded metal felt heat pipe wicks. Int. J. Heat Mass Transf., Vol. 48, pp 293–305, 2005.
[75] Figus, C., Ounougha, L., Bonzom P., Supper, W. and Puillet, C. Capillary fluid loop developments in Astrium. Appl. Therm. Eng., Vol. 23, pp 1085–1098, 2003.
[76] Ottenstein, L., Butler, D. and Ku, J. Flight testing of the capillary pumped loop 3 experiment. Proceedings of the 2002 International Two-phase Thermal Control Technology Workshop, Mitchellville, MA, 24–26 September 2002.
[77] Swanson, T.D. and Birur, G.C. NASA thermal control technologies for robotic spacecraft. Appl. Therm. Eng., Vol. 23, pp 1055–1065, 2003.
[78] Kempers, R., Ewing, D. and Ching, C.Y. Effect of number of mesh lasers and fluid loading on the performance of screen mesh wicked heat pipes. Appl. Therm. Eng, In Press, Available on line on Science Direct.
[79] Kar, K. and Dybbs, A. Effective thermal conductivity of fully and partially saturated metal wicks. Proceedings of the 6th International Heat Transfer Conference, Vol. 3, pp 91–97, Toronto, Canada, 1978.
[80] Van Oost, S. and Aalders, B. Cryogenic heat pipe ageing. Paper J-6, Proceedings of the 10th International Heat Pipe Conference, Stuttgart, September 21–25, 1997.
[81] Anon. On the influence of the overshoot of oxygen on corrosion of carbon steels in heat supply systems. Teploenergetika, No. 12, pp 36–38, December 1992.
[82] Rassamakin, B.M., Gomelya, N.D., Khairnasov, N.D. and Rassamakina, N.V. Choice of the effective inhibitors of corrosion and the results of the resources tests of steel and aluminium thermosyphon with water. Paper J-1, Proceedings of the 10th International Heat Pipe Conference, Stuttgart, 21–25 September 1997.
[83] Zhao Rong Di et al. Experimental investigation of the compatibility of mild carbon steel and water heat pipes. Proceedings of the 6th International Heat Pipe Conference, Grenoble, 1988.
[84] Zhang, H. and Zhuang, J. Research, development and industrial application of heat pipe technology in China. Appl. Therm. Engi., Vol. 23, pp 1067–1083, 2003.

4

DESIGN GUIDE

4.1 INTRODUCTION

The design of a heat pipe or thermosyphon to fulfil a particular duty involves four broad processes:

Selection of appropriate type and geometry
Selection of candidate materials
Evaluation of performance limits
Evaluation of the actual performance

The background to each of these stages is covered in Chapters 2 and 3. In this chapter, the theoretical and practical aspects are discussed with reference to sample design calculations.

4.2 HEAT PIPES

The design procedure for a heat pipe is outlined in Fig. 4.1.

As with any design process, many of the decisions that must be taken are interrelated and the process is iterative. For example, choice of the wick and case material eliminates many candidate working fluids (often including water) due to compatibility constraints. If the design then proves inadequate with the available fluids, it is necessary to reconsider the choice of construction materials.

Two aspects of practical design, which must also be taken into consideration, are the fluid inventory and start-up of the heat pipe.

4.2.1 Fluid inventory

A feature of heat pipe design, which is important when considering small heat pipes and units for space use is the working fluid inventory. It is a common practice to include a slight excess of working fluid over and above that required to saturate

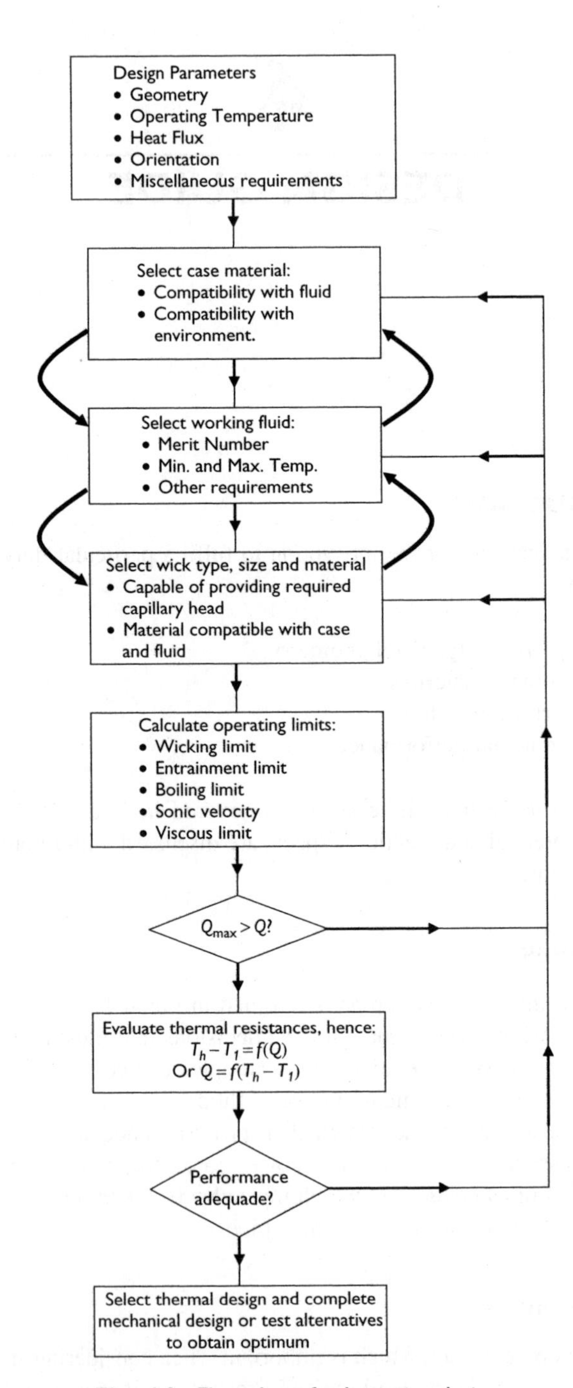

Fig. 4.1 Flow sheet for heat pipe design.

the wick, but when the vapour space is of small volume a noticeable temperature gradient can exist at the condenser, similar to that indicating the presence of noncondensable gas. This reduces the effective length of the condenser, hence impairing heat pipe performance. Another drawback of excess fluid is peculiar to heat pipes in space, where in zero gravity the fluid can move about the vapour space, affecting the dynamics of the spacecraft.

If there is a deficiency of working fluid, the heat pipe may fail because of the inability of any artery to fill. This is not so critical with homogeneous wicks as some of the pores will still be able to generate capillary action. Marcus [1] discusses in detail these effects and the difficulties encountered in ensuring whether the correct amount of working fluid is injected into the heat pipe. One way of overcoming the problem is to provide an excess fluid reservoir, which behaves as a sponge, absorbing working fluid that is not required by the primary wick structure.

4.2.2 Priming

With heat pipes having some form of arterial wick, it is necessary to ensure that should an artery become depleted of working fluid, it should be able to refill automatically. It is possible to calculate the maximum diameter of an artery to ensure that it will be able to reprime. The maximum priming height that can be achieved by a capillary is given by the equation.

$$h + h_c = \frac{\sigma_l \cos\theta}{(\rho_l - \rho_v)g} \times \left(\frac{1}{r_{p1}} + \frac{1}{r_{p2}}\right) \tag{4.1}$$

where h is the vertical height to the base of the artery, h_c the vertical height to the top of the artery, r_{p1} the first principal radius of curvature of the priming meniscus, and r_{p2} is the second principal radius of curvature of the priming meniscus.

For the purpose of priming, the second principal radius of curvature of the meniscus is extremely large (approximately $1 \sin\phi$). For a cylindrical artery

$$h_c = d_a$$

and

$$r_{p1} = \frac{d_a}{2}$$

where d_a is the artery diameter.

Hence the above equation becomes:

$$h + d_a = \frac{2\sigma_l \cos\theta}{(\rho_l - \rho_v)g \times d_a} \tag{4.2}$$

which produces a quadratic in d_a that may be solved as:

$$d_a = \frac{1}{2}\left[\left(\sqrt{h^2 + \frac{8\sigma_l \cos\theta}{(\rho_l - \rho_v)g}}\right) - h\right] \tag{4.3}$$

An artery can deprime when vapour bubbles become trapped in it. It may be necessary to reduce the heat load in such circumstances, to enable the artery to reprime. It is possible to design a heat pipe incorporating a tapered artery, effectively a derivative of the monogroove wick system illustrated in Fig. 3.4 (Chapter 3) Use of this design [2] facilitated bubble venting into the vapour space and achieved significant performance improvements.

4.3 DESIGN EXAMPLE 1

4.3.1 Specification

A heat pipe is required which will be capable of transferring a minimum of 15 W at a vapour temperature between 0 °C and 80 °C over a distance of 1 m in zero gravity (a satellite application). Restraints on the design are such that the evaporator and condenser sections are each 8 cm long, located at each end of the heat pipe and the maximum permissible temperature drop between the outside wall of the evaporator and the outside wall of the condenser is 6 °C. Because of weight and volume limitations, the cross-sectional area of the vapour space should not exceed 0.197 cm^2. The heat pipe must also withstand bonding temperatures.

Design a heat pipe to meet this specification.

4.3.2 Selection of materials and working fluid

The operating conditions are contained within the specification. The selection of wick and wall materials is based on the criteria discussed in Chapter 3. As this is an aerospace application, low mass is an important factor.

On this basis, aluminium alloy 6061 (HT30) is chosen for the wall and stainless steel for the wick.

If it is assumed that the heat pipe will be of a circular cross section, the maximum vapour space area of 0.197 cm^2 yields a radius of 2.5 mm.

Working fluids compatible with these materials, based on available data, include

Freon 11
Freon 113
Acetone
Ammonia

Water must be dismissed at this stage, both on compatibility grounds and because of the requirement for operation at 0 °C with the associated risk of freezing. Note that the freon refrigerants are CFCs and are no longer available, but are included in the evaluation in order to demonstrate the properties of a range of fluids.

The operating limits for each fluid must now be examined.

4.3.2.1 Sonic limit

The minimum axial heat flux due to the sonic limitation will occur at the minimum operating temperature, 0 °C, and can be calculated from the equation 2.58 with the Mach number set to unity.

$$\dot{q}_s = \rho_v L \sqrt{\frac{\gamma R T_v}{2(\gamma+1)}}$$

The gas constant for each fluid may be obtained from

$$R = \frac{R_o}{Molecular\ weight} = \frac{8315}{M_w}\ \mathrm{J/kgK}$$

For ammonia, $\gamma = 1.4$ [3] and its molecular weight is 31, the latent heat and density may be obtained from Appendix 1, therefore $\dot{q}_s$ is given by

$$\dot{q}_s = 3.48 \times 1263 \times \sqrt{\frac{1.4}{2(1.4+1)} \frac{8315 \times 273}{31}} = 84 \times 10^7\ \mathrm{W/m^2}$$
$$= 84\ \mathrm{kW/cm^2}$$

Similar calculations may be carried out for the other candidate fluids, yielding

Freon 11	$0.69\ \mathrm{kW/cm^2}$
Freon 113	$3.1\ \mathrm{kW/cm^2}$
Acetone	$1.3\ \mathrm{kW/cm^2}$
Ammonia	$84\ \mathrm{kW/cm^2}$

Since the required axial heat flux is $15/0.197\ \mathrm{W/cm^2} \equiv 0.076\ \mathrm{kW/cm^2}$, the sonic limit would not be encountered for any of the candidate fluids.

It is worth noting that the term $\sqrt{\gamma/\gamma+1}$ varies from 0.72 to 0.79 for values of γ from 1.1 to 1.66; therefore, when the sonic limit is an order of magnitude above the required heat flux it is nor essential that γ be known precisely for the fluid.

4.3.2.2 Entrainment limit

The maximum heat transport due to the entrainment limit may be determined from equation 2.62

$$\dot{q} = \sqrt{\frac{2\pi \rho_v L^2 \sigma_l}{z}}$$

which may be written

$$\dot{Q}_{ent} = \pi r_v^2 L \sqrt{\frac{2\pi \rho_v \sigma_l}{z}}$$

where z is the characteristic dimension of the liquid–vapour interface for a fine mesh that may be taken as 0.036 mm. The entrainment limit is evaluated at the highest operating temperature.

The fluid properties of the fluids and the resulting entrainment limits are given in Table 4.1.

A sample calculation for acetone is reproduced below

Particular care must be taken to ensure that the units used are consistent.

$$L = \mathrm{J/kg} \qquad \sigma_1 = \mathrm{N/m} \qquad \rho_v = \mathrm{kg/s} \qquad r_v = \mathrm{m}\ (or\ A = \mathrm{m}^2)\ z = \mathrm{m}$$

Some useful conversion factors are given in Appendix 1

Then we have

$$\mathrm{m}^2 \times \frac{\mathrm{J}}{\mathrm{kg}} \times \sqrt{\frac{\mathrm{kg}}{\mathrm{m}^3}\frac{\mathrm{N}}{\mathrm{m}}\frac{1}{\mathrm{m}}} \equiv \mathrm{m}^2 \times \frac{\mathrm{J}}{\mathrm{kg}} \times \sqrt{\frac{\mathrm{kg}}{\mathrm{m}^3}\frac{\mathrm{kg}\times\mathrm{m}}{\mathrm{m}\times\mathrm{s}^2}\frac{1}{\mathrm{m}}} \equiv \frac{\mathrm{J}}{\mathrm{s}} = \mathrm{W}$$

$$\dot{Q}_{\mathrm{ent,acetone}} = \pi r_v^2 L \sqrt{\frac{2\pi\rho_v\sigma_1}{z}} = \pi \times (2.5 \times 10^{-3})^2 \times 495$$

$$\times 10^3 \sqrt{\frac{2\pi \times 4.05 \times 0.0162}{0.036 \times 10^{-3}}} = 1040\,\mathrm{W}$$

4.3.2.3 Wicking limit

At this stage the wick has still to be specified, but a qualitative comparison of the potential performance of the four fluids can be obtained by evaluating the Merit number

$$\rho_1 \frac{\sigma_1 L}{\mu_1}$$

for each fluid over the temperature range (Fig. 4.2).

4.3.2.4 Radial heat flux

Boiling in the wick may result in the vapour blocking the supply of liquid to all parts of the evaporator. In arterial heat pipes, bubbles in the artery itself can create

Table 4.1 Properties of candidate fluids at 80 °C

Fluid	L (kJ/kg)	σ_1 (mN/m)	ρ_v (kg/m^3)	$\dot{Q}_{\mathrm{ent}}$ (kW)
Freon 11	221	10.7	27.6	0.98
Freon 113	132	10.6	18.5	0.48
Acetone	495	16.2	4.05	1.04
Ammonia	891	7.67	34	3.75

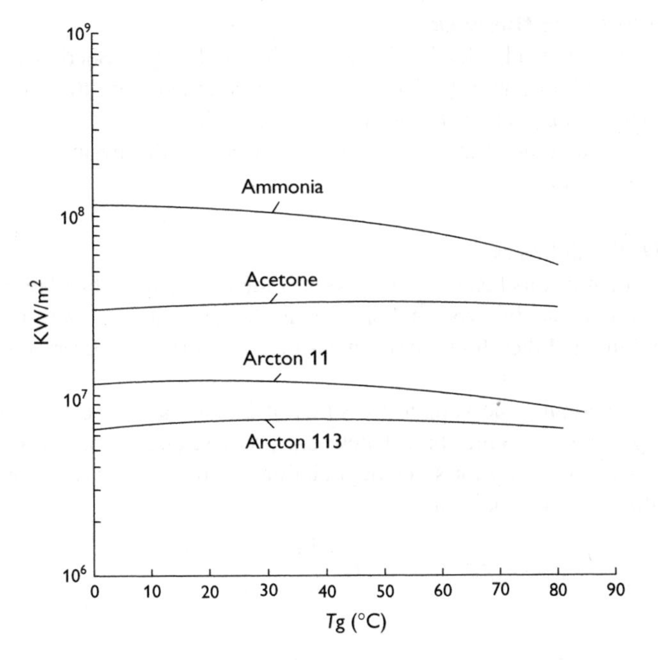

Fig. 4.2 Merit number for candidate fluids.

even more serious problems. It is therefore desirable to have a working fluid with a high superheat ΔT to reduce the chance of nucleation. The degree of superheat to cause nucleation is given by

$$\Delta T = \frac{3.06\sigma_l T_{sat}}{\rho_v L\delta}$$

where ö is the thermal layer thickness, and, taking a representative value of 15 μm allows comparison of the fluids. ΔT is evaluated at 80 °C, as the lowest permissible degree of superheat will occur at the maximum operating temperature.
These are

Freon 11	0.13 K
Freon 113	0.31 K
Acetone	0.58 K
Ammonia	0.02 K

These figures suggest that the freons and ammonia require only very small superheat temperatures at 80 °C to cause boiling. Acetone is the best fluid from this point of view.

4.3.2.5 Priming of the wick

A further factor in fluid selection is the priming ability (see Section 3.7). A comparison of the priming ability of fluids may be obtained from the ratio ρ_l/ρ_v and this is plotted against vapour temperature in Fig. 4.3.

Acetone and ammonia are shown to be superior to the freons over the whole operating temperature range.

4.3.2.6 Wall thickness

The requirement of this heat pipe necessitates the ability to be bonded to a radiator plate. Depending on the type of bonding used, the heat pipe may reach 170 °C during bonding, and therefore vapour pressure is important in determining the wall thickness.

At this temperature, the vapour pressures of ammonia and acetone are 113 and 17 bar, respectively. Taking the 0.1 per cent proof stress, Ω, of HT30 aluminium as $46.3\,\mathrm{MN/m^2}$ (allowing for some degradation of properties in weld regions), and using the thin cylinder formula,

$$t = \frac{\mathrm{Pr}}{\Omega}$$

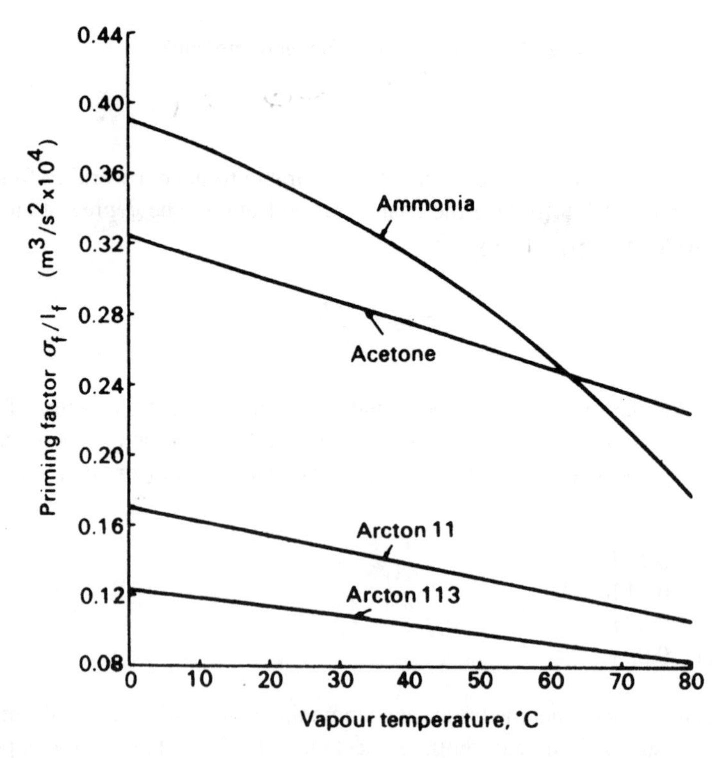

Fig. 4.3 Priming factor for selected fluids.

the minimum wall thickness for ammonia is 0.65 mm and 0.1 mm for acetone. There is therefore a mass penalty attached to the use of ammonia.

4.3.2.7 Conclusions on selection of working fluid

Acetone and ammonia both meet the heat transport requirements, ammonia being superior to acetone in this respect. Nucleation occurs more readily in an ammonia heat pipe, and the pipe may also be heavier. The handling of ammonia to obtain high purity is difficult, and the presence of any water in the working fluid may lead to long-term degradation in performance.

Acetone is therefore selected in spite of the somewhat inferior thermal performance.

4.3.3 Detail design

4.3.3.1 Wick selection

Two types of wick structure are proposed for this heat pipe, homogeneous and arterial types. A homogeneous wick may be a mesh, twill or felt, and arterial types normally incorporate a mesh to distribute liquid circumferentially.

Homogeneous meshes are easy to form but have inferior properties to arterial types. The first question is, therefore, will a homogeneous wick transport the required amount of fluid over 1m to meet the heat transport specification?

To determine the minimum flow area to transport 15 W, one can equate the maximum capillary pressure to the sum of the liquid and gravitational pressure drops (neglecting vapour ΔP).

$$\Delta P_{\mathrm{l}} + \Delta P_{\mathrm{g}} = \Delta P_{\mathrm{c}}$$

where

$$\Delta P_{\mathrm{c}} = \frac{2\sigma_{\mathrm{l}} \cos\theta}{r_{\mathrm{e}}}$$

$$\Delta P_{\mathrm{l}} = \frac{\mu_{\mathrm{l}} \dot{Q} l_{\mathrm{eff}}}{\rho_{\mathrm{l}} L A_{\mathrm{w}} K}$$

$$\Delta P_{\mathrm{g}} = \rho_{\mathrm{l}} g h$$

The gravitational effect is zero in this application, but is included to permit testing of the heat pipe on the ground. The value of h may be taken as 1 cm based on end-to-end tilt plus the tube diameter. The effective length l_{eff} is 1 m and $\cos\theta$ is taken to be unity.

The effective capillary radius for the wick is 0.029 mm; therefore, using the surface tension of acetone at 80 °C,

$$\Delta P_{\mathrm{c}} = \frac{2\sigma_{\mathrm{l}} \cos\theta}{r_{\mathrm{e}}} = \frac{2 \times 0.0162 \times 1}{0.029 \times 10^{-3}} = 1120\,\mathrm{N/m^2}$$

The wick permeability, K, is calculated using the Blake–Koseny equation

$$K = \frac{d_w^2 (1-\varepsilon)^3}{66.6\varepsilon^2}$$

where ε is the volume fraction of the solid phase (0.314) and d_W the wire diameter (0.025 mm).

$$K = \frac{(2.5 \times 10^{-5})^2 (1-0.314)^3}{66.6 \times 0.314^2} = 3 \times 10^{-11}$$

Therefore, considering the properties of acetone at 80 °C

$$\rho_l = 719 \, \text{kg/m}^3$$

$$\mu_l = 0.192 \, \text{cP} = 192 \times 10^{-6} \, \text{kg/ms or Ns/m}^2$$

$$\Delta P_l = \frac{\mu_l \dot{Q} l_{\text{eff}}}{\rho_l L A_w K} = \frac{192 \times 10^{-6} \times 15 \times 1}{719 \times 495 \times 10^3 \times 3 \times 10^{-11} A_w}$$

$$= \frac{0.027}{A_w} \, \text{N/m}^2$$

The gravitational pressure is

$$\Delta P_g = \rho_l g h = 719 \times 9.81 \times 0.01 = 70 \, \text{N/m}^2$$

Equating the three terms gives

$$\Delta P_l + \Delta P_g = \Delta P_c$$

$$\frac{0.027}{A_w} + 70 = 1120$$

$$A_w = \frac{0.027}{1120 - 70} = 26 \times 10^{-6} \, \text{m}^2 \equiv 0.26 \, \text{cm}^2$$

Since the required wick area ($0.26 \, \text{cm}^2$) is greater than the available vapour space area ($0.197 \, \text{cm}^2$), it can be concluded that the homogeneous type of wick is not acceptable. An arterial wick must be used.

4.3.3.2 Arterial diameter

Equation 3.12 in Chapter 3 describes the artery priming capability, setting a maximum value on the size of any arteries

$$d_a = \frac{1}{2}\left[\sqrt{\left(h^2 + \frac{8\sigma_l \cos\theta}{(\rho_l - \rho_v) g}\right)} - h\right]$$

Using this equation, d_a is evaluated at a vapour temperature of 20 °C (for convenience priming ability may be demonstrated at room temperature), h is taken as 1 cm to cater for arteries near the top of the vapour space.

$$d_a = \frac{1}{2}\left[\sqrt{\left(0.01^2 + \frac{8 \times 0.0237 \times 2}{(790 - 0.64) \times 9.81}\right)} - 0.01\right] = 0.58 \times 10^{-3}\,\text{m}$$

Thus, the maximum permitted value is 0.58 mm. To allow for uncertainties in fluid properties, wetting (θ assumed 0°) and manufacturing tolerances, a practical limit is 0.5 mm.

4.3.3.3 Circumferential liquid distribution and temperature difference

The circumferential wick is the most significant thermal resistance in this heat pipe, and its thickness is limited by the fact that the temperature drop between the vapour space and the outside surface of the heat pipe and vice versa should be 3 °C maximum. Assuming that the temperature drop through the aluminium wall is negligible, the thermal conductivity of the wick may be determined and used in the steady state conduction equation.

$$k_{\text{wick}} = \left(\frac{\beta - \varepsilon}{\beta + \varepsilon}\right) k_l$$

where

$$\beta = \left(1 - \frac{k_s}{k_l}\right) / \left(1 - \frac{k_s}{k_l}\right)$$

$$k_s = 16\,\text{W/m °C (steel)}$$

$$k_l = 0.165\,\text{W/m °C (acetone)}$$

$$\therefore \beta = \frac{1 + 97}{1 - 97} = -1.02$$

The volume fraction ε of the solid phase is approximately 0.3

$$\therefore k_{\text{wick}} = \left(\frac{-1.02 - 0.3}{-1.02 + 0.3}\right) \times 0.165$$

$$= 0.3\,\text{W/m °C}$$

Using the basic conduction equation

$$\dot{Q} = kA_e \frac{\Delta T}{t}$$

$$t = \frac{kA_e \Delta T}{\dot{Q}}$$

where A_e is the area of the evaporator (8 cm long, 0.5 cm diameter) and $\dot{Q}$ is the required heat load.

$$t = \frac{0.3 \times 3 \times \pi \times 5 \times 10^{-3} \times 80 \times 10^{-3}}{15} = 75 \times 10^{-6}\,\text{m} \equiv 0.075\,\text{mm}$$

Thus, the circumferential wick must be 400 mesh, which has a thickness of 0.05 mm. Coarser meshes are too thick, resulting in unacceptable temperature differences across the wick in the condenser and evaporator.

4.3.3.4 Arterial wick

Returning to the artery, the penultimate section revealed that the maximum artery depth permissible was 0.5 mm. In order to prevent nucleation in the arteries, they should be kept away from the heat pipe wall and formed of low conductivity material. It is also necessary to cover the arteries with a fine pore structure and 400-mesh stainless steel is selected. It is desirable to have several arteries to give a degree of redundancy, and two proposed configurations are considered, one having six arteries as shown in Fig. 4.4, and the other having four arteries. In the former case, each groove is nominally 1.0-mm wide, and in the latter case, 1.5 mm.

4.3.3.5 Final analysis

It is now possible to predict the overall capability of the heat pipe, to check that it meets the specification.

We have already shown that entrainment and sonic limitations will not be met and that the radial heat flux is acceptable. The heat pipe should also meet the overall temperature drop requirement, and the arteries are sufficiently small to allow

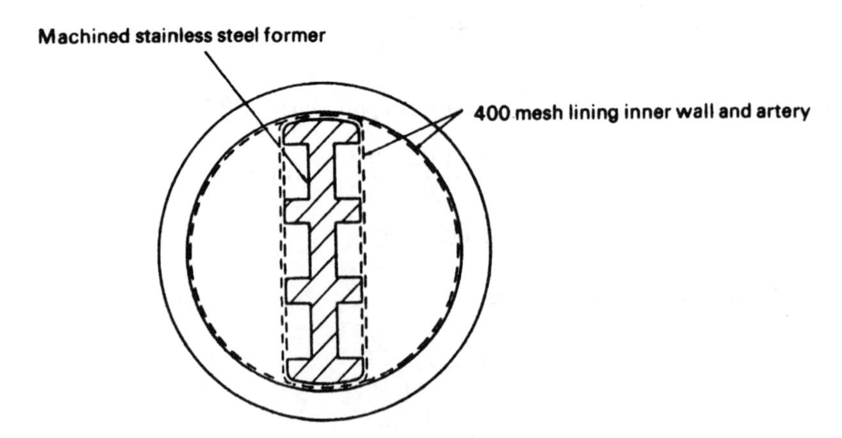

Fig. 4.4 Design of wick.

repriming at 20 °C. The wall thickness requirement for structural integrity (0.1 mm minimum) can easily be satisfied. The wicking limitation will therefore determine the maximum performance.

$$\text{i.e.} \qquad \Delta P_{la} + \Delta P_{lm} + \Delta P_g + \Delta P_v = \Delta P_c$$

where ΔP_{la} is the pressure drop in the artery, ΔP_{lm} the loss in the circumferential wick.

The axial flow in the mesh will have little effect and can be neglected. McAdams [4] presents an equation for the pressure loss, assuming laminar flow, in a rectangular duct, and shows that the equation is in good agreement with experiment for streamline flow in rectangular ducts having depth/width ratios.
a_a/b_a of 0.05 – 1.0.

This equation may be written as

$$\Delta P_{la} = \frac{4K_l \times l_{eff} Q}{a_a^2 b_a^2 \theta_c N}$$

where N is the number of channels, θ_c a function of channel aspect ratio and is given in Fig. 4.5.

and $$K_l = \frac{\mu_l}{\pi_l L}$$

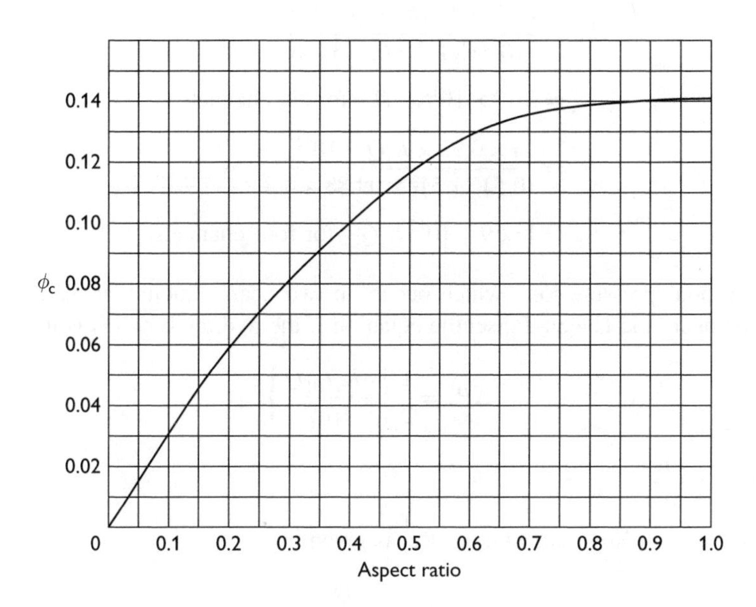

Fig. 4.5 Channel aspect ratio factor.

The summed pressure loss in the condenser and evaporator is given by

$$\Delta P_{\mathrm{la}} = \frac{K_1 \times l_{\mathrm{eff\,c}} Q}{2KA_{\mathrm{c}}}$$

where $l_{\mathrm{eff\,c}}$ is the effective circumferential flow length approximately equal to

$$\frac{\pi r_{\mathrm{w}}}{4}$$

$$\text{carrying} \quad \frac{\dot{m}}{4}$$

where $\dot{m}$ is the liquid mass flow, A_{c} the circumferential flow area (mesh thickness × cond. or evap. length) and K the permeability of 400 mesh.

The circumferential flow area for 400 mesh with two layers is

$$A_{\mathrm{c}} = 8 \times 10^{-2} \times 0.1 \times 10^{-3}\,\mathrm{m}^2 = 8 \times 10^{-6}\,\mathrm{m}^2$$

A resistance occurs in both the evaporator and the condenser, therefore; substituting in the above equation

$$\Delta P_{\mathrm{lm}} = \frac{\pi \times 2.5 \times 10^{-3}}{2 \times 4} \frac{1}{8 \times 10^{-6}} \frac{K_1 Q}{0.314 \times 10^{-10}}$$

$$= 4.00 \times 10^{12}\, K_1 Q \quad \text{for each section}$$

$$\Delta P_{\mathrm{la}} = \frac{4 \times 0.92 \times K_1 \times Q \times 10^{12}}{(0.5)^2 (1)^2 \times 0.115 \times 6}$$

$$= 21.3 \times 10^{12}\, K_1 Q \quad \text{for six channels}$$

$$= \frac{4 \times 0.92 \times K_1 Q \times 10^{12}}{(0.5)^2 (1.5)^2 \times 0.088 \times 4}$$

$$= 18.59 \times 10^{12}\, K_1 Q \quad \text{for four channels}$$

The vapour pressure loss, which occurs in two near-semicircular ducts, can be obtained using the Hagen–Poiseuille equation if the hydraulic radius is used

$$\Delta P_{\mathrm{v}} = \frac{1}{2} \left\{ \frac{8 K_{\mathrm{v}} l_{\mathrm{eff}} Q}{\pi r_{\mathrm{H}}^4} \right\}$$

where $K_{\mathrm{v}} = \dfrac{\mu_{\mathrm{v}}}{\rho_{\mathrm{v}} L}$

Now the axial Reynolds number Re_z is given by

$$\mathrm{Re}_z = \frac{Q}{\pi r_{\mathrm{H}} \mu_1 L}$$

The transitional heat load at which the flow becomes turbulent can be calculated assuming that transition from laminar to turbulent flow occurs at $\mathrm{Re}_z = 1000$, based on hydraulic radius (corresponding to 2000 if the Reynolds number is based on the diameter). Restricting the width of the stainless steel former in Fig. 4.4 to 1.5 mm, $r_H = 1.07\,\mathrm{mm}$.

$\dot{Q}$ may be evaluated at the transition point using values of μ_v and L at several temperatures between 0 °C and 80 °C, giving

Vapour temperatures (° C)	Transitional load (W)
0	31.1
20	31.2
40	30.6
60	30.2
80	30.0

The transitional load is always greater than the design load of 15 W, but as the heat pipe may be capable of operating in excess of the design load, it is necessary to investigate the turbulent regime.

For $\mathrm{Re}_z > 100$ and for two ducts

$$\Delta P_v = \frac{0.00896\,\mu_v^{0.25} Q^{1.75} l_{\mathrm{eff}}}{2\rho_v r_H^{4.75} L^{1.75}}$$

This is the empirical Blasius equation.

$$\Delta P_v \text{ (laminar)} = \frac{1}{2}\left(\frac{8 \times 0.92\, K_v \times Q \times 10^{12}}{p \times 1.07^4}\right)$$

$$= 0.9 \times 10^{12}\, K_v \times Q$$

$$\Delta P_v \text{ (turbulent)} = \frac{0.00896 \times 0.92}{2 \times (1.07 \times 10^{-3})^{4.75}}\; Q^{1.75} \left(\frac{m_v^{0.25}}{L^{1.75}\; r_v}\right)$$

$$= 0.53 \times 10^{12}\; Q^{1.75} \left(\frac{m_v^{0.25}}{L^{1.75}\; r_v}\right)$$

The gravitational pressure drop is

$$\Delta P_g = \rho_l\, gl \sin\phi$$

$$= 0.0981 \rho_{l'}$$

taking $l \sin\phi = 1\,\mathrm{cm}$

The capillary pressure generated by the arteries is given by

$$\Delta P_c = \frac{2\sigma_l \cos\theta}{r_c}$$
$$= \frac{2}{0.003 \times 10^{-2}} \sigma_l \cos\theta$$
$$= 0.667 \times 10^{-5} \sigma_l$$

Summarising

$$\Delta P_c = 0.667 \times 10^5 \sigma_l$$
$$\Delta P_g = 0.0981 \rho_l$$
$$\Delta P_{vl} = 0.9 \times 10^{12} K_v Q$$
$$\Delta P_{vt} = 0.53 \times 10^{12} Q^{1.75} \left(\frac{\mu_v^{0.25}}{L^{1.75} \rho_v} \right)$$
$$\Delta P_{lm} = 4 \times 10^{12} K_l Q$$
$$\Delta P_{la} = 21.3 \times 10^{12} K_l Q \quad \text{(six channels)}$$
$$= 18.59 \times 10^{12} K_l Q \quad \text{(four channels)}$$

These equations involve $\dot{Q}$ and the properties of the working fluid. Using properties at each temperature (in 20 °C increments) over the operating range, the total capability can be determined:

$$\Delta P_c = \Delta P_{lm} + \Delta P_{la} \begin{Bmatrix} \text{six channels} \\ \text{four channels} \end{Bmatrix} + \Delta P_g + \Delta P_v \begin{Bmatrix} \text{laminar} \\ \text{turbulent} \end{Bmatrix}$$

This yields the following results

Vapour temperatures (°C)	Q(W)			
	Laminar four channels	Laminar six channels	Turbulent four channels	Turbulent six channels
0	21.6	20.9	–	–
20	34.0	32.5	22.6	22.0
40	42.6	40.2	27.9	27.0
60	49.1	45.8	33.0	32.0
80	51.4	47.6	36.4	35.0

In this example, it is assumed that the maximum resistance to heat transfer is across the wick, the other thermal resistances have not been calculated explicitly.

This heat pipe was constructed with six grooves in the artery structure and met the specification.

4.4 DESIGN EXAMPLE 2

4.4.1 Problem

Estimate the liquid flow rate and heat transport capability of a simple water heat pipe operating at 100 °C having a wick of two layers of 250 mesh against the inside wall. The heat pipe is 30 cm long and has a bore of 1 cm diameter. It is operating at an inclination to the horizontal of 30°, with the evaporator above the condenser.

It will be shown that the capability of the above heat pipe is low. What improvement will be made if two layers of 100 mesh are added to the 250-mesh wick to increase liquid flow capability?

4.4.2 Solution – original design

The maximum heat transport in a heat pipe at a given vapour temperature may be obtained from the equation

$$Q_{max} = \dot{m}_{max} L$$

where $\dot{m}_{max}$ is the maximum liquid flow rate in the wick.

Using the standard pressure balance equation

$$\Delta P_c = \Delta P_v + \Delta P_l + \Delta P_g$$

and neglecting, for the purposes of a first approximation, the vapour pressure drop ΔP_v, we can substitute for the pressure terms and obtain

$$\frac{2\sigma_l \cos\theta}{r_c} = \frac{\mu_l}{\rho_l L} \times \frac{Q l_{eff}}{A_w K} + \rho_l\, g l \sin\phi$$

Rearranging and substituting for $\dot{m}$, we obtain

$$\dot{m} = \frac{\rho_l K A_w}{\mu_l l_{eff}} \left\{ \frac{2\sigma_l}{r_c} \cos\theta - \rho_l g l_{eff} \sin\phi \right\}$$

The wire diameter of 250 mesh is typically 0.0045 cm, and therefore the thickness of two layers of 250 mesh is 4×0.0045 cm or 0.0180 cm.

The bore of the heat pipe is 1 cm

$$\begin{aligned} \therefore A_{wick} &= 0.018 \times \pi \times 1 \\ &= 0.057\,\text{cm}^2 \\ &= 0.057 \times 10^{-4}\,\text{m}^2 \end{aligned}$$

From Table 3.4, the pore radius r and permeability K of 250 mesh are 0.002 cm and $0.302 \times 10^{-10}\,\text{m}^2$, respectively. Assuming perfect wetting ($\theta = 0°$), the mass flow $\dot{m}$ may be calculated using the properties of water at 100 °C.

$$L = 2258\,\mathrm{J/kg}$$
$$\rho_l = 958\,\mathrm{kg/m^3}$$
$$\mu_l = 0.283\,\mathrm{mNs/m^2}$$
$$\tau_l = 58.9\,\mathrm{mN/m}$$

Converting all terms to the base units (kg, J, N, m and s)

$$\dot{m}_{max} = \frac{958 \times 0.302 \times 10^{-10} \times 0.057 \times 10^{-4}}{0.283 \times 10^{-3} \times 0.3} \times \left(\frac{2 \times 58.9 \times 10^{-3}}{0.2 \times 10^{-4}} - 958 \times 9.810 \times 0.3 \times 0.5 \right)$$
$$= 1.95 \times 10^{-9}\,(5885 - 1410)$$
$$= 8.636 \times 10^{-6}\,\mathrm{kg/s}$$
$$\dot{Q}_{max} = \dot{m}_{max} \times L$$
$$= 8.636 \times 10^{-6} \times 2.258 \times 10^{6}$$
$$= 19.5\,\mathrm{W}$$

4.4.3 Solution – revised design

Consider the addition of two layers of 100-mesh wick below the original 250 mesh.

The wire diameter of 100 mesh is 0.010 cm
∴ thickness of two layers is 0.040 cm
Total wick thickness = 0.040 + 0.018 cm
= 0.058 cm

$$\therefore A_{wick} \approx 0.058 \times \pi \times 1$$
$$\approx 0.182\,\mathrm{cm^2}$$

The capillary pressure is still governed by the 250 mesh, and $r_c = 0.002$ cm. The permeability of 100 mesh is used, Langston and Kunz giving a value of $1.52 \times 10^{-10}\,\mathrm{m^2}$. The mass flow may now be calculated.

$$\dot{m}_{max} = \frac{958 \times 1.52 \times 10^{-10} \times 0.057 \times 10^{-4}}{0.283 \times 10^{-3} \times 0.3} \times \left(\frac{2 \times 58.9 \times 10^{-3}}{0.2 \times 10^{-4}} - 958 \times 9.810 \times 0.3 \times 0.5 \right)$$
$$= 31 \times 10^{-9}\,(5885 - 1410)$$
$$= 139 \times 10^{-6}\,\mathrm{kg/s}$$
$$\dot{Q}_{max} = \dot{m}_{max} \times L$$
$$= 139 \times 10^{-6} \times 2.258 \times 10^{6}$$
$$= 314\,\mathrm{W}$$

The modified wick structure has resulted in an estimated increase in capacity from 19.5 to 314 W, that is the limiting performance has been improved by more than an order of magnitude. A disadvantage of the additional layers lies in the additional thermal resistance introduced at the evaporator and the condenser.

4.5 THERMOSYPHONS

The design process for a thermosyphon is similar to that for a wicked heat pipe in that it requires identification of suitable case material and working fluid, followed by evaluation of the performance of the unit using the techniques discussed in Chapter 2. The wicking limitation is not relevant to thermosyphon performance. The thermal resistance of any wick used to distribute liquid through the evaporator must, however, be taken into account.

4.5.1 Fluid inventory

In the case of thermosyphons, the fluid inventory is based on different considerations from wicked heat pipes. [5]. The amount of liquid is governed by two considerations; too small a quantity can lead to dryout, while an excess of liquid can lead to quantities being carried up to the condenser, where blockage of surface for condensation can result. Bezrodnyi and Alekseenko [6] recommended that the liquid fill should be at least 50 per cent of the volume of the evaporator and also that the volume of liquid, V_l should be related to the thermosyphon dimensions as follows:

$$V_l > 0.001D(l_e + l_a + l_c)$$

where D is the pipe internal diameter.

When a wick is fitted to the evaporator section, ESDU [7] recommends

$$V_l > 0.001D(l_a + l_c) + A_w l_e \varepsilon$$

where ε is the wick porosity.

Comprehensive work by Groll et al. at IKE, Stuttgart, investigated the effect of fluid inventory on the performance of thermosyphons incorporating a variety of novel liquid–vapour flow separators (to minimise interaction between these components). It was found, for several working fluids, [8], that the fluid inventory, expressed as a percentage of the evaporator volume occupied by the liquid working fluid, had a rather flat optimum between about 20 and 80 per cent.

The constraints on filling for a closed two-phase thermosyphon are illustrated by El-Genk and Saber [9]. Low initial filling ratio results in dryout of the evaporator while excess working fluid results in liquid filling the entire evaporator when bubbles form due to boiling.

4.5.2 Entrainment limit

The third limit shown in Fig. 4.6 is the entrainment or counter current flooding limit discussed in Section 2.4.2.

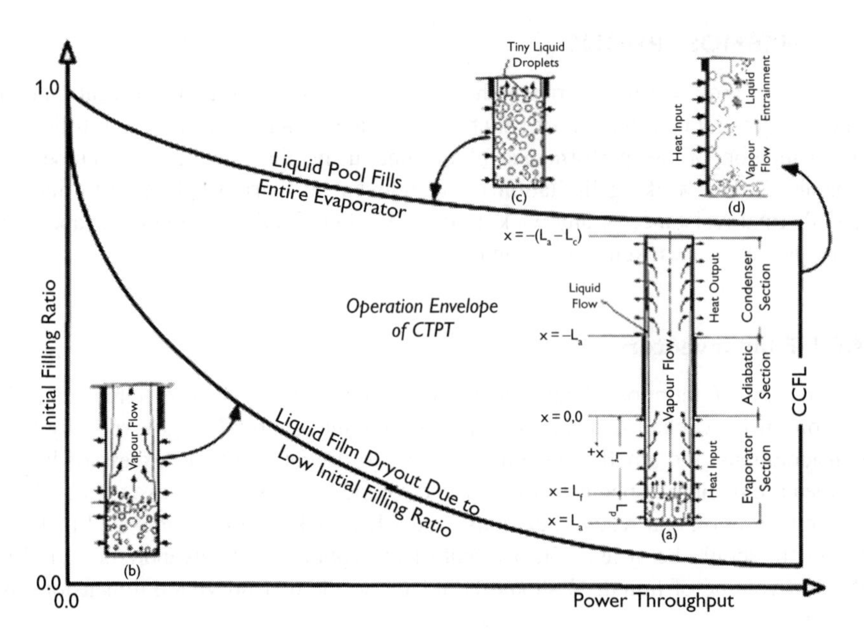

Fig. 4.6 Operating envelope of closed two-phase thermosyphons (CTPT) [9].

4.6 SUMMARY

In this chapter, the theory presented in Chapter 2 and the practical considerations of material and working fluid choice described in Chapter 3 have been used to illustrate the process of designing simple wicked heat pipes. The design of thermosyphons uses similar techniques, but the important differences have been highlighted.

REFERENCES

[1] Marcus, B.D. Theory and design of variable conductance heat pipes. TRW Systems Group, NASA CR-2018, April 1972

[2] Holmes, H.R. and Field, A.R. The gas-tolerant high capacity tapered artery heat pipe. AIAA/ASME 4th Joint Thermophysics and Heat Transfer Conference. Paper AIAA-86-1342, Boston, 1986.

[3] Anon. ASHRAE Handbook of fundamentals, p 20.35. ASHRAE, 2005

[4] McAdams, W.H. Heat Transmission, 3rd Edn. McGraw-Hill, 1954

[5] Anon. Heat pipes – performance of two-phase closed thermosyphons. Data – Item No. 81038, Engineering Sciences Data Unit, London, 1981.

[6] Bezrodnyi, M.K. and Alekseenko, D.B. Investigation of the critical region of heat and mass transfer in low temperature wickless heat pipes. High Temp., Vol.15, pp 309–313, 1977

[7] Anon. Heat pipes – general information on their use, operation and design. Data Item No. 80013, Engineering Sciences Data Unit, London, 1980

[8] Groll, M. et al. Heat recovery units employing reflux heat pipes as components. Final Report, Contract EE-81-133D(B). Commission of the European Communities Report EUR9166EN, 1984.
[9] El-Genk M.S. and Saber H.H. Determination of operation envelopes for closed/two-phase thermosyphons. Int. J Heat Mass Transf. Vol. 42, pp 889–903, 1999.

5

HEAT PIPE MANUFACTURE AND TESTING

The manufacture of conventional capillary driven heat pipes involves a number of comparatively simple operations, particularly when the unit is designed for operation at temperatures of the order of, say 50–200 °C. It embraces skills such as welding, machining, chemical cleaning and nondestructive testing, and can be carried out following a relatively small outlay on capital equipment. The most expensive item is likely to be the leak detection equipment. (Note that many procedures described are equally applicable to thermosyphons.)

With all heat pipes, however, cleanliness is of prime importance to ensure that no incompatibilities exist (assuming that the materials selected for the wick, wall and working fluid are themselves compatible), and to make certain that the wick and wall will be wetted by the working fluid. As well as affecting the life of the heat pipe, negligence in assembly procedures can lead to inferior performance, due, for example, to poor wetting. Atmospheric contaminants, in addition to those likely to be present in the raw working fluid, must be avoided. Above all, the heat pipe must be leak-tight to a very high degree. This can involve outgassing of the metal used for the heat pipe wall, end caps, etc., although this is not essential for simple low-temperature operations.

Quality control cannot be overemphasised in heat pipe manufacture, and in the following discussion of assembly methods, this will be frequently stressed.

A substantial part of this chapter is allocated to a review of life test procedures for heat pipes. The life of a heat pipe often requires careful assessment in view of the many factors that can affect long-term performance, and most establishments seriously involved in heat pipe design and manufacture have extensive life test programmes in progress. As discussed later, data available from the literature can indicate satisfactory wall/wick/working fluid combinations, but the assembly procedures used differ from one manufacturer to another, and this may introduce an unknown factor that will necessitate investigation.

Measuring the performance of heat pipes is also a necessary part of the work leading to an acceptable product, and the interpretation of the results may prove difficult. Test procedures for heat pipes destined for use in orbiting satellites have their own special requirements brought about by the need to predict performance in zero gravity by testing in earth gravity.

While the vast majority of heat pipes manufactured today are conventional wicked circular (or near circular) cross-section units, there is an increasing trend towards miniaturisation and the use of microgroove type structures as wicks. The implications of these trends for manufacturing procedures are highlighted in appropriate parts of this chapter.

Although the manufacture of special types such as loop or micro-heat pipes are outside the scope of this book, again there are features unique to these types that require close attention during manufacture and assembly, and these are identified and referenced, where it is believed appropriate.

5.1 MANUFACTURE AND ASSEMBLY

5.1.1 Container materials

The heat pipe container, including the end caps and filling tube, is selected on the basis of several properties of the material used and these are listed in Chapter 3. (Unless stated otherwise, the discussion in this chapter assumes that the heat pipes are tubular in geometry.) However, the practical implications of the selection are numerous.

Of the many materials available for the container, three are by far the most common in use, namely copper, aluminium and stainless steel. Copper is eminently satisfactory for heat pipes operating between 0 °C and 200 °C in applications such as electronics cooling. While commercially pure copper tube is suitable, the oxygen-free high-conductivity type is preferable. Like aluminium and stainless steel, the material is readily available and can be obtained in a wide variety of diameters and wall thicknesses in its tubular form.

Aluminium is less common as a material in commercially available heat pipes but has received a great deal of attention in aerospace applications, because of its obvious weight advantages. It is generally used in alloy form, typically 6061-T6, the nearest British equivalent being aluminium alloy HT30. Again this is readily available and can be drawn to suit by the heat pipe manufacturer, or extruded to incorporate, for example, a grooved wick.

Stainless steel unfortunately cannot generally be used as a container material with water where a long life is required, owing to gas generation problems, but it is perfectly acceptable with many other working fluids, and is in many cases the only suitable container, as for example, with liquid metals such as mercury, sodium and potassium. Types of stainless steel regularly used for heat pipes include 302, 316 and 321. (Comments on compatibility with water are made in Chapter 3.) Mild steel may be used with organic fluids, and, again, reference to Chapter 3 can be made for a discussion on its possible compatibility with water.

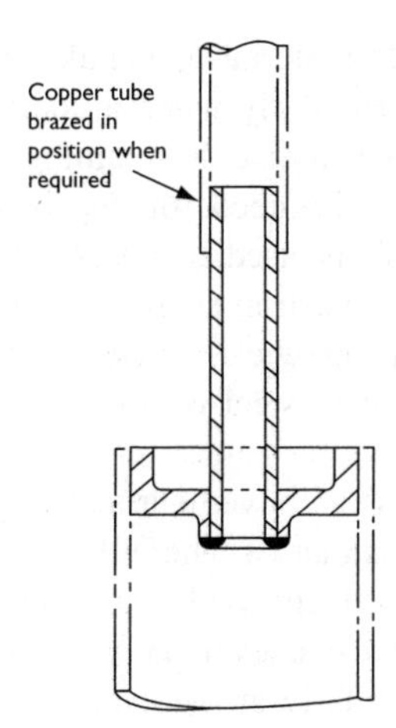

Fig. 5.1 End cap and filling tube.

In the assembly of heat pipes, provision must be made for filling, and the most common procedure involves the use of an end cap with a small diameter tube attached to it, as shown in Fig. 5.1. The other end of the heat pipe contains a blank end cap. End cap and filling tube materials are generally identical to those of the heat pipe case, although for convenience a copper extension may be added to a stainless steel filling tube for cold welding (see Section 5.1.9). It may be desirable to add a valve to the filling tube where, for example, gas analysis may be carried out following life tests (see Section 5.2). The valve material must, of course, be compatible with the working fluid.

If the heat pipe is to operate at high vapour pressures, a pressure test should be carried out to check the integrity of the vessel.

5.1.2 Wick materials and form

The number and form of materials that have been tested as wicks in heat pipes is very large. Reference has already been made to some of these in analysis of the liquid pressure drop, presented in Chapter 2, and in the discussion on selection criteria in Chapter 3.

5.1.2.1 Wire mesh

The most common form of wick is a woven wire mesh or twill which can be made in many metals. Stainless steel, monel and copper are woven to produce meshes having very small pore sizes (see Table 3.4) and 400 mesh stainless steel is available 'off the shelf' from several manufacturers. Aluminium is available, but because of difficulties in producing and weaving fine aluminium wires, the requirements of small pore wicks cannot be met.

Stainless steel is the easiest material to handle in mesh form. It can be rolled and retains its shape well, particularly when a coarse mesh is used. The inherent springiness in the coarse meshes assists in retaining the wick against the heat pipe wall, in some cases obviating the need for any other form of wick location. In heat pipes where a 400 mesh is used, a coarse 100 mesh layer located at the inner radius can hold the finer mesh in shape. Stainless steel can also be diffusion bonded, giving strong permanent wick structures attached to the heat pipe wall. The diffusion bonding of stainless steel is best carried out in vacuum furnace at a temperature of 1150–1200 °C.

The spot welding of wicks is a convenient technique for preserving shape or for attaching the wick to the wall in cases where the heat pipe diameter is sufficiently large to permit insertion of an electrode. Failing this, a coil spring can be used.

It is important to ensure that whatever the wick form it is in very close contact with the heat pipe wall, particularly at the evaporator section, otherwise local hot spots will occur. With mesh the best way of making certain that this is the case is to diffusion bond the assembly.

The manufacture of heat pipes for thermal control of the chips in laptop computers and the like conventionally involves wicked copper heat pipes. Here copper or nickel mesh may be employed instead of stainless steel, depending upon the choice of the working fluid.

5.1.2.2 Sintering

A similar structure having an intimate contact with the heat pipe wall is a sintered wick. Sintering is often used to produce metallic filters, and many components of machines are now produced by this process as opposed to die casting or moulding.

The process involves bonding together a large number of particles in the form of a packed metal powder. The pore size of the wick thus formed can be arranged to suit by selecting powders having a particular size. The powder, which is normally spherical, is placed in containers giving the shape required and then either sintered without being further compacted or, if a temporary binder is used, a small amount of pressure may be applied. Sintering is normally carried out at a temperature 100–200 °C below the melting point of the sintering material.

The simplest way of making wicks by this method is to sinter the powder in the tube that will form the final heat pipe. This has the advantage that the wick is also sintered to the tube wall and thus makes a stronger structure. In order to leave the central vapour channel open, a temporary mandrel has to be inserted in the tube. The powder is then placed in the annulus between mandrel and tube. In the case of copper powder, a stainless steel mandrel is satisfactory as the copper will not bond to stainless steel and thus the bar can easily be removed after sintering. The bar is held in a central position at each end of the tube by a stainless steel collar.

A typical sintering process is described below. Copper was selected as the powder material and also as the heat pipe wall. The particle size chosen was $-150+300$ grade, giving particles of 0.05–0.11 mm diameter. The tube was fitted with the mandrel and a collar at one end. The powder was then poured in from the other end.

No attempt was made to compact the powder apart from tapping the tube to make sure there were no gross cavities left. When the tube was full the other collar was put in place and pushed up against the powder. The complete assembly was then sintered by heating in hydrogen at 850 °C for 1/2 h. After the tube was cooled and removed and the tube, without the mandrel, was then resintered. (The reason for this was that when the mandrel was in place the hydrogen could not flow easily through the powder and as a result sintering may not have been completely successful since hydrogen is necessary to reduce the oxide film that hinders the process.) After this operation, the tube was ready for use. Figure 5.2 shows a cross section of a completed tube and Figure 5.3 shows a magnified view of the structure of the copper wick. The porosity of the finished wick is of the order of 40–50 per cent.

A second type of sintering may be carried out to increase the porosity. This necessitates the incorporation of inert filler material to act as pore formers. This is subsequently removed during the sintering process, thus leaving a very porous structure. The filler used was a perspex powder that is available as small spheres. This powder was sieved to remove the −150+300 (0.050–0.100 mm) fraction. This

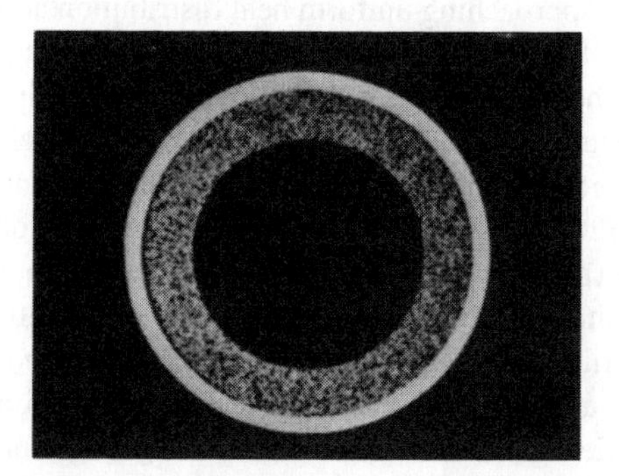

Fig. 5.2 Sintered wick cross section (Copper).

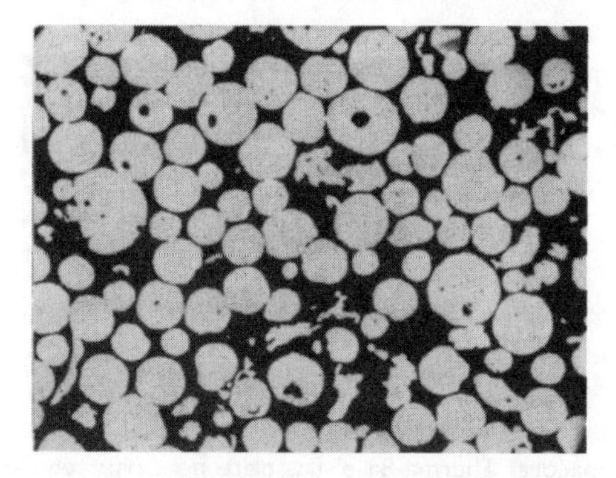

Fig. 5.3 Magnified view of sinter structure.

was mixed with an equal volume of very fine copper powder ($-200\,\mu$m). On mixing, the copper uniformly coats the plastic spheres. This composite powder then shows no tendency to separate into its components.

The wick is now made up exactly as the previous tube with the exception that more compaction is required in order to combat the very high shrinkage that takes place during sintering. During the initial stages of the sintering, the plastic is vapourised and diffused out of the copper compact, thus leaving a skeletal structure of fine copper powder with large interconnected pores. The final porosity is probably of the order of 75–85 per cent.

It is obvious that there are many possible variations of the wicks made by sintering methods. Porosity, capillary rise and volume flow can all be optimised by the correct choice of metal powder size, filler size, filler proportion and by incorporation of channel forming fillers.

Flat plate heat pipes, sometimes called vapour chambers (where the role of the wick may be solely for liquid distribution across the evaporator), most commonly use sintered wick structures. The Thermacore 'Therma-Base' unit [19] uses a copper sintered wick, the unit being illustrated in Fig. 5.4. An alternative method for achieving something approaching uniform heat distribution across plate is to embed heat pipes of tubular form within it, as shown in the Thermacore unit in Fig. 5.5.

Not all flat plate heat pipes are designed to operate against gravity, hence the fact that the equipment can operate with relatively simple wick structures. One heat pipe that was mounted vertically was used in the thermal control of thermoelectric refrigeration units [20]. Although subject to further development, the use of a sintered copper (40 μm particles) wick with longitudinal channels – a graded wick structure or composite structure – was interesting. As highlighted elsewhere, the impact of the fluid fill on the heat pipe performance was shown to be significant. Results showed that as the fluid loads was varied between 5 and 40 g, the thermal resistance initially decreased from 0.25 K/W at 5 g inventory to a minimum of 0.15 K/W at 20 g, before rising to over 0.3 K/W at 40 g. Careful assessment for fluid inventory in wicked heat pipes of all types is essential.

Fig. 5.4 The Thermacore 'Therma-Base' flat plate heat pipe, employing a sintered wick (Courtesy Thermacore International Inc.).

Fig. 5.5 An alternative approach to 'spreading the heat' over a flat surface – the Thermacore heat spreader (Courtesy Thermacore International Inc.).

5.1.2.3 Vapour deposition

Sintering is not the only technique whereby a porous layer can be formed which is in intimate contact with the inner wall of the heat pipe. Other processes include vapour coating, cathode sputtering and flame spraying. Brown Boveri, in UK Patent 1313525, describe a process known as 'vapour plating' that has been successfully used in heat pipe wick construction. This involves plating the internal surface of the heat pipe structure with a tungsten layer by reacting tungsten hexafluoride vapour with hydrogen, the porosity of the layer being governed by the surface temperature, nozzle movement and distance of the nozzle from the surface to be coated.

5.1.2.4 Microlithography and other techniques

The trend towards micro-heat pipes (see also Chapter 6) has led to the use of manufacturing techniques for such units that copies some of the methods used in the microelectronics area that these devices are targeting. Sandia National Laboratory is one of a number of laboratories using photolithographic methods, or similar techniques, for making heat pipes, or more specifically the wick. Workers at Sandia [21] wanted to make a micro-heat pipe that could cool multiple heat sources, keep the favourable permeability characteristics of longitudinally grooved wicks and allow fabrication using photolithography.

Full arguments behind the selection of the anisotropic wick concept, based upon longitudinal liquid flow while minimising transverse movement, are given in the referenced paper, but the wick made was of rectangular channels of width $40\,\mu m$ and height $60\,\mu m$, formed using an electroplating process that plates the copper wick material onto the flat 150 mm substrate wafers.

Final assembly comprised cutting the wick parts, putting in spacers and the fill tube and electron beam welding around the periphery. (Support posts inside were resistance welded.)

Even in the 2-year period since this work was reported, the progress in microfabrication technology, including direct writing methods such as selective laser melting (SLM) is such that a three-dimensional wick structure of almost any form could be fabricated by a rapid prototyping method in several laboratories around the world.

5.1.2.5 Grooves

A type of wick, which is widely used in spacecraft applications but which is unable to support significant capillary heads in earth gravity, is a grooved system. The simplest way of producing longitudinal grooves in the wall of a heat pipe is by extrusion or by broaching. Aluminium is the most satisfactory material for extruding, where grooves may be comparatively narrow in width, but possess a greater depth. An example of a copper grooved heat pipe wick is shown in Fig. 5.6. The external cross section of the heat pipe can also be adapted for a particular application. If the heat pipe is to be mounted on a plate, a flat surface may be incorporated on the wall of the heat pipe to give better thermal contact with the plate.

An alternative groove arrangement involves 'threading' the inside wall of the heat pipe using taps or a single-point cutting tool to give a thread pitch of up to 40 threads per cm. Threaded arteries are attractive for circumferential liquid distribution, and may be used in conjunction with a different artery system for axial liquid transport.

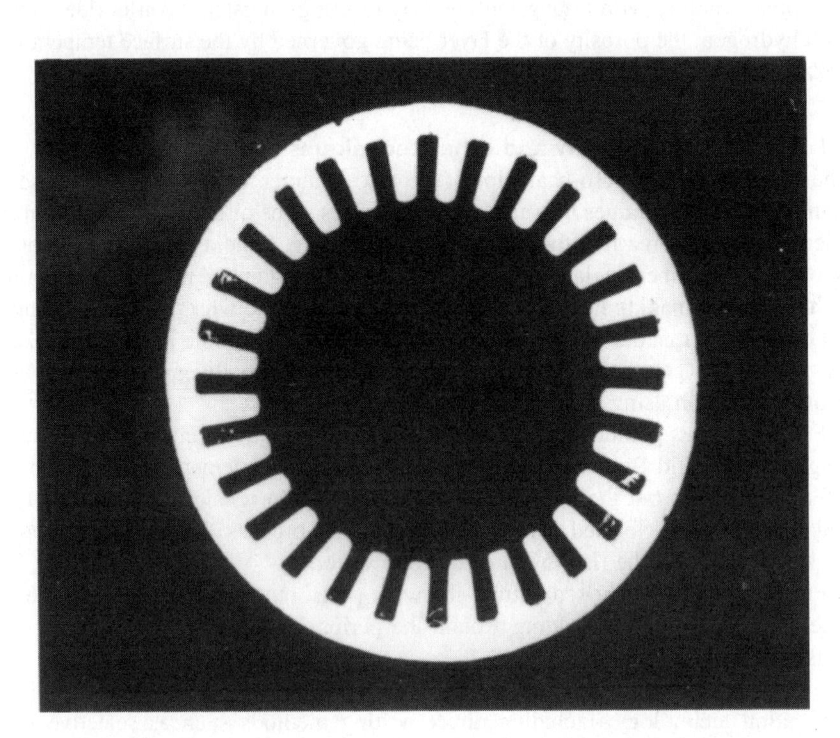

Fig. 5.6 Grooved wick in a copper tube, obtained by mandrel drawing.

Longitudinal grooves may be used as arteries (see Chapter 4 for example) and Fig. 5.7a shows such an artery set, machined in the form of six grooves in a former for insertion down the centre of a heat pipe. Prior to insertion, the arteries are completed by covering them with diffusion-bonded mesh to the outer surface. Fig. 5.7b shows diffusion-bonded mesh.

Triangular-shaped grooves have been fabricated in silicon for electronics cooling duties. Work at INSA in France [22, 23] used two processes – anisotropic chemical etching followed by direct silicon wafer bonding – for micro-heat pipe fabrication. In one unit, 55 triangular parallel micro-heat pipes were constructed (230 μm wide and 170 μm deep) and ethanol was used as the working fluid. A section through the array is shown in Fig. 5.8a, while a second unit, shown in Fig. 5.8b, uses arteries, fabricated in an identical manner, in a third layer of silicon.

5.1.2.6 Felts and foams

Several companies are now producing metal and ceramic felts and metal foam which can be effectively used as heat pipe wicks, particularly where units of noncircular

(a)

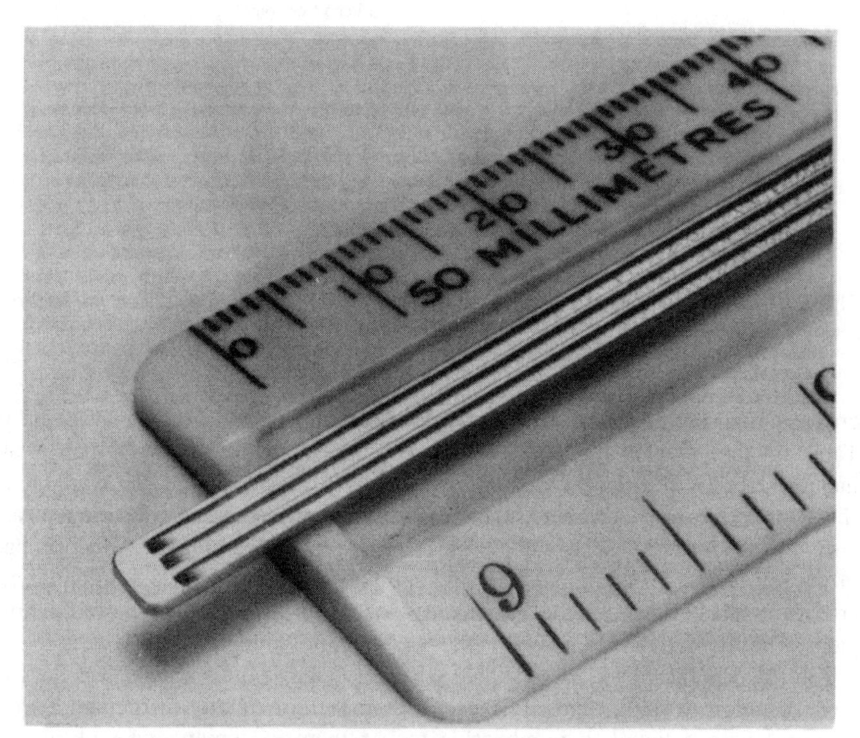

Fig. 5.7a Artery set prior to covering with mesh.

(b)

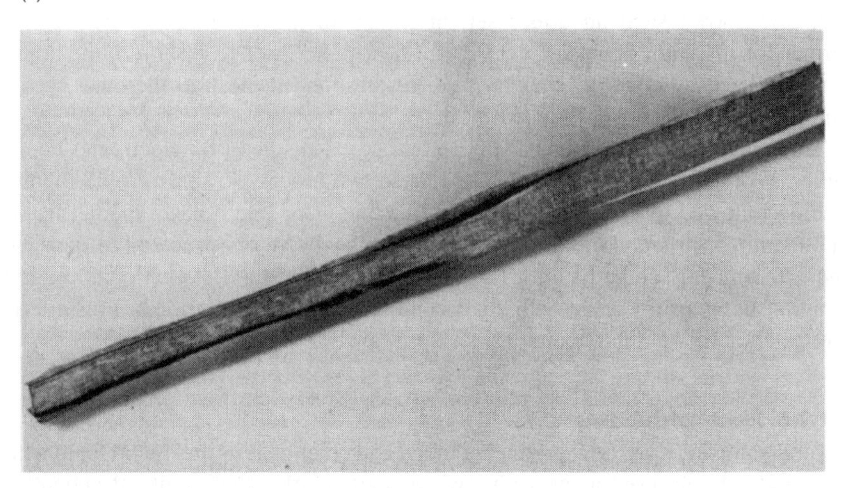

Fig. 5.7b Diffusion-bonded mesh as used to cover arteries.

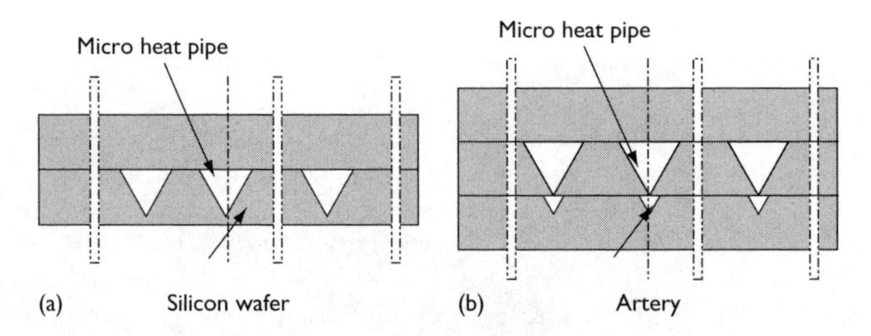

Fig. 5.8 Cross sections of two micro-heat pipe arrays fabricated in silicon: (a) without artery and (b) with artery [23].

cross section are required. The properties of some of these materials are given in Table 3.3. Foams are available in nickel, stainless steel and copper, and felt materials include stainless steel and woven ceramic fibres (Refrasil). Foams are available in sheet and rod forms, and can be supplied in a variety of pore sizes. Metallic felts are normally produced in sheets and are much more pliable than foams. An advantage of the felt is that by using mandrels and applying a sintering process, longitudinal arteries could be incorporated inside the structure, providing low-resistance flow paths. The foam, however, may double as a structural component.

Knitted ceramic fibres are available with very small pore sizes and are inert to most common working fluids. Because of their lack of rigidity, particularly when saturated with a liquid, it is advisable to use them in conjunction with a wire mesh wick to retain their shape and desired location. The ceramic structure can be obtained in the form of multilayer sleeves, ideal for immediate use as a wick, and a range of diameters of sleeves is available. Some stretching of the sleeve can be applied to reduce the diameter, should the exact size is not available.

5.1.3 Cleaning of container and wick

All the materials used in a heat pipe must be clean. Cleanliness achieves two objectives. It ensures that the working fluid will wet the materials and that no foreign matter is present which could hinder capillary action or create incompatibilities.

The cleaning procedure depends upon the material used, the process undergone in manufacturing and locating the wick, and the requirements of the working fluid, some of which wet more readily than others. In the case of wick/wall assembles produced by processes such as sintering or diffusion bonding, carried out under an inert gas or vacuum, the components are cleaned during the bonding process, and provided that the time between this process and final assembly is short, no further cleaning may be necessary.

If the working fluid is a solvent, such as acetone, no extreme precautions are necessary to ensure good wetting, and an acid pickle followed by a rinse in the working fluid appears to be satisfactory. However, cleaning procedures become more rigorous as one moves up the operating temperature range to incorporate liquid metals as working fluids.

The pickling process for stainless steel involves immersing the components in a solution of 50 per cent nitric acid and 5 per cent hydrofluoric acid. This is followed by a rinse in demineralised water. If the units are to be used in conjunction with water, the wick should then be placed in an electric furnace and heated in air to 400 °C for 1 h. At this temperature, grease is either volatised or decomposed, and the resulting carbon burnt off to form carbon dioxide. Since an oxide coating is required on the stainless steel, it is not necessary to use an inert gas blanket in the furnace.

Nickel may undergo a similar process to that described above for stainless steel but pickling should be carried out in a 25 per cent nitric acid solution. Pickling of copper demands a 50 per cent phosphoric acid and 50 per cent nitric acid mixture.

Cleanliness is difficult to quantify, and the best test is to add a drop of demineralised water to the cleaned surface. If the drop immediately spreads across the surface, or is completely absorbed into the wick, good wetting has occurred, and satisfactory cleanliness has been achieved.

Stainless steel wicks in long heat pipes sometimes create problems in that furnaces of sufficient size, which contain the complete wick may not be readily available. In this case, a flame cleaning procedure may be used, whereby the wick is passed through a Bunsen flame as it is fed into the container.

An ultrasonic cleaning bath is a useful addition for speeding up the cleaning process but is by no means essential for low-temperature heat pipes. As with this process or any other associated with immersion of the components in a liquid to removed contaminants, the debris will float to the top of the bath and must be skimmed off before removing the parts being treated. If this is not done, the parts could be recontaminated as they are removed through this layer. Electropolishing may also be used to aid cleaning of metallic components.

Ceramic wick materials are generally exceptionally clean when received from the manufacturer, owing to the production process used to form them, and therefore need no treatment, provided that the handling during assembly of the heat pipe is under clean conditions.

It is important, particularly when water is used as the working fluid, to avoid skin contact with the heat pipe components. Slight grease contamination can prevent wetting, and the use of surgical gloves for handling is advisable. Wetting can be aided by additives (wetting agents) [1] applied to the working fluid, but this can introduce compatibility problems and also affect surface tension.

5.1.4 Material outgassing

When the wick or wall material is under vacuum, gases will be drawn out, particularly if the components are metallic. If not removed prior to sealing of the heat pipe, these gases could collect in the heat pipe vapour space. The process is known as outgassing.

While outgassing does not appear to be a problem in low-temperature heat pipes for applications that are not too arduous, high-temperature units (>400 °C) and pipes for space use should be outgassed in the laboratory prior to filling with working fluid and sealing.

The outgassing rate is strongly dependent on temperature, increasing rapidly as the component temperature is raised. It is advisable to outgas components following cleaning, under vacuum at a baking temperature of about 400 °C. Following baking the system should be vented with dry nitrogen. The rate of outgassing depends on the heat pipe operating vapour pressure, and if this is high the outgassing rate will be restricted.

If the heat pipe has been partially assembled prior to outgassing, and the end caps fitted, it is necessary to make sure that no welds, etc., leak, as these could produce misleading results to outgassing rate. It will generally be found that analysis of gases escaping through a leak will show a very large air content, whereas those brought out by outgassing will contain a substantial water vapour content. A mass spectrometer can be used to analyse these gases. Leak detection is covered in Section 5.1.6. below.

The outgassing characteristics of metals can differ considerably. The removal of hydrogen from stainless steel, for example, is much easier to effect than its removal from aluminium. Aluminium is particularly difficult to outgas and can hold comparatively large quantities of noncondensables. In one test, it was found [2] that gas was suddenly released from the aluminium when it approached red heat under vacuum. Two hundred grams of metal gave 89.5 cc of gas at NTP, 88 cc being hydrogen and the remainder carbon dioxide. It is also believed that aluminium surfaces can retain water vapour even when heated to 500 °C or dried over phosphorous pentoxide. This could be particularly significant because of the known incompatibility of water with aluminium. (See Section 5.1.12 for high-temperature heat pipes.)

5.1.5 Fitting of wick and end caps

Cleaning of the heat pipe components is best carried out before insertion of the wick, as it is easy to test the wick for wettability. Outgassing may be implemented before assembly or while the heat pipe is on the filling rig (see Section 5.1.8).

In cases where the wick is an integral part of the heat pipe wall, as in the case of grooves, sintered powders, or diffusion bonded meshes, cleaning of the heat pipe by flushing through with the appropriate liquid is convenient, prior to the welding of the end caps.

If a mesh wick is used, and the mesh layers are not bonded to one another or to the heat pipe wall, particularly when only a fine mesh is used, a coiled spring must be inserted to retain the wick against the wall. This is readily done by coiling the spring tightly around a mandrel giving a good internal clearance in the heat pipe. The mandrel is inserted into the pipe, the spring tension released and the mandrel then removed. The spring will now be holding the wick against the wall. Typically the spring pitch is about 1 cm. (In instances where two mesh sizes may be used in the heat pipe, say two layers of 200 mesh and one layer of 100 mesh, the liquid–vapour interface must always be in the 200 mesh to achieve maximum capillary rise. It is therefore advisable to wrap the 200 mesh over the end of the 100 mesh, as shown in Fig. 5.9. It is possible to locate the fine mesh against the wall, where it will suppress boiling.)

The fitting of end caps is normally carried out by argon arc welding. This need not be done in a glove box and is applicable to copper, stainless steel and aluminium heat pipes. The advantage of welding over brazing or soldering is that no flux is required, therefore the inside of cleaned pipes do not suffer from possible contamination. However, possible inadequacies of the argon shield, in conjunction

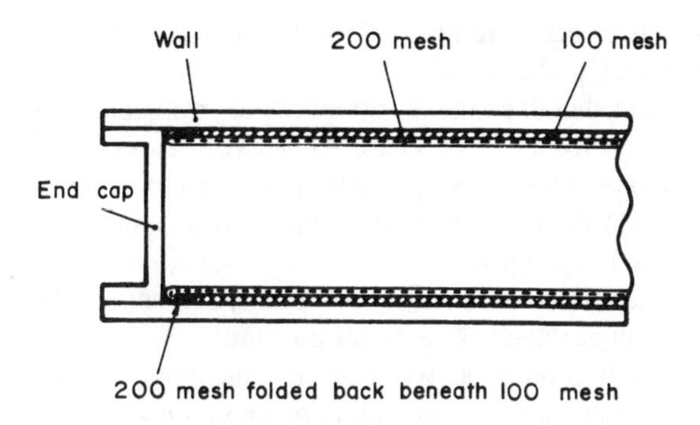

Fig. 5.9 Sealing of mesh at the end of heat pipe.

with the high temperatures involved can lead to local material oxidation which may be difficult to remove from the heat pipe interior. Assembly in a glove box filled with argon would overcome this but would be expensive. The use of a thermal absorbant paste such as Rocol HS to surround the area of heat pipe local to the weld can considerably reduce the amount of oxide formed.

Electron beam welding may also be used for heat pipe assembly, but this added expense cannot be justified in most applications.

5.1.6 Leak detection

All welds on heat pipes should be checked for leaks. If quality control is to be maintained, a rigorous leak check procedure is necessary because a small leak that may not affect heat pipe performance initially could make itself felt over a period of months.

The best way to test a heat pipe for leaks is to use a mass spectrometer that can be used to evacuate the heat pipe to a very high vacuum, better than 10^{-5}torr, using a diffusion pump. The weld area is then tested by directing a small jet of helium gas onto it. If a leak is present, the gauge head on the mass spectrometer will sense the presence of helium once it enters the heat pipe. After an investigation of the weld areas and location of the general leak area(s), if present, a hypodermic needle can then be attached to the helium line and careful traversing of the suspected region can lead to very accurate identification of the leak position, possibly necessitating only a very local rewelding procedure to seal it.

Obviously, if a very large leak is present, the pump on the mass spectrometer may not even manage to obtain a vacuum better than 10^{-2} or 10^{-3} torr. Porosity in weld regions can create conditions leading to this, and may point to impure argon or an unsuitable welding filler rod.

It is possible, if the leak is very small, for water vapour from the breath to condense and block, albeit temporarily, the leak. It is therefore important to keep the pipe dry during leak detection.

5.1.7 Preparation of the working fluid

It is necessary to treat the working fluid used in a heat pipe with the same care as that given to the wick and container.

The working fluid should be the most highly pure available, and further purification may be necessary following purchase. This may be carried out by distillation. In the case of low-temperature working fluids such as acetone, methanol, and ammonia the presence of water can lead to incompatibilities, and the minimum possible water content should be achieved.

Some brief quotations from a treatise on organic solvents [3] highlight the problems associated with acetone and its water content:

'Acetone is much more reactive than is generally supposed. Such mildly basic materials as alumina gel cause aldol condensation to 4-hydroxy-4-methyl-2-pentanone, (diacetone alcohol), and an appreciable quantity is formed in a short time if the acetone is warm. Small amounts of acidic material, even as mild as anhydrous magnesium sulphate, cause acetone to condense'.

'Silica gel and alumina increased the water content of the acetone, presumably through the aldol condensation and subsequent dehydration. The water content of acetone was increased from 0.24 to 0.46% by one pass over alumina. All other drying agents tried ...... caused some condensation'.

Ammonia has a very great affinity for water, and it has been found that a water content of the order of <10 ppm is necessary to obtain satisfactory performance. Several chemical companies are able to supply high-purity ammonia but exposure to air during heat pipe filling must be avoided.

The above examples are extreme but serve to illustrate the problems that can arise when the handling procedures are relaxed.

A procedure that is recommended for all heat pipe working fluids used up to 200 °C is freeze-degassing. This process removes all dissolved gases from the working fluid, and if the gases are not removed they could be released during heat pipe operation and collected in the condenser section. Freeze-degassing may be carried out on the heat pipe filling rig described in Section 5.1.8 and is a simple process. The fluid is placed in a container in the rig directly connected to the vaccum system and is frozen by surrounding the container with a flask containing liquid nitrogen. When the working fluid is completely frozen the container is evacuated and resealed and the liquid nitrogen flask removed. The working fluid is then allowed to thaw and dissolved gases will be seen to bubble out of the liquid. The working fluid is then refrozen and the process repeated. All gases will be removed after three or four freezing cycles.

The liquid will now be in a sufficiently pure state for insertion into the heat pipe.

5.1.8 Heat pipe filling

A flow diagram for a rig that may be used for heat pipe filling is shown in Fig. 5.10. The rig may also be used to carry out the following processes:

Working fluid degassing
Working fluid metering
Heat pipe degassing
Heat pipe filling with inert gas.

Before describing the rig and its operation, it is worth mentioning the general requirements when designing vacuum rigs. The material of construction for pipework is generally either glass or stainless steel. Glass has advantages when handling liquids in that the presence of liquid droplets in the ductwork can be observed and their vapourisation under vacuum noted. Stainless steel has obvious strength benefits and must be used for all high-temperature work, together with high-temperature packless valves such as Hoke bellows valves. The rig described below is for low-temperature heat pipe manufacture.

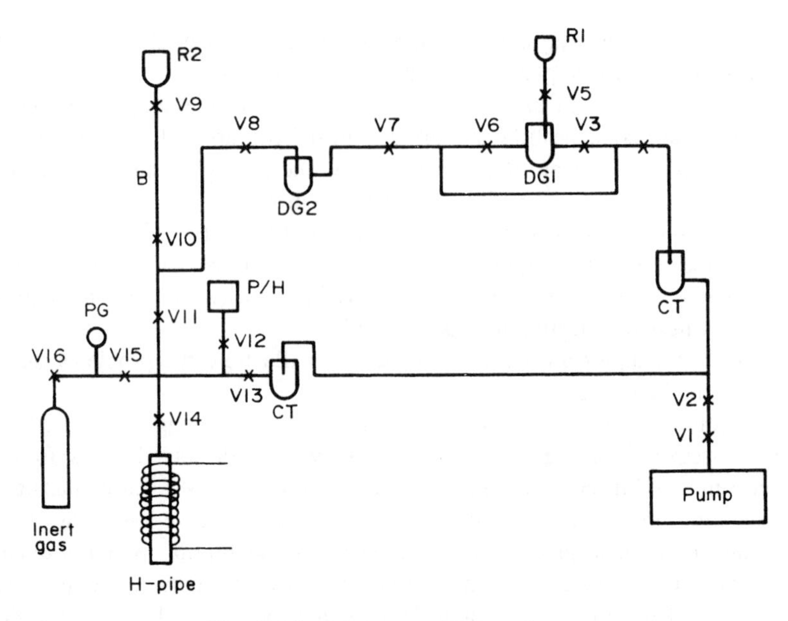

Fig. 5.10 A heat pipe filling rig layout.

Valves used in vacuum rigs should preferably have 'O' ring seals, and it is important to ensure that the ductwork is not too long or has a small diameter, as this can greatly increase evacuation times.

The vacuum pump may be the diffusion type or a sorption pump containing a molecular seive that produce vacuums as low as 10^{-4} torr. It is, of course, advisable to refer to experts in the field of high-vacuum technology when considering designing a filling rig.

5.1.8.1 Description of rig

The heat pipe filling rig described below is made using glass for most of the pipework. Commencing from the right-hand side, the pump is of the sorption type, which is surrounded by a polystyrene container of liquid nitrogen when a vacuum is desired. Two valves are fitted above the pump, the lower one being used to disconnect the pump when it becomes saturated. (The pump may be cleaned by baking out in a furnace for a few hours.) Above the valve V2, a glass-to-metal seal is located and the rest of the pipework is glass. Two limbs lead from this point, both interrupted by cold traps, in the form of small glass flasks, which are used to trap stray liquid and any impurities which could affect other parts of the rig or contaminate the pump. The cold traps are formed by surrounding each flask with a container of liquid nitrogen.

The upper limb includes provision for adding working fluid to the rig and two flasks are included (DG1 and DG2) for degassing the fluid. The section of the rig used for adding fluid can be isolated once a sufficient quantity of fluid has been passed to flask DG2 and thence to the burette between valves V9 and V10.

The lower limb incorporates a Pirani head that is used to measure the degree of vacuum in the rig. The heat pipe to be filled is fitted below the burette, and provision is also made to electrically heat the pipe to enable outgassing of the unit to be carried out on the rig (see also Section 5.1.4). An optional connection can be made via valve V15 to permit the loading of inert gas into the heat pipe for variable conductance types.

5.1.8.2 Procedure for filling a heat pipe

The following procedure may be followed using this rig for filling, for example, a copper/ethanol heat pipe.

(i) Close all valves linking rig to atmosphere (V5, V9, V14, V15).
(ii) Attach sorption pump to rig via valves V1 and V2, both of which should be closed.
(iii) Surround the pump with liquid nitrogen, and also top up the liquid nitrogen containers around the cold traps. (It will be found that the liquid N_2 evaporates quickly initially, and regular topping up of the pump and traps will be necessary.
(iv) After approximately 30 min, open valves V1 and V2, commencing rig evacuation. Evacuate to about 0.010 mmHg, the time to achieve this depending on the pump capacity, rig cleanliness and rig volume.
(v) Close valves V4 and V6, and top up reservoir R1 with ethanol.
(vi) Slowly crack valve V5 to allow ethanol into flask DG1. Reclose V5 and freeze the ethanol using a flask of liquid N_2 around DG1.
(vii) When all the ethanol is frozen, open V4 and evacuate. Close V4 and allow ethanol to melt. All gas will bubble out of the ethanol as it melts. The ethanol is then refrozen.
(viii) Open V4 to remove gas.
(ix) Close V4, V3 and V8; open V6 and V7. Place liquid N_2 container around flask DG2.
(x) Melt the ethanol in DG1 and drive it into DG2. (This is best carried out by carefully heating the frozen mass using a hair dryer. Warming of the ductwork between DG1 and DG2 and up to V4, will assist.)
(xi) The degassing process may be repeated in DG2 until no more bubbles are released. V4 and V6 are now closed, isolating DG1.
(xii) Close V7 and V11; open V8 and V10, and drive the ethanol into the burette as in (x). Close V10 and V8 and open V11. The lower limb and upper limb back to V8 are now brought to a high vacuum ($\approx$0.005 mm/Hg).

The heat pipe to be filled should now be attached to the rig. In cases where the heat pipe does not have its own valve, the filling tube may be connected to the rig below V14 using thick-walled rubber tubing, or in cases where this may be attacked by the working fluid, another flexible tube material or a metal compression or 'O' ring coupling. If a soft tube material is used, the joints should be covered with a silicone-based vacuum grease to ensure no leaks.

The heat pipe may be evacuated by opening valve V14. Following evacuation that should take only a few minutes, depending on the diameter of the filling tube, the heat pipe may be outgassed by heating. This can be done by surrounding the heat pipe with electric heating tape, and applying heat until the Pirani gauge returns to the maximum vacuum obtained before heating commenced. (It is worth emphasising the fact that, depending on the diameter of the heat pipe and filling tube, the pressure recorded by the Pirani is likely to be less than that in the heat pipe. It is preferable from this point of view to have a large diameter filling tube.)

To prepare the heat pipe for filling, the lower end is immersed in liquid nitrogen so that the working fluid, which flows towards the coldest region, will readily flow to the heat pipe base. Valve V10 is then cracked and the correct fluid inventory (in most cases enough to saturate the wick plus a small excess) allowed to flow down into the heat pipe. Should fluid stray into valve seats or other parts of the rig, local heating of these areas using the hot air blower will evaporate any liquid, which should then condense and freeze in the heat pipe. A further freeze-degassing process may be carried out with the fluid in the heat pipe, allowing it to thaw with V14 closed, refreezing and then opening V14 to evacuate any gas. The heat pipe may then be sealed.

5.1.9 Heat pipe sealing

Unless the heat pipe is to be used as a demonstration unit, or for life testing, in which case, a valve may be retained on one end, the filling tube must be permanently sealed.

With copper, this is conveniently carried out using a tool that will crimp and cold weld the filling tube. A typical crimp obtained with this type of tool is shown in Fig. 5.11 and the force to operate this is applied manually. The tool is illustrated in Fig. 5.12.

If stainless steel or aluminium are used as the heat pipe filling tube material, crimping followed by argon arc welding is a more satisfactory technique. Once the desired vacuum has been attained and the fluid injected, two 0.5 in. (12.7 mm) jaws are brought into contact with the evacuating tube and the latter is flattened. The heat pipe is then placed between two 0.25 in. thick jaws located at the lower half

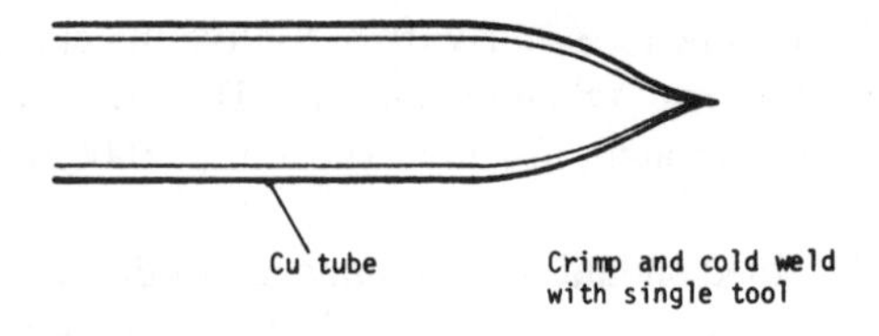

Fig. 5.11 Crimped and cold welded seal.

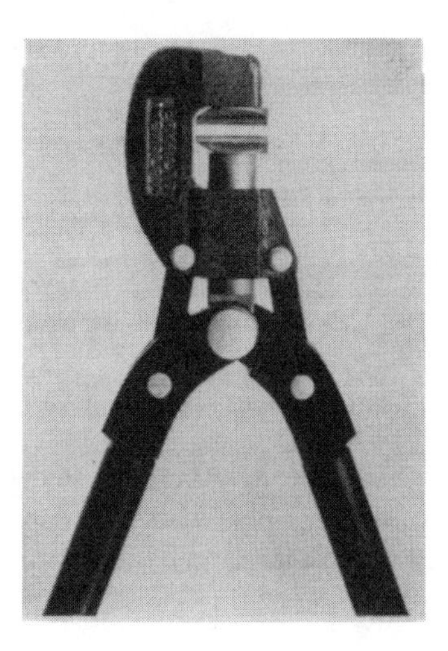

Fig. 5.12 Crimp and cold welding tool.

of the 0.5 in. flattened section. Sufficient load is placed on the evacuating tube to temporarily form a vacuum-tight seal and the remaining 0.25 in. flattened section is simultaneously cut through and welded using an argon arc torch. The 0.25 in. (6.3 mm) crimping tool, which fits between the jaws of a standard vice, is shown in Fig. 5.13. Results obtained are shown in Fig. 5.14.

Fig. 5.13 Jaws for crimping prior to welding.

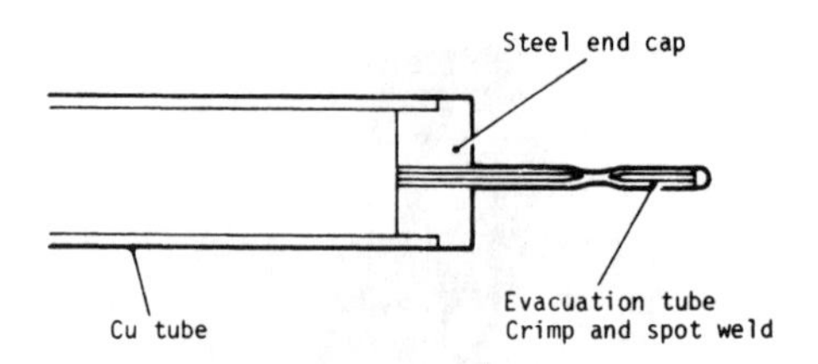

Fig. 5.14 Crimped and argon-arc welded end.

Following sealing the filling tube may be protected by a cap having an outer diameter the same as that of the heat pipe wall. The cap may be filled with solder, a metal-loaded resin or any other suitable material.

5.1.10 Summary of assembly procedures

The following is a list of the procedures described above, which should be followed during heat pipe assembly:

(i) Select container material
(ii) Select wick material and form
(iii) Fabricate wick and end caps, etc.
(iv) Clean wick, container and end caps
(v) Outgas metal components
(vi) Insert wick and locate
(vii) Weld end caps
(viii) Leak check welds
(ix) Select working fluid
(x) Purify working fluid (if necessary)
(xi) Degas working fluid
(xii) Evacuate and fill heat pipe
(xiii) Seal heat pipe.

It may be convenient to weld the blank end cap before wick insertion, and in cases of sintered and diffusion bonded wicks, the outgassing may be done with the wick in place in the container.

For the manufacturer considering the production of a considerable number of identical heat pipes, for example 50 or more units following prototype trials, a number of the manufacturing stages may be omitted. Outgasing of metal components may be unnecessary, and it may be found that, depending upon the filling and evacuation procedure used, the fluid degasing may be eliminated as a separate activity.

5.1.11 Heat pipes containing inert gas

Heat pipes of the variable conductance type (see Section 6.1 in Chapter 6) contain an inert gas in addition to the normal working fluid, and an additional step in the filling process must be carried out. The additional features on the filling rig to cater for inert gas metering are shown in Fig. 5.10.

The working fluid is inserted into the heat pipe in the normal way, and then the system is isolated and the line connecting the heat pipe to the inert gas bottle is opened and the inert gas bled into the heat pipe. The pressure increases as the inert gas quantity in the heat pipe is raised, as indicated by the pressure gauge in the gas line. The pressure appropriate to the correct gas inventory may be calculated, taking into account the partial pressure of the working fluid vapour in the heat pipe (see earlier Editions for mass calculation) and when this is reached, the heat pipe is sealed in the normal manner.

Two aspects of variable conductance (gas-loaded) heat pipes need to be accounted for in manufacture and test procedures. Firstly, most theories for variable conductance heat pipes are based on the assumption that the gas/vapour interface is sharp and diffusion between the two regions is not present. In practice this is not the case, and in some designs it is necessary to take diffusion into account.

A second more serious phenomen resulting from the introduction of inert gas control into a heat pipe occurs when gas bubbles enter the wick structure via the working fluid.

These two features are briefly discussed below.

5.1.11.1 Diffusion at the vapour/gas interface

It has been demonstrated that in some gas-buffered heat pipes the energy and mass diffusion between the vapour and the noncondensable gas could have an appreciable effect on heat transfer in the interface region and the temperature distribution along the heat pipe.

The diffusion coefficient of the inert gas has an effect on the extent of the diffuse region, gases having higher diffusion coefficients being less desirable, reducing the maximum heat transport capability of the heat pipe by reducing local condenser temperature. It must be noted that the diffusion coefficient is inversely proportional to density, and therefore at lower operating temperatures, particularly during start-up of the heat pipe, the diffuse region may be extensive and of even greater significance. It is therefore important to cater for this during any transient performance analysis, and it should be noted during inert gas selection.

5.1.11.2 Gas bubbles in arterial wick structures

Although in simple heat pipes containing only the working fluid, freeze-degassing of the liquid can remove any dissolved gases; in a variable conductance heat pipe inert gas is always present. If the gas dissolves in the working fluid, or finds its way in bubble form into arteries carrying liquid, the performance of the heat pipe can be adversely affected.

Saaski [24] carried out theoretical and experimental work on the isothermal dissolution of gas in arterial heat pipes, examining effects of solubility and diffusivity of helium and Argon in ammonia, and methanol.

One of the significant factors determined by Saaski was the venting time of bubbles in working fluids (the time for a bubble to disappear).

The venting time t_v may be calculated from the equation:

$$t_v = \frac{R_o^2}{3\alpha D}$$

where R_o is the bubble radius (initial); α the Ostwald coefficient, given by the ratio of the solute concentration in the liquid phase to the concentration in the gaseous phase [25] and D the diffusion coefficient.

Predicted values of t_v are given in Table 5.1.

Table 5.1 shows that the venting times can be considerable when the working fluid is at a low temperature, but in general argon is more easily vented than helium.

The equation above is not valid when noncondensable gas pressure is significant compared to the value of $2\sigma_l/R_o$, where σ_l is the surface tension of the working fluid. Saaski stated that the venting time increases linearly with noncondensable gas pressure, other factors being equal, and showed that if, as in a typical gas-controlled heat pipe, the helium pressure is about equal to the ammonia working fluid vapour pressure, the vent time can be 9 days. This is a very long time when compared with the transients to be expected in a variable conductance heat pipe (VCHP); by changing the working fluid and/or control gas, relatively long venting times may still be obtained.

Having established venting times for spherical bubbles, Saaski developed a theory to cater for elongated bubbles, the type most likely to form in an arteries. He obtained the results in Table 5.2 for the half lives of elongated arterial bubbles in a VCHP at 20 °C (artery radius 0.05 cm, noncondensable gas partial pressure equal to vapour pressure).

Table 5.1 Venting time of gas bubbles in working fluids ($R_o = 0.05$ cm)

Fluid	Temperature	t_v (s)	
		Helium	Argon
Ammonia	−40	1200	107
	20	63	6.7
	60	7	1.6
Methanol	−40	1030	154
	20	133	55
	60	50	26
Water	22	1481	1215

Table 5.2 Half lives of arterial bubbles in various working fluids

Fluid	$t_{1/2}$ (Helium)	$t_{1/2}$ (Argon)
Ammonia	7 days	17 h
Methanol	4.8 h	1.7 h
Water	3 h	2.5 h

The models used to calculate these values were confirmed experimentally, and it was concluded that the venting times are of sufficient length that repriming of an arterial heat pipe containing gas may be possible only if some assistance in releasing gas occlusions can be given during start-up or steady state operation, either by internal phenomena or by external interference.

Kosson et al. [26] introduced another factor affecting VCHP's, namely variations in pressure within the pipe due to oscillations in the diffusion zone. These pressure variations are of the same order as the capillary pressure and can cause vapour flashing within the artery, with accompanying displacement of liquid from the artery.

In order to overcome this and occlusion problems, subcooling of the liquid in the artery was carried out by routing the fluid to the condenser wall so that it experienced sink conditions before returning to the evaporator. As shown in Saaski's results, lowering the liquid temperature improved venting time. It was also found to reduce the sensitivity of the artery to vapour formation caused by the pressure oscillations described above.

5.1.12 Liquid metal heat pipes

The early work on liquid metal heat pipes was concerned with the application to thermionic generators. For this application, there are two temperature ranges of interest, the emitter range of 1400–2000 °C and the collector range of 500–900 °C. In both temperature ranges, liquid metal working fluids are required and there is a considerable body of information on the fabrication and performance of such heat pipes. More recently, heat pipes operating in the lower temperature range have been used to transport heat from the heater to the multiple cylinders of a Stirling engine and for industrial ovens. A large range of material combinations have been found suitable in this temperature range and compatibility and other problems are well understood. The alkali metals are used with containment materials such as stainless steel, nickel, niobium–zirconium alloys and other refractory metals. Lifetimes of greater than 20 000 h are reported [4]. Grover [5] reports on the use of a light-weight pipe made from beryllium and using potassium as the working fluid. The beryllium was inserted between the wick and wall of a pipe both made from niobium–1 per cent zirconium. The pipe operated at 750 °C for 1200 h with no signs of attack, alloying or mass transport.

The use of Inconel 600 as a container material with sodium working fluid was investigated by Japanese researchers, with a view to show its long-term durability compared to stainless steel type 316. Standard assembly procedures were used, and results of life tests showed some superiority of Inconel 600 over 316 stainless steel, but grain boundary examination suggested that after 60 000 h of operation, pitting would lead to pin holes of corrosion [27].

High-performance, long life, liquid-metal heat pipes can be constructed with some confidence; they are, however, expensive. Hence, before commencing the design of a liquid metal heat pipe, it is important to decide what is to be required from it. It frequently happens that an application does not require the pipe to pump against a gravity head so that a thermal syphon will be adequate. This greatly reduces the importance of working fluid purity. Again short operation life at low rating will enable cheaper and less time-consuming fabrication methods to be adopted. If gas buffering is possible a simpler crimped seal arrangement can be used.

Two examples of heat pipes using liquid sodium as the working fluid are shown in Figs 5.15 and 5.16. These heat pipes, manufactured by Transterm in Romania., are tubular and annular units, respectively. The unit in Fig. 5.15 is destined for a chemical reactor, while the second unit, an isothermal oven, may be used for crystal growing.

5.1.13 Liquid metal heat pipes for the temperature range 500–1100°C

In this temperature range potassium and sodium are the most suitable working fluids and stainles steel is selected for the container. The construction and fabrication of a sodium heat pipe [6] will be described to indicate the processes involved. The

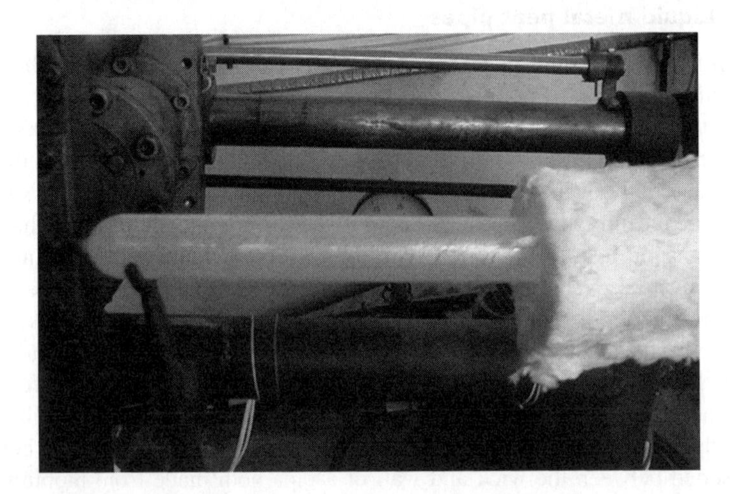

Fig. 5.15 Sodium heat pipe during manufacturing process. Dimensions: length, 2800 mm; outside diameter, 38 mm; inside diameter, 32 mm; destination: catalytic reactors.

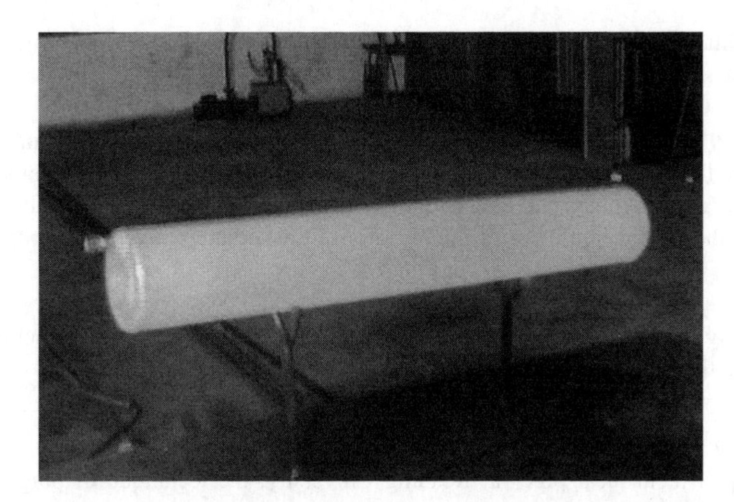

Fig. 5.16 Isothermal oven for growing crystals experiments. Working fluid: sodium; dimensions: length, 500 mm; outside diameter, 51 mm; inner working space diameter 25 mm.

heat pipe container was made from type 321 (EN58B) stainless steel tube 2.5 cm diameter and 0.9 mm wall thickness. The capillary structure was two layers of 100-mesh stainless steel having a wire diameter of 0.1016 mm and an aperture size of 0.152 mm. The pipe was 0.9 m in length and the wick welded by spot welds using a special tool built for the purpose.

5.1.13.1 Cleaning and filling

The following cleaning process was followed:

(i) Wash with water and detergent
(ii) Rinse with demineralised water
(iii) Soak for 30 min in 1:1 mixture of hydrochloric acid and water
(iv) Rinse with demineralised water
(v) Soak for 20 min in an ultrasonic bath filled with acetone and repeat with a clean fluid.

After completion of the welds and brazes this procedure was repeated. Argon arc welding was used throughout, and after leak testing, the pipe was outgassed at a temperature of 900 °C and a pressure of 10^{-5} torr for several hours in order to remove gases and vapours.

Various methods may be used to fill the pipe with liquid metal including

(i) distillation, sometimes from a getter sponge to remove oxygen;
(ii) breaking an ampoule contained in the filler pipe by distortion of the filler pipe.

Distillation is essential if a long life is required. The method adopted for the pipe being described was as follows:

(iii) 99.9 per cent industrial sodium was placed in a glass filter tube attached to the filling tube of the heat pipe. A bypass to the filter allowed the pipe to be initially evacuated and outgassed. The filling pipe and heat pipe were immersed in the heated liquid paraffin bath to raise the sodium above its melting point. The arrangement is shown in Fig. 5.17.

Finally the bypass valve is closed and a pressure applied by means of helium gas to force the molten sodium through the filter and into the heat pipe.

5.1.13.2 Sealing

For liquid metal heat pipes at Reading University, the technique of plug sealing was adopted, as shown in Fig. 5.18.

A special rig has been constructed which allows for outgassing of an open-ended tube and sodium filling by the filtering method described above. On completion of the filling process, the end sealing plug, supported by a swivel arm within the filling chamber, is swung into position and placed within the heat pipe. The plug is then induction heated to effect a brazed vacuum seal. The apparatus and sequence of operation is illustrated in Fig. 5.17. The end sealing plug is finally argon arc welded after removal of the heat pipe from the filling apparatus.

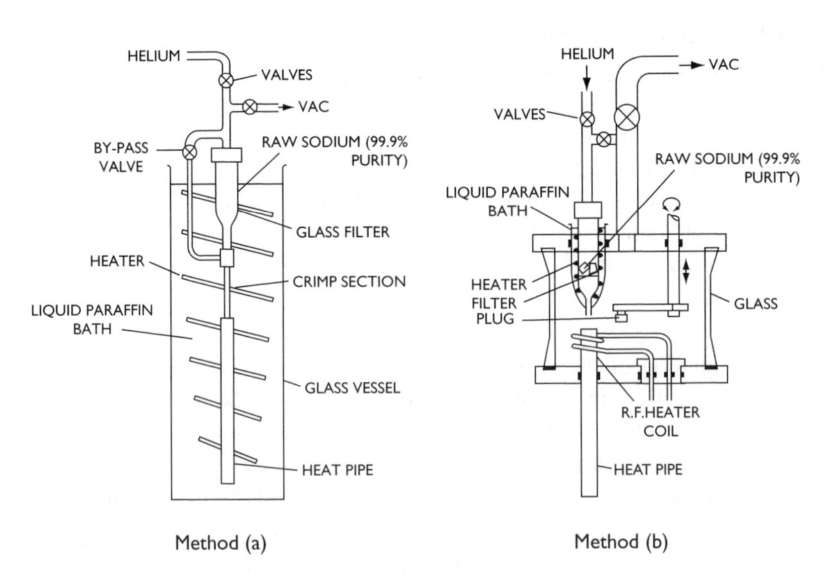

Fig. 5.17 Liquid metal heat pipe filling (Courtesy Reading University).

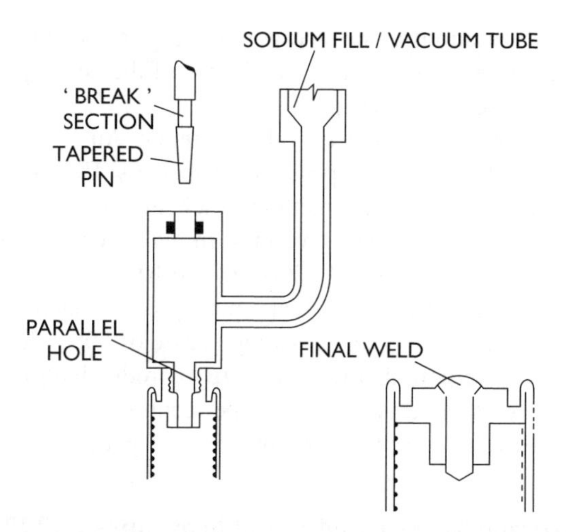

Fig. 5.18 Plug sealing technique for sealing liquid metal heat pipes (Courtesy Reading University).

5.1.13.3 Operation

It has been found that wetting of the wick structure does not occur immediately, and it was necessary to heat the pipe as a thermal syphon for several hours at 650 °C. Heating was by an R.F. induction heater over a length of 10 cm. Temperature profiles are given in Fig. 5.19 for heat inputs of 1.2 and 1.4 kW. Before sealing, the heat pipe was filled with helium at a pressure of 20 torr to protect the copper crimp by the resulting gas buffer. It is seen that the gas buffer length is approximately proportional to the power input as might be expected.

The start-up of the heat pipe after conditioning was interesting. In the thermal syphon mode, that is with the pipe vertical and heated at the bottom, there were

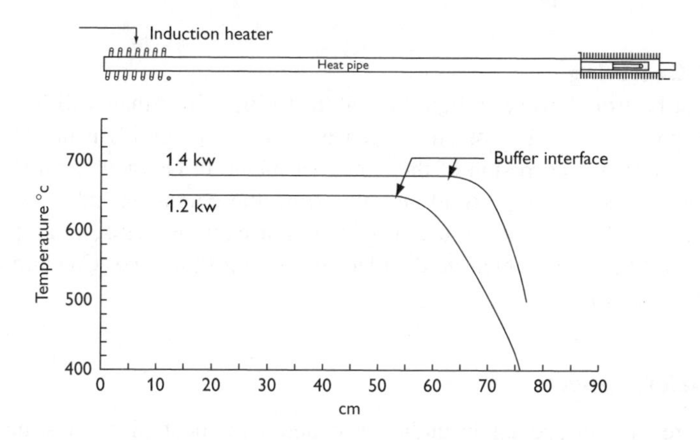

Fig. 5.19 Temperature profiles along a sodium heat pipe (Courtesy Reading University).

violent temperature variations associated with boiling in the evaporator zone. This was not experienced when the heater was at the top of the heat pipe. Further sodium work at Reading is described in [16].

Similar work has been reported by other authors. An interesting method for making rigid thin-walled wicked pipes is described by Vinz et al. [7]. Previous work on mesh wicks has included methods such as spot welding, drawing and sintering. The first method does not give uniform adhesion, and drawing methods cannot be used for very fine wicks (<200–400 mesh) because of damage. Vinz's method consists of winding a screen strip spirally on a mandrel and sintering it under simultaneous axial pulling and twisting. Gauze of 508×3600 mesh has been used successfully to give pore diameters of $10\,\mu$ reproducible to ± 10 per cent and with a free surface for evaporation of 15–20 per cent.

Broached grooves can be used either alone or with gauze wicks.

5.1.13.4 High-temperature liquid metal heat pipes >1200 °C

At the lower end of the range, lithium is preferred as the working fluid and niobium–zirconium or tantalum as the container material. At higher temperatures silver may be used as the working fluid with tungsten or rhenium as the container material. Data on the compatibility and lifetime of heat pipes made from these materials are given in Chapter 3. Such refractory materials have a high affinity for oxygen and must be operated in a vacuum or inert gas.

Busse and his collaborators have carried out a considerable programme on lithium and silver working fluid heat pipes, and the techniques used for cleaning, filling, fabrication and sealing are described in Refs [8, 9].

More recently [28], a lithium heat pipe system has been studied by Advanced Cooling Technologies in the United States, on behalf of Lockheed Martin and the US Air Force Research Laboratory. The system, operating at slightly lower temperatures (to 1100 °C), is directed at cooling the wings of spacecraft on re-entry into the Earth's atmosphere.

5.1.13.5 Gettering

Oxides can be troublesome in liquid metal heat pipes since they will be deposited in the evaporator area. Dissolved oxygen is a particular problem in lithium heat pipes since it causes corrosion of the container material. Oxygen can arise both as an impurity in the heat pipe fluid and also from the container and wick material. A number of authors report the use of getters. For example Busse et al. [9] used a zirconium sponge from which he distilled lithium into the pipe. Calcium can also be used for gettering.

5.1.14 Safety aspects

While there are no special hazards associated with heat pipe construction and operation, there are a number of aspects that should be borne in mind.

Where liquid metals are employed, standard handling procedures should be adopted. The affinity of alkali metals for water can give rise to problems; a fire was started in one laboratory when a sodium in stainless steel pipe distorted releasing the sodium and at the same time fracturing a water pipe.

Mercury is a highly toxic material and its saturated vapour density at atmospheric pressure is many times the recommended maximum tolerance.

One danger that is sometimes overlooked is the high pressure which may occur in a heat pipe when it is accidentally raised to a higher temperature than its design value. Water is particularly dangerous in this respect. The critical pressure of water is 220 bar and occurs at a temperature of 374 °C. When a water in copper heat pipe sealed by a soldered plug was inadvertently overheated, both the 30 cm long heat pipe and the plug were ejected from the clamps at very high velocity and could well have had fatal results. It is imperative that a release mechanism such as a crimp seal be incorporated in such heat pipes.

Cryogenic heat pipes employing fluids such as liquid air should have special provision for pressure release or be of sufficient strength since they are frequently allowed to rise to room temperature when not in use. The critical pressure of nitrogen is 34 bar.

Organisations using specific ‘Health and Safety at Work’ documentation and procedures will find that many aspects of heat pipe manufacture and use may need bringing to the attention of personnel, such as the toxicity/flammability of some working fluids, the high temperature of some surfaces and the need to keep within internal pressure guidelines. (Note that pressure will not in most cases be monitored and can only be assessed from knowledge of the heat pipe temperature and the fluid used within it.)

5.2 HEAT PIPE LIFE TEST PROCEDURES

Life testing and performance measurements on heat pipes, in particular when accelerated testing is required, are the most important factors in their selection.

In spacecraft, for example, the European Space Agency (ESA) stipulate [10] that heat pipes should be suitable for operation for 7–10 years in space after 5 years of ground storage and testing during spacecraft development. It is specifically stated that evidence of long-term compatibility of materials and working fluids must be available before spacecraft ‘qualification’ can be granted. (Requirements are presented in more detail later.)

Life tests on heat pipes are commonly regarded as being primarily concerned with the identification of any incompatibilities that may occur between the working fluid and wick and wall materials. The ultimate life test, however, would be in the form of a long-term performance test under conditions appropriate to those in the particular application. If this is done, it is difficult, however, in cases where the wick is pumping against gravity, to accelerate the life test by increasing, say the evaporator heat flux, as this could well cause heat pipe failure owing to the fact that it is likely to be operating well in excess of its design capabilities. This, therefore, necessitates operation in the reflux mode. (Compatibility data are presented in Chapter 3.)

There are many factors to be taken into account when setting up a full life test programme, and the relative merits of the alternative techniques are discussed below.

5.2.1 Variables to be taken into account during life tests

The number of variables to be considered when examining the procedure for life tests on a particular working fluid/wick/wall combination is very extensive and would require a large number of heat pipes to be fully comprehensive.

Several of these may be discounted because of existing available data on particular aspects, but one important point which must be emphasised is the fact that quality control and assembly techniques inevitably vary from one laboratory to another, and these differences can be manifested in differing compatibility data and performance.

5.2.1.1 The working fluid

The selection of the working fluid must take into account the following factors which can all be investigated by experiments:

(i) Purity – the working fluid must be free of dissolved gases and other liquids, for example water. Such techniques as freeze-degassing and distillation are available to purify the working fluid. It is important to ensure that the handling of the working fluid following purification does not expose it to contaminants.

(ii) Temperature – some working fluids are sensitive to operating temperature. If such behaviour is suspected, the safe temperature band must be identified.

(iii) Heat flux – high heat fluxes can create vigorous boiling action in the wick, which can lead to errosion.

(iv) Compatibility with wall and wick – the working fluid must not react with the wall and wick. This can also be a function of temperature and heat flux, the tendency for reactions to occur generally increases with increasing temperature or flux.

(v) Noncondensable gas – in the case of variable conductance heat pipes, where a noncondensable gas is used in conjunction with the working fluid, the selection of the two fluids must be based on compatibility and also on the solubility of the gas in the working fluid. (In general these data are available from the literature, but in specific arterial design the effect of solubility may only be apparent after experimentation.)

5.2.1.2 The heat pipe wall

In addition to the interface with the heat pipe working fluid, as discussed above, the wall and associated components such as end caps have their own particular requirements with regard to life, and also interface with the wick. The successful operation of the heat pipe must take into account the following:

(i) Vibration and acceleration – the structure must be able to withstand any likely vibrations and accelerations, and any qualification procedures designed to ensure that the units meet these specifications should be regarded as an integral part of any life test programme.
(ii) Quality assurance – the selection of the outer case material should be based on the purity or at least the known alloy specification of the metal used.
(iii) External environment – the external environment could affect the case material properties or cause degradation of the outer surface. This should also be the subject of life test investigation if any deterioration is suspected.
(iv) Interface corrosion – it is possible that some corrosion could occur at metallic interfaces, particularly where dissimilar metals are used, in the presence of the working fluid.

5.2.1.3 The wick

The heat pipe wick is subjected to the same potential hazards as the heat pipe wall, with the exception of external attack. Vibration is much more critical however, and the wick itself contains, in most cases, many interfaces where corrosion could occur (Fig. 5.20).

5.2.2 Life test procedures

There are many ways of carrying out life tests, all having the same aim, namely to demonstrate that the heat pipe can be expected to last for its design life with an excellent degree of certainty.

The most difficult part of any life test programme is the interpretation of the results and the extrapolation of these results to predict long-term performance. (One technique used for extrapolating results obtained from gas generation measurements is described in Section 5.2.3.)

The main disadvantage of carrying out life tests of one particular combination of materials, be the test accelerated or at design load, is the fact that if any reaction does occur, insufficient data are probably available to enable one to explain the main causes of the degradation. For example, in some life tests carried out at IRD, diacetone alcohol was formed as a result of acetone degradation. It was not possible without further testing over a considerable period, however, to state whether this phenomenon was a function of operating temperature, as life tests on identical units operating at several different vapour temperatures will have to be carried out. It is even possible that comprehensive life test programmes may never provide the complete answer to some questions, new aspects being found during each study.

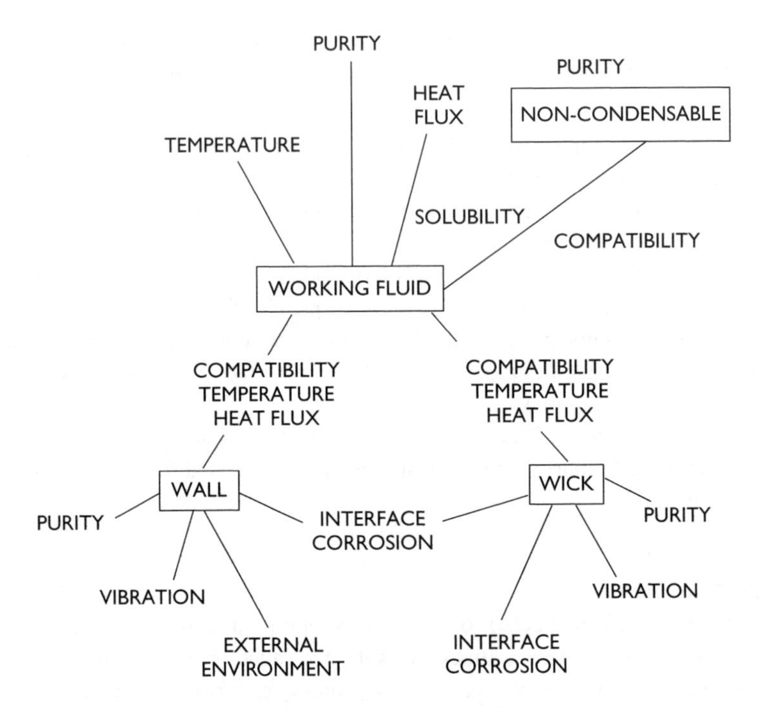

Fig. 5.20 Heat pipe life test factors.

5.2.2.1 Effect of heat flux

The effect of heat flux on heat pipe lives and performance can only really be investigated using units in the reflux mode, where fluxes well in excess of design values may be applied.

By setting up experiments involving a number of heat pipes operating at the same vapour temperature but with differing evaporator heat fluxes, one can later examine the inner surface of the evaporator for corrosion, etc.

If carried out in a representative heat pipe, performance tests could be carried out at regular intervals during the life tests.

5.2.2.2 Effect of temperature

Compatibility and working fluid make-up can both be affected by the operating temperature of the heat pipe. It is therefore important to be able to discriminate between any effects resulting from temperature levels.

5.2.2.3 Compatibility

As opposed to the effect of heat flux or temperature on the working fluid alone, it is necessary to investigate the compatibility of the working fluid with the wall and wick materials.

Here one is looking for reactions between the materials which could change the surface structure in the heat pipe, generate noncondensable gas or produce impurities in the form of deposits that could affect evaporator performance. Of course, all three phenomena could occur at the same time, at differing degrees, and this can make the analysis of the degradation much more complex.

Compatibility tests can be carried out at design conditions on a heat pipe operating horizontally or under tilt against gravity. To be meaningful, such tests should continue for years, but if compatibility is shown to be satisfactory over, say, a three-year period, some conclusions can be made concerning the likely behaviour over a much longer life. Accelerated compatibility tests could also be performed, with occasional tests in the heat pipe mode to check on the design performance.

5.2.2.4 Other factors

The life of a heat pipe can be affected by assembly and cleaning procedures, and it is important to ensure that life test pipes are fully representative as far as assembly techniques are concerned. The working fluid used must, of course, be of the highest purity.

Another feature of life testing is the desirability of incorporating valves on the pipes so that samples of gas, etc., can be taken out without necessarily causing the unit to cease functioning. One disadvantage of valves is the introduction of a possibly new incompatibility that of the working fluid and valve material, although this can be ruled out with modern stainless steel valves.

When testing in the heat pipe mode, a valve body can be filled with working fluid that may be difficult to remove. This should be taken into account when carrying out such tests, in case depletion of the wick or artery system occurs.

5.2.3 Prediction of long-term performance from accelerated life tests

One of the major drawbacks of accelerated life tests has been the uncertainty associated with the extrapolation of the results to estimate performance over a considerably longer period of time. Baker [11] has correlated data on the generation of hydrogen in stainless steel heat pipes, using an Arrhenius plot, with some success, and this has been used to predict noncondensable gas generation over a 20-year period.

The data were based on life tests carried out at different vapour temperatures over a period of 2 years, the mass of hydrogen generated being periodically measured. Vapour temperatures of 100, 200 and 300 °F were used, five heat pipes being tested at each temperature.

Baker applied Arrhenius plots to these results, which were obtained at the Jet Propulsion Laboratory, in the following way.

The Arrhenius model is applicable to activation processes, including corrosion, oxidation, creep and diffusion. Where the Arrhenius plot is valid, the plot of the log of the response parameter (F) against the reciprocal of absolute temperature is a straight line.

The response parameter is defined by the equation:

$$F = \text{Const.} \times \exp. - A/kT \tag{5.1}$$

where A is the reaction activation energy; k the Boltzmann constant (1.38×10^{-23} J/K) and T the absolute temperature.

For the case of the heat pipe, Baker described the gas generation process as

$$\dot{m}(t, T) = f(t)F(T) \tag{5.2}$$

where $\dot{m}$ is the mass generation rate; t denotes time and $F(t)$ is given in equation (5.1).

By plotting the mass of hydrogen generation in each heat pipe against time, with results at different temperatures, one can use these figures to obtain a universal curve, presenting the mass of hydrogen generated as a function of time × shift factor, which will be a straight line on logarithmic paper. Finally, the shift factors are plotted against the reciprocal of absolute temperature for each temperature examined, and the slope of this curve gives the activation energy A in equation (5.1).

The mass of hydrogen generated at any particular operating temperature can then be determined using the appropriate value of shift factor. Baker concluded that stainless steel/water heat pipes could operate for many years at temperatures of the order of 60 °F, but at 200 °F the gas generation would be excessive.

It is probable that this model could be applied to other wall/wick/working fluid combinations, the only drawback being the large number of test units needed for accurate predictions. The minimum is of the order of 12, results being obtained at three vapour temperatures, four heat pipes being tested at each temperature.

Another study was concerned with the evolution of hydrogen in nickel/water heat pipes. Anderson used a corrosion model to enable him to predit the behaviour of heat pipes over extended periods, based on accelerated life tests, following Baker's method [12].

He argued that oxidation theory predicts that passivating film growth occurs with a parabolic time dependence and an exponential temperature dependence.

Anderson gives the following values for A, the reaction activation energy:

Stainless steel (304)/water	8.29×10^{-20} J
Nickel/water	10.3×10^{-20} J

and confirms Baker's model.

Later, work in Japan [17, 18] concentrated on a statistical treatment of life test data from accelerated tests on copper/water heat pipes. This has been directed in part at investigating the formation of small quantities of noncondensable gas (CO_2) in such pipes where lifetimes of 20 years or more are required.

The investigations were carried out on axially grooved heat pipes, some of which used commercial phosphorous, deoxidised copper and others oxygen-free copper.

High-temperature ageing was done for periods of 20, 40 and 150 days. Analysis was by x-ray microscopy and infrared spectroscopy. The infrared absorption spectra showed absorption caused by benzene rings, phenyl groups, an O–H link and a C–O–C link. It was therefore concluded that the products were organic. After the full ageing period, corrosion was observed on the inside of the commercial copper, while the OFC copper showed only slight corrosion. It was concluded that phosphorous used during refining of the copper had a profound effect on its corrosion properties.

With regard to the generation of CO_2, on the basis of the criterion that the active–inactive (i.e. buffered) boundary in the vapour space is where the temperature drop becomes one half of the total temperature drop, it was possible to estimate the gas column length and temperature drop achieved after 20 years of use of the heat pipe.

After 1000 days at 160 °C yielded a temperature difference of about 2.5 °C. Ageing for 470 days at 393 K corresponds, according to data in [17] to a 20-year use at 333 K, and it was thus concluded that commercially available heat pipes using this phosphorous-containing material could be used satisfactorily if a temperature drop of 3 °C is acceptable.

5.2.4 A life test programme

A life test programme must provide detailed data on the effects of temperature, heat flux and assembly techniques on the working fluid, and the working fluid/wall and wick material compatibility.

The alternative techniques for testing have been discussed in Section 5.2.2 and it now remains to formulate a programme that will enable sufficient data to be accumulated to enable the life of a particular design of heat pipe to be predicted accurately.

Each procedure may be given a degree of priority (numbered 1–3, in decreasing order of importance) and these are presented in Table 5.3 together with the MINIMUM number of units required for each test. The table is self-explanatory.

This programme should provide sufficient data to enable one to confidently predict the long-term performance of a heat pipe, based on an Arrhenius plot, and the maximum allowable operating temperature, based on fluid stability. The programme involves a considerable amount of testing, but the cost should be weighed against satisfactory heat pipe performance over its life in applications where reliability is of prime importance.

5.2.5 Spacecraft qualification plan

The use of heat pipes in spacecrafts, as discussed earlier, makes heavy demands on heat pipe technology. The qualification plan set out by the ESA, and illustrated in Fig. 5.21, involves the construction of 15 sample heat pipes, the majority of which are eventually opened up for detailed physical analysis. Of the heat pipes included in the qualification programme, it is stipulated that at least 10 per cent, but not less than 3 units, should be subjected to life tests of 8000 h. The qualification procedure

Table 5.3 Heat pipe life test priorities

Priority	Minimum no. units to be tested	Number with valves	Test specification
1	–	–	Cleanliness of materials
1	–	–	Purity of working fluid
1	–	–	Sealing of case
1	–	–	Outgassing
2	2	2	Refluxing – vapour temperature at maximum design
1	4 at each temp.	all	Refluxing – temperature range up to maximum design (to include) bonding temperatures)
3	2	2	Refluxing – heat flux at maximum design
2	2	1	Heat pipe mode – intermittent tests between refluxing
1	2	1	Heat pipe mode – long-term continuous performance test
1[1]	2	0	Heat pipe mode – vibration test with intermittent performance tests
1	2	2	Variable conductance heat pipe – solubility of gas in working fluid and effect on artery

[1] where applicable.

guidelines set out the temperature and heat load requirements of these 'ageing' tests and it is pointed out that these tests, by themselves, do not necessarily prove the longevity of the heat pipes for 7–10 years of operation in space.

An example of the qualification procedures for a specific ESA heat pipe type is given in [29]. Alcatel Space [30] present data of value to those developing heat pipes for spacecraft in the referenced paper summarising a 'roadmap'. This sets out the objectives and the technical challenges of a development programme, in particular for aluminium/ammonia heat pipes, but applicable more broadly.

The Company, interestingly, outlines the facilities that enable it to produce typically 2000 heat pipes per annum. These include

Filling stations
Automatic welding machine
Testing benches – proof pressure, ageing and thermal cycling
Performance test benches
Bending machines

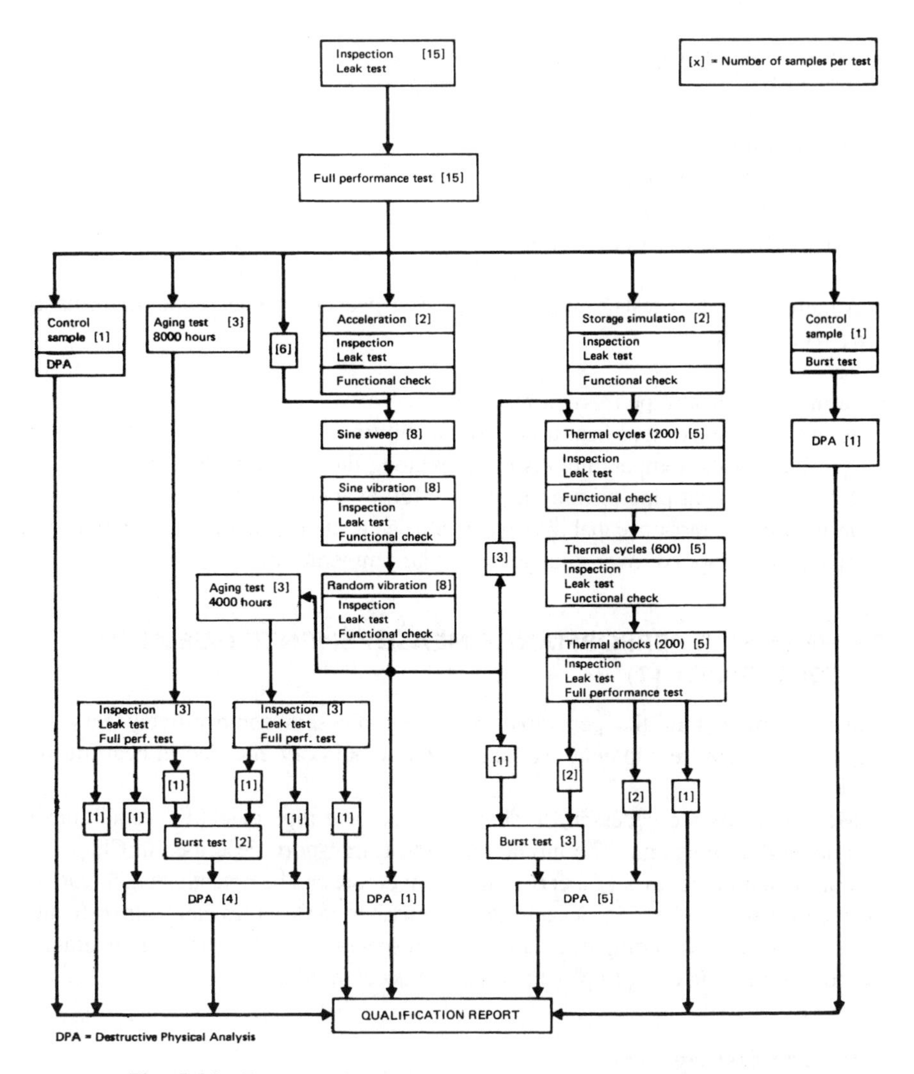

Fig. 5.21 European Space Agency heat pipe qualification plan [10].

The thermal tests carried out on the bench include burnout (maximum heat transport capability), heat flux tests and precise measurement of the thermal conductance. The ability to incline the pipe (a necessity for space qualification procedures) is also incorporated in the test bench.

Other countries use ESA criteria for space qualification. In the Ukraine [31], this led to the following tests being proposed:

- Inspection and physical measurement
- Proof pressure testing – leak test

- Performance testing
- Burst test
- Random vibration
- Storage simulation test
- Thermal cycles/shock test
- Ageing test (life test)
- Noncondensable gas definition test

A number of other tests to measure various aspects of the heat pipe performance were recommended:

- Definition of heat pipe thermal resistance
- Definition of maximum heat transport capability
- Definition of the temperature distribution along the length of the heat pipe
- Definition of heat pipe priming time after full evaporator dryout
- Definition of start-up capability when 80 per cent of maximum heat transport capability is applied, over a range of vapour temperatures.

5.3 HEAT PIPE PERFORMANCE MEASUREMENTS (SEE ALSO SECTION 5.1.12)

The measurement of the performance of heat pipes is comparatively easy and requires in general equipment available in any laboratory engaged in heat transfer work.

Measurements are necessary to show that the heat pipe meets the requirements laid down during design. The limitations to heat transport, described in Chapter 2, and presented in the form of a performance envelope, can be investigated, as can the degree of isothermalisation. A considerable number of variables can be investigated by bench testing, including orientation with respect to gravity, vapour temperature, evaporator heat flux, start-up, vibrations and accelerations.

5.3.1 The test rig

A typical test rig is shown diagrammatically in Fig. 5.22. The rig has the following features and facilities:

(i) Heater for evaporator section
(ii) Wattmeter for power input measurement
(iii) Variac for power control
(iv) Condenser for heat removal
(v) Provision for measuring flow and temperature rise of condenser coolant
(vi) Provision for tilting heat pipe
(vii) Thermocouples for temperature measurement and associated readout system
(viii) Thermal insulation.

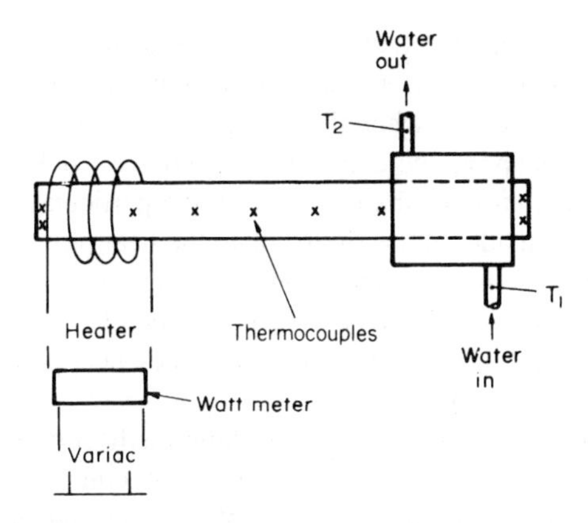

Fig. 5.22 Heat pipe performance test rig.

The heater may take several forms, as long as heat is applied uniformly and the thermal resistance between the heater and the evaporator section is low. This can be achieved using rod heaters mounted in a split copper block clamped around the heat pipe or by wrapping insulated heater wire directly on the heat pipe. For many purposes eddy current heating is convenient, using the condenser as a calorimeter. Heat losses by radiation and convection to the surroundings should be minimised by applying thermal insulation to the outside of the heater. An accurate wattmeter covering the anticipated power range, and a variac for close control of power, should be incorporated in the heater circuit. Where orientation may be varied, long leads between the heater and the instruments should be used for convenience.

An effective technique for measuring the power output of heat pipes operating at vapour temperatures appropriate to most organic fluids and water is to use a condenser jacket through which a liquid is passed. For many cases this can be water. The heat given up to the water can be obtained if the temperature rise between the condenser inlet and outlet is known, together with the flow rate. The temperature of the liquid flowing through the jacket may be varied to vary the heat pipe vapour temperature. Where performance measurements are required at vapour temperatures of about 0 °C, a cryostat may be used.

Cryogenic heat pipes should be tested in a vacuum chamber. This prevents convective heat exchange and a cold wall may be used to keep the environment at the required temperature. As a protection against radiation heat input, the heat pipe, fluid lines and cold wall should all be covered with superinsulation. If the heat pipe is mounted such that the mounting points are all at the same temperature (cold wall and heat sink) it can be assumed that all heat put into the evaporator will be transported by the heat pipe as there will be no heat path to the environment. Further data on cryogenic heat pipe testing can be obtaining from [13, 14].

An important factor in many heat pipe applications is the effect of orientation on performance. The heat transport capability of a heat pipe operating with the evaporator below the condenser (thermosyphon or reflux mode) can be up to an order of magnitude higher than that of a heat pipe using the wick to return liquid to the evaporator from a condenser at a lower height. In many cases, the wick may prove incapable of functioning when the heat pipe is tilted so that the evaporator is only a few centimetres above the condenser. Of course, wick selection is based in part on the likely orientation of the heat pipe in the particular application.

Provision should be made on the rig to rotate the heat pipe through 180 °C while keeping heater and condenser in operation. The angle of the heat pipe should be accurately set and measured. In testing of heat pipes for satellites, a tilt of only 0.5 cm over a length of 1 m may be required to check heat pipe operation, and this requires very accurate rig alignment.

The measurement of temperature profiles along the heat pipe is normally carried out using thermocouples attached to the heat pipe outer wall. If it is required to investigate transient behaviour, for example during start-up, burnout or on a VCHP, automatic electronic data collection is required. For steady state operation a switching box connected to a digital voltmeter or a multichannel chart recorder should suffice, but most laboratories now possess computer-aided data collection and real-time presentation of such data.

5.3.2 Test procedures

Once the heat pipe is fully instrumented and set up in the rig, the condenser jacket flow may be started and heat applied to the evaporator section. Preferably heat input should be applied at first in steps, building up to design capability and allowing the temperatures along the heat pipe to achieve a steady state before adding more power. When the steady state condition is reached, power input, power output (i.e. condenser flow rate and ΔT) and the temperature profile along the heat pipe should be noted.

If temperature profiles as shown in Fig. 5.23 are achieved, the heat pipe is operating satisfactorily. However, several modes of failure can occur, all being recognisable by temperature changes at the evaporator or condenser.

The most common failure is burnout, created by excessive power input at the evaporator section. It is brought about by the inability of the wick to feed sufficient liquid to the evaporator, and is characterised by a rapid rise in evaporator temperature compared to other regions of the heat pipe. Typically, the early states of burnout are represented by the upper curve in Fig. 5.23.

Once burnout has occurred, the wick has to be reprimed and this is best achieved by cutting off the power input completely. When the temperature difference along the pipe drops to 1–2 °C, the power may be reapplied. The wick must reprime, i.e. be rewetted and saturated with working fluid along its complete length, if operation against gravity or zero gravity is envisaged. If this is the case, the recovery after burnout must be demonstrated in the tilted condition. In other cases, the recovery may be aided by gravity assistance.

A second failure mechanism recognisable by an increased evaporator temperature, and known as overheating, occurs at elevated temperatures. As explained in

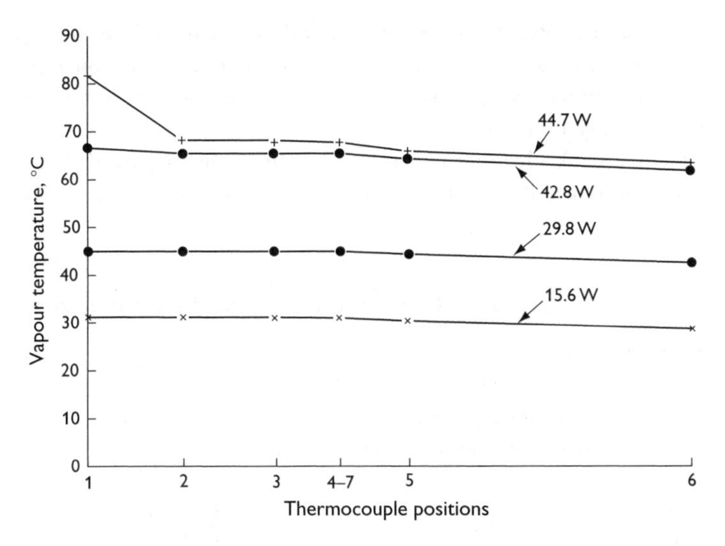

Fig. 5.23 Typical temperature profiles along a heat pipe under test.

Chapters 2 and 3, each working fluid has an operating temperature range characterised by the Merit number, which achieves an optimum value at a particular temperature and then decreases as this temperature is exceeded. This means that the fluid is able to transport less heat. Thus, the temperature of the evaporator becomes higher than the rest of the pipe. In general, the evaporator temperature does not increase as quickly as in a burnout condition, but these two phenomena are difficult to distinguish.

Temperature changes at the condenser section can also point to failure mechanisms or a decrease in performance. A sudden drop in temperature at the end of the heat pipe downstream of the cooling jacket occurring at high powers can be attributed to the collection of working fluid in that region, insulating the wall and creating a cold spot. This has been called 'coolout' [15]. Complete failure need not necessarily occur when this happens, but the overall ΔT will be substantially increased and the effective heat pipe length reduced.

A similar drop in temperature downstream of the condenser jacket can occur in pipes of small diameter (<6 mm bore) when the fluid inventory is greater than that needed to completely saturate the wick. The vapour tends to push the excess fluid to the cooler end of the heat pipe, where, because of the small vapour space volume, a small excess of fluid will create a long cold region. This can occur at low powers and adjustments in fluid inventory may be made if a valve is incorporated in the heat pipe. One way is to use an excess fluid reservoir, which acts as a sponge but has pores sufficiently large to prevent it from sucking fluid out of the wick. This technique is used in heat pipes for space use and the reservoir may be located at any convenient part of the vapour space.

Failure can be brought about by incompatibilities of materials, generally in the form of the generation of noncondensable gases that collect in the condenser section. Unlike liquid accumulation, the gas volume is a function of vapour temperature and its presence is easily identified.

Unsatisfactory wick cleaning can inhibit wetting, and if partial wetting occurs the heat pipe will burnout very quickly after the application of even small amounts of power.

5.3.3 Evaluation of a copper heat pipe and typical performance

5.3.3.1 Capabilities

A copper heat pipe using water as the working fluid was manufactured and tested to determine the temperature profiles and the maximum capability.

The design parameters of the pipe were as follows:

Length	320 mm
Outside dia.	12.75 mm
Inside dia.	10.75 mm
Material of case	Copper
Wick form	4 layers 400 mesh
Wick wire dia.	0.025 mm
Effective pore radius	0.031 mm
Calculated porosity	0.686
Wick Material	Stainless steel
Locating spring length	320 mm
Pitch	7 mm
Wire dia.	1 mm
Material	Stainless steel
Working fluid	Water (10^6 Ω resistivity)
Quantity	2 ml
End fittings	Copper
Instrumentation thermocouples	7

5.3.3.2 Test procedure

The evaporator section was fitted into the 100-mm long heater block in the test rig, and the condenser section covered by a 150-mm long water jacket. The whole system was then lagged.

First tests were carried out with the heat pipe operating vertically with gravity assistance. The power was applied and on achievement of a steady state condition the thermocouple readings and temperature rise through the water jacket were noted, as was the flow rate.

Power to the heaters was increased incrementally and the steady state readings noted until dryout was seen to occur. (This was characterised by a sudden increase in the potential of the thermocouple at the evaporator section relative to the readings of the other thermocouples.)

The above procedure was performed for various vapour temperatures and heat pipe orientations with respect to gravity.

5.3.3.3 Test results

Typical results obtained are shown in Fig. 5.23, showing the vapour temperature profile along the pipe when operating with the evaporator 10 mm above the condenser.

The table below gives power capabilities for a 9.5-mm-outside diameter copper heat pipe of length 30 cm, with a composite wick of 100 and 400 mesh, operating at an elevation (evaporator above condenser) of 18 cm.

Vapour temp (°C)	Power out (W)
84	17
121	30.5
162.5	54
197	89

The working fluid was again water, and a capability of 165 W was measured with horizontal operation (290 W with gravity assistance).

REFERENCES

[1] Brown-Boveri & Cie Ag. UK Patent No. 1281272, April 1969.

[2] Evans, U.R. The Corrosion And Oxidation of Metals. First Supplementary Volume. St. Martin's Press, Inc., 1968.

[3] Organic Solvents, Weissberger, Proskauer, Riddick & Toops (Ed.). Inter-science, 1955.

[4] Birnbreier and Gammal, G. Long time tests of Nb 1% Zr heat pipes filled with sodium and caesium. International Heat Pipe Conference, Stuttgart, October, 1973.

[5] Grover, G.M., Kemme, J.E. and Keddy, E.S. Advances in heat pipe technology. International Symposium on Thermionic Electrical Power Generation, Stresa, Italy, May 1968.

[6] Rice, G.R. and Jennings, J.D. Heat pipe filling. International Heat Pipe Conference, Stuttgart, October 1973.

[7] Vinz, P., Cappelletti, C. and Geiger, F. Development of capillary structures for high performance sodium heat pipes. International Heat Pipe Conference, Stuttgart, October 1973.

[8] Quataert, D., Busse, C.A. and Geiger, F. Long time behaviour of high temperature tungsten-rhenium heat pipes with lithium and silver as the working fluid. International Heat Pipe Conference, Stuttgart, October 1973.

[9] Busse, C.A., Geiger, F. and Strub, H. High temperature lithium heat pipes. International Symposium on Thermionic Electrical Power Generation, Stresa, Italy, May 1968.
[10] Anon. Heat pipe qualification requirements. European Space Agency. Report ESA PSS-49 (TST-OL). Issue No.1. January 1979.
[11] Baker, E. Prediction of long term heat pipe performance from accelerated life tests. AIAA J., Vol. LL, No. 9, 1973.
[12] Anderson, W.T. Hydrogen evolution in nickel-water heat pipes. AIAA Paper 73–726, 1973.
[13] Kissner, G.L. Development of a cryogenic heat pipe. 1st International Heat Pipe Conference, Paper 10–2, Stuttgart, October 1973.
[14] Nelson, B.E. and Petrie, W. Experimental evaluation of a cryogenic heat pipe/radiator in a vacuum chamber. 1st International Heat Pipe Conference, Paper 10-2a, Stuttgart, October 1973.
[15] Marshburn, J.P. Heat pipe investigations. NASA TN-D-7219, August 1973.
[16] Rice, G. and Fulford, D. Capillary pumping in sodium heat pipes. Proceedings of 7th International Heat Pipe Conference, Minsk, 1990. Hemisphere, New York, 1991.
[17] Murakami, M. et al. Statistical prediction of long term reliability of copper-water heat pipes from accelerated test data. Proceedings of 6th International Heat Pipe Conference, Grenoble, 1987.
[18] Kojima, Y. and Murakami, M. A statistical treatment of accelerated heat pipe life test data. Proceedings of 7th International Heat Pipe Conference, Minsk, 1990. Hemisphere, New York, 1991.
[19] Mehl, D. Therma-base vapor chamber heat sinks eliminate semiconductor hot spots. See *www.thermacore.com*
[20] Esarte, J. and Domiguez, M. Experimental analysis of a flat heat pipe working against gravity. Appl. Therm. Eng., Vol. 23, pp 1619–1627, 2003.
[21] Rightley, M.J., Tigges, C.P., Givler, R.C., Robino, C.V., Mulhall, J.J. and Smith, P.M. Innovative wick design for multi-source, flat plate heat pipes. Microelectron. J., Vol. 34, pp 187–194, 2003.
[22] Le Berre, M., Launay, S., Sartre, V. and Lallemand, M. Fabrication and experimental investigation of silicon micro heat pipes for cooling electronics. J. Micromech. Microeng., Vol. 13, pp 436–441, 2003.
[23] Launay, S., Sartre, V. and Lallemand, M. Experimental study on silicon micro-heat pipe arrays. Appl. Therm. Eng., Vol. 24, pp 233–243, 2004.
[24] Saaski, E.W. Gas occlusions in arterial heat pipes. AIAA Paper 73–724, AIAA, New York, 1973.
[25] Saaski, E.W. Investigation of bubbles in arterial heat pipes. NASA CR-114531, 1973.
[26] Kosson, R. et al. Development of a high capacity variable conductance heat pipe. AIAA Paper 73–728, AIAA, New York, 1973.
[27] Matsumoto, S., Yamamoto, T. and Katsuta, M. Heat transfer characteristic change and mass transfer under long-term operation in sodium heat pipe. Proceedings of 10th International Heat Pipe Conference, Paper I-4, Stuttgart, 21–25 September 1997.
[28] Coppinger, R. Lithium capillary system to cool wings on re-entry. Flight Int., Vol. 168, No. 4998, p 26, 16–22 August 2005.
[29] Dubois, M., Mullender, B. and Supper, W. Development and space qualification of high capacity grooved heat pipes. Proceedings of 10th International Heat Pipe Conference, Stuttgart, 21–25 September 1997.

[30] Hoa, C., Demolder, B. and Alexandre, A. Roadmap for developing heat pipes for ALCATEL SPACE's satellites. Appl. Therm. Eng., Vol. 23, pp 1099–1108, 2003.

[31] Baturkin, V., Zhuk, S., Olefirenko, D. and Rudenko, A. Thermal qualification tests of longitudinal ammonia heat pipes for using in thermal control systems of small satellites. Proceedings of IV Minsk International Seminar 'Heat Pipes, Heat Pumps, Refrigerators', Minsk, Belarus, 4–7 September 2000.

[illegible]

6

SPECIAL TYPES OF HEAT PIPE

The variety of heat pipes, in terms of their geometry, function and/or the methods used to transport the liquid from the condenser to the evaporator, is great. Some types of heat pipe have become less popular over the decades or have ceased to be regarded as 'special', for example flat plate heat pipes and vapour chambers (see Chapters 5 and 8 for examples of the latter types). Other types have seen a reawakening of interest, brought about by demands for increasing miniaturisation and the need to enhance performance, in particular liquid return flow rates. The use of electrokinetic forces is one technique in which the 'wheel has gone full circle'.

Where heat pipe types have been omitted from the discussion in this Chapter, but were present in earlier editions, some suggestions for further reading are retained in the Bibliography section in the Appendices.

The following types will be described below:

Variable conductance heat pipes
Thermal diodes
Pulsating (oscillating) heat pipes
Loop heat pipes (LHPs) and capillary pumped loops (CPLs)
Micro-heat pipes
Use of electrokinetic forces
Rotating heat pipes
Miscellaneous types – sorption heat pipe (SHP); magnetic fluid heat pipes

6.1 VARIABLE CONDUCTANCE HEAT PIPES

The variable conductance heat pipe (VCHP), sometimes called in a specific variant the gas-controlled or gas-loaded heat pipe, has a unique feature that sets it apart from other types of heat pipe. This is its ability to maintain a device mounted at the evaporator at a near constant temperature, independent of the amount of power being generated by the device.

Variable conductance heat pipes are now routinely used in many applications (see also Chapter 7). These applications range from thermal control of components and systems on satellites to precise temperature calibration duties and conventional electronics temperature control.

The temperature control functions of a gas-buffered heat pipe were first examined as a result of noncondensable gas generation within a sodium/stainless steel basic heat pipe. It was observed [1] that as heat was put into the evaporator section of the heat pipe, the hydrogen generated was swept to the condenser section. An equilibrium situation shown in Fig. 6.1 was reached.

Subsequent visual observation of high-temperature heat pipes, and temperature measurements, indicated that the working fluid vapour and the noncondensable gas were segregated, that a sharp interface existed between the working fluid and the noncondensable gas and that the noncondensable gas effectively blocked off the condenser section it occupied, stopping any local heat transfer.

Significantly, it was also observed that the noncondensable gas interface moved along the pipe as a function of the thermal energy being transported by the working fluid vapour, and it was concluded that suitable positioning of the gas interface could be used to control the temperature of the heat input section within close limits.

Much of the subsequent work on heat pipes containing noncondensable or inert gases has been in developing means for controlling the positioning of the gas front, and in ensuring that the degree of temperature control achievable is sufficient to enable components adjacent to the evaporator section to be operated at essentially constant temperatures, independent of their heat dissipation rates, over a wide range of powers.

The first extension of the simple form of gas-buffered pipe shown in Fig. 6.1 was the addition of a reservoir downstream of the condenser section (Fig. 6.2). This was added to allow all the heat pipe length to be effective when the pipe was operating at maximum capability and to provide more precise control of the vapour temperature. The reservoir could also be conveniently sealed using a valve.

The early workers in the field of cold-reservoir VCHP's were troubled by vapour diffusion into the reservoir, followed by condensation, even if liquid flow into the gas area had been arrested. It is necessary to wick the reservoir of a cold-reservoir unit in order to enable the condensate to be removed. The partial pressure of the vapour in the reservoir will then be at the vapour pressure corresponding to its temperature (Fig. 6.3).

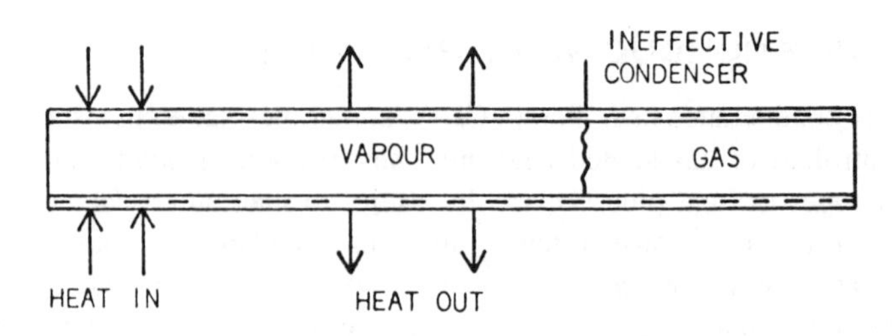

Fig. 6.1 Equilibrium state of a gas-loaded heat pipe.

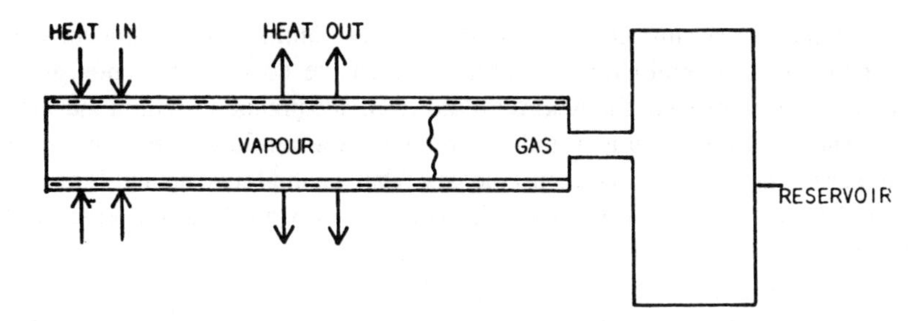

Fig. 6.2 Cold-reservoir variable conductance heat pipe.

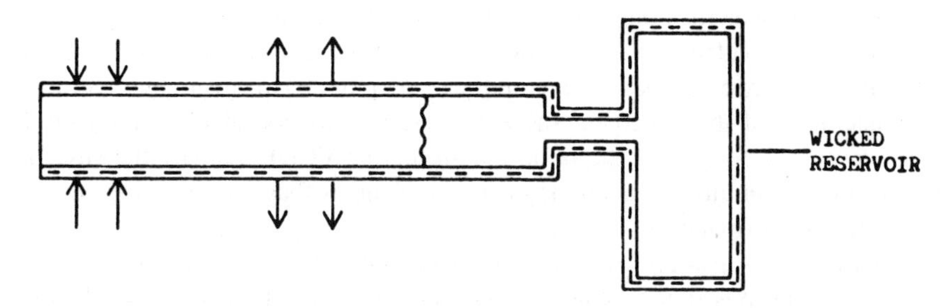

Fig. 6.3 Cold wicked reservoir variable conductance heat pipe.

The type of VCHP described above is of the passively controlled type. The active condenser length varies in accordance with temperature changes in various parts of the system. An increase in evaporator temperature causes an increase in vapour pressure of the working fluid, which causes the gas to compress into a smaller volume, releasing a larger amount of active condenser length for heat rejection. Conversely, a drop in evaporator temperature results in a lower vapour surface area. The net effect is to provide a passively controlled variable condenser area that increases or decreases heat transfer in response to the heat pipe vapour temperature.

A successful cold-reservoir VCHP was that constructed by Kosson et al. [2]. An arterial wick system of high liquid transport capability was used in conjunction with ammonia to carry up to 1200 W. The nominal length of the heat pipe, including reservoir, approached 2 m and the diameter was 25 mm. Nitrogen was the control gas.

An important feature of this heat pipe was provision for subcooling the liquid in the artery. This was to reduce inert gas bubble sizes and help the liquid to absorb any gas in the bubbles. This phenomenon is discussed in Section 5.1.11.1 in Chapter 5.

6.1.1 Passive control using bellows

Wyatt [1] proposed as early as 1965 that a bellows might be used to control the inert gas volume, but he was not specific in suggesting ways in which the bellows volume might be adjusted.

It is impractical to insert a wick into a bellows unit, and consequently it is necessary to have a semipermeable plug between the condenser section and the bellows in order to prevent the working fluid from accumulating within the storage volume. The plug must be impervious to both the working fluid vapour and the liquid, but permeable to the noncondensable gas. Marcus and Fleischman [3] proposed and tested a perforated Teflon plug with success, preventing liquid entering the reservoir during vibration tests.

However, Wyatt did put forward proposals that would have the effect of overcoming one of the major problems associated with the 'cold-reservoir' VCHP, although he did not appreciate the significance at the time. His proposal was to electrically heat the bellows that would be thermally insulated from the environment. His argument for doing this was to ensure that stray molecules of working fluid would not condense in the bellows if the bellows temperature was maintained at about 1 °C above the heat pipe vapour temperature. However, by controlling the temperature of the noncondensable gas reservoir, one is able to eliminate the most undesirable feature of the basic cold-reservoir VCHP, namely the susceptibility of the gas to environmental temperature changes that can upset the constant temperature performance.

Turner [4] investigated the use of bellows to change the reservoir volume and/or pressure. He proposed a mechanical positioning device to control the bellows between two precisely determined points but listed several disadvantages of this type of control, including the fact that mechanical devices require electrical energy for their activation and are also subject to failure due to jamming and friction. In proposing ammonia as the working fluid, he also felt that the associated pressure might add considerably to the weight of the bellows for containment reasons and also fatigue failure was a possibility.

6.1.2 Hot-reservoir variable conductance heat pipes

The cold-reservoir VCHP is particularly sensitive to variations in sink temperature which could affect the reservoir pressure and temperature. In an attempt to overcome this drawback, the hot-reservoir unit was developed.

One attractive layout for a hot-reservoir system is to locate the gas reservoir adjacent to, or even within, the evaporator section of the heat pipe. Thermal coupling of the reservoir to the evaporator minimises gas temperature fluctuations that limit the controllability. Figure 6.4 illustrates one form that this concept might take.

It has previously been stated that it is undesirable to have working fluid inside the hot reservoir, and the semi-permeable plug has been put forward as one way of preventing diffusion of large amounts of vapour or liquid into the gas volume. If the reservoir has a wick, and contains working fluid, there will be within it a vapour pressure corresponding to its temperature, which in the case of the hot-reservoir pipe having a gas reservoir within the evaporator, would be essentially the same as the temperature throughout the whole pipe interior, and there would not be gas in the reservoir.

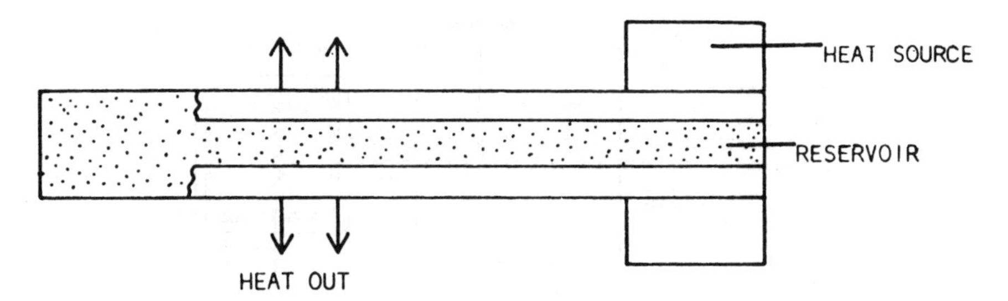

Fig. 6.4 Hot-reservoir variable conductance heat pipe.

An alternative technique [5] applied to the hot-reservoir VCHP is to provide a cold trap between the condenser and the reservoir. This effectively reduces the partial pressure of the working fluid vapour in the gas, and the system provides temperature control which is relatively independent of ambient radiation environments. This system is not applicable to a hot reservoir located inside the evaporator.

6.1.3 Feedback control applied to the variable conductance heat pipe

The cold-reservoir VCHP is particularly sensitive to variations in sink temperature which could effect the reservoir pressure and temperature. In an attempt to overcome this drawback, the hot-reservoir unit was developed, as described above.

Ideally, each of these forms of heat pipe is at best capable of maintaining its own temperature constant, and this is true only if an infinite storage volume is used. Thus, if the thermal impedance of the heat source is large, or if the power required to be dissipated by the component is liable to fluctuations over a range, the temperature of the sources would not be kept constant and severe fluctuations could occur, making the system unacceptable.

The development of feedback-controlled VCHPs has enabled absolute temperature control to be obtained, and this has been demonstrated experimentally [6]. These heat pipes are representative of the third generation of thermal control devices incorporating the heat pipe principle.

Two forms of feedback control are feasible, active (electrical) and passive (mechanical).

6.1.3.1 Electrical feedback control (active)

An active feedback-controlled VCHP is shown diagrammatically in Fig. 6.5. A temperature sensor, electronic controller and a heated reservoir (internal or external heaters) are used to adjust the position of the gas–vapour interface such that the source temperature remains constant. As in the cold-reservoir system, the wick is continuous and extends into the storage volume. Consequently, saturated working fluid is always present in the reservoir. The partial pressure of the vapour in equilibrium with the liquid in the reservoir is determined by the reservoir temperature that can be varied by the auxiliary heater.

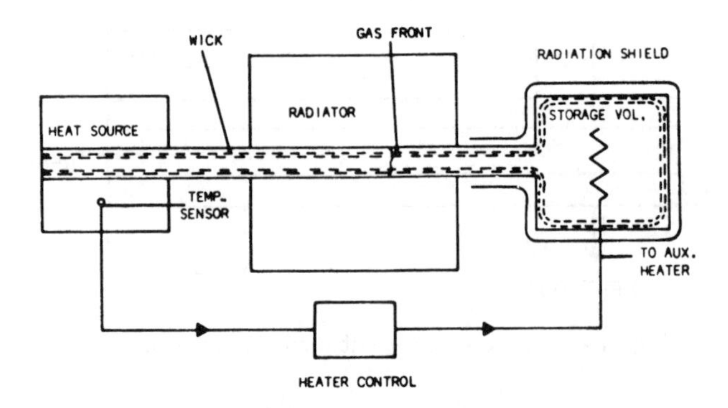

Fig. 6.5 Active feedback-controlled variable conductance heat pipe.

The two extremes of control required in the system are represented by the high-power/high-sink and low-power/low-sink conditions. The former case necessitates operation at maximum conductance, conversely the low-power/low-sink condition is appropriate for operation with minimum power dissipation.

By using a temperature-sensing device at the heat source, and connecting this via a controller to a heater at the reservoir, the auxiliary power and thus the reservoir temperature can be regulated so that precise control of the gas–buffer interface occurs, maintaining the desired source temperature at a fixed level.

6.1.3.2 Mechanical feedback control (passive)

Most of the work on mechanical feedback control has involved the use of a bellows reservoir, as advocated in several earlier proposals on nonfeedback-controlled passive VCHPs. A proposed passive feedback system utilising bellows was designed by Bienert et al. [7, 8] and is illustrated in Fig. 6.6. The control system consists of two bellows and a sensing bulb located adjacent to the heat source. The inner bellows contains an auxiliary fluid, generally an incompressible liquid, and is connected to the sensing bulb by a capillary tube.

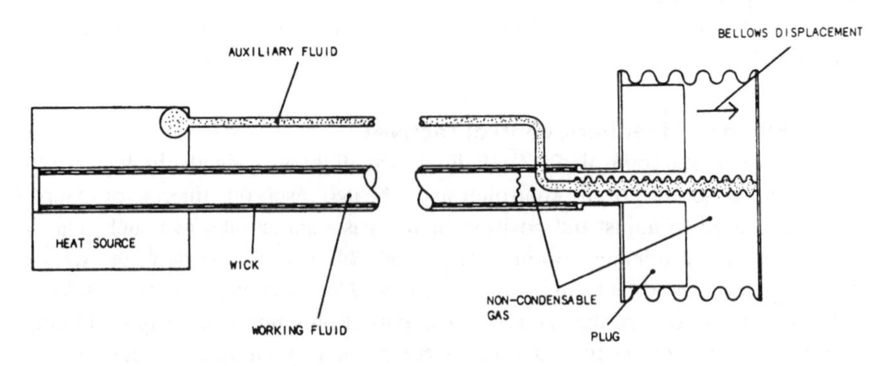

Fig. 6.6 Passive feedback-controlled variable conductance heat pipes.

Variations in source temperature will cause a change in the pressure of the auxiliary fluid, resulting in a displacement of the inner bellows. This displacement causes a movement of the main reservoir bellows. By relating the displacement of the bellows system, and therefore that of the vapour/gas interface in the heat pipe, to the heat source, a feedback-controlled system that regulates the source temperature is obtained.

The construction and testing of a passive control VCHP using methanol as the working fluid and nitrogen as the gas has been reported by Depew et al. [9]. The system was run over a power input range of 2–30 W, with the heat source at ambient temperature and the heat sink at a nominal value of 0 °C. Control was obtained with a metal bellows gas reservoir that was actuated by an internal liquid-filled bellows. The liquid bellows was pressurised by expanding the liquid methanol in an auxiliary reservoir in the evaporator heater block. Temperature variation of the heat source was restricted to ±4 °C using this design.

The use of VCHPs for primary temperature measurement has proved attractive because of the ability to set temperatures within ±0.5 mK, an accuracy difficult to achieve with other temperature control systems. The CNR, *Istituto di Metrologia 'G. Colonnetti'*, (IMGC) in Turin, Italy, has had considerable success in applying liquid metal VCHPs for accurate temperature measurement [10]. Used for temperature measurements in the range from 100 °C to 962 °C (sodium being used as the working fluid in the latter case), a mercury unit was also employed to study the vapour pressure relationship for this fluid [11].

Illustrated in Fig. 6.7, the heat pipe is constructed using stainless steel and is 400 mm long and 30 mm in diameter. The capillary structure is helical knurling, covering all of the inner surfaces, and the inert gas reservoir is connected to the heat pipe via a water-cooled limb. Mercury was filled to about 280 g.

Following initial tests, it was found that temperature response times after a change in pressure were unsatisfactory, so modifications were made to the capillary structure that resulted in excellent temperature stability (within less than 1 mK) and a rapid response to pressure changes. Further units were constructed using sodium as the working fluid (for high-temperature tests) and the 3 M fluid 'Fluorinert FC-43' for operation around 177 °C.

With the reawakening of interest in nuclear power generation, spurred by concerns about global warming and CO_2 emissions, researchers have been studying methods for controlling heat removal from reactors, in particular the modular high-temperature gas-cooled reactor. Should a loss of forced circulation through the core occur, or the primary system suffers depressurisation, passive decay heat removal becomes necessary, and one way of implementing this is to use VCHPs [12].

Experiments on a laboratory-scale simulation of the decay showed that the VCHP could effectively compensate for a threefold increase in decay heat rate by allowing only a modest rise in system temperature. In the tests, nitrogen was used as the control gas and water as the working fluid. It was concluded that the VCHP system could potentially control the reactor vessel temperature within safety limits, should

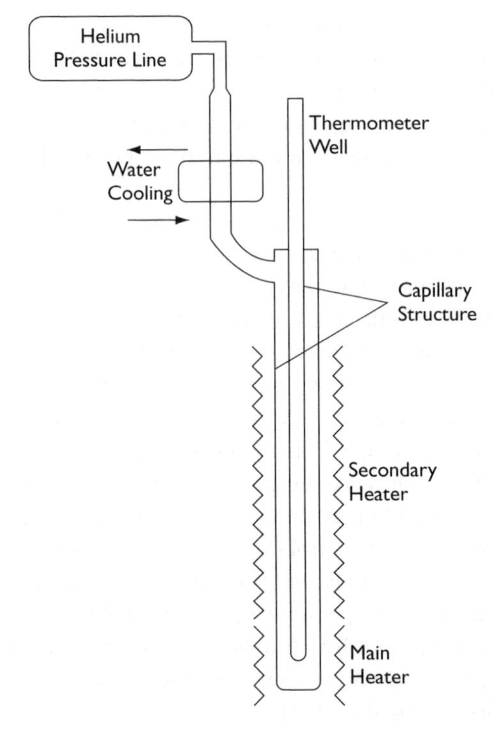

Fig. 6.7 The stainless steel mercury gas-controlled heat pipe developed at CNR-IMGC, Italy [10].

an accident occur. The passive nature of the system of course has added safety attractions.

An analogy may be made with the work of the Tokyo Electric Power Company and others [13] where a VCHP has been studied for load levelling on sodium–sulphur (NAS) batteries. Such batteries need to be kept above 300 °C during operation, but if they overheat by more than 10 °C, their life is reduced. The VCHP was shown to provide a satisfactory level of heat control, thus contributing to improved charge/discharge efficiency. (See information on aerospace VCHPs in Chapter 7.)

6.1.3.3 Comparison of Systems

The most common VCHP systems are compared in Table 6.1. When comparing the two types of feedback control (active and passive), better temperature control is obtained using an active system. In the active system, all the noncondensable gas will be in the condenser when the low-power sink condition is attained. However, in the passive arrangement, noncondensable gas will be present in the reservoir, regardless of the use of a plug, and the excess gas must be accommodated when the bellows is at maximum size (high-power/sink condition). Hence, storage requirements will be generally greater for a passive system having the same degree of temperature

Table 6.1 Comparison of temperature-controlled heat pipe systems

System	Advantages	Disadvantages
Wicked cold	Reliable without moving parts. No auxiliary power needed. Sensitivity is governed by vapour within the reservoir.	Very sensitive to sink conditions. Large storage volume. Only heat pipe temperature controlled.
Non-wicked hot reservoir. Passive control.	Reliable without moving parts. Less sensitive to sink conditions than cold reservoir. No auxiliary power needed.	Sensitive to heat carrier diffusion into the reservoir. Only heat pipe temperature controlled.
Passive feedback-controlled bellows system.	Heat source control. Slight sensitivity to sink conditions. No auxiliary power needed.	Complex and expensive system. Sensitive to heat carrier diffusion into the reservoir. Moving parts used.
Active electrical feedback-controlled system.	Heat source control. Best adjustability to various set points. Sensitivity is governed by vapour pressure within the reservoir. Minimum storage volume of all concepts. Relative insensitivity to gas generation. No moving parts.	Auxiliary power needed.

control as an active system. Better temperature control can be achieved with an active system than with the equivalent volume passive system. The necessity to incorporate a semi-permeable plug and the use of moving parts in the form of a bellows also adds to the complexity of passive systems.

In some applications, the use of additional electrical power to heat the reservoir, the increased complication of a bellows system, may be unacceptable. In this case, the choice lies between simple hot- and cold-reservoir variable conductance heat pipes.

6.2 HEAT PIPE THERMAL DIODES AND SWITCHES

6.2.1 The thermal diode

Here we are concerned with on/off rather than proportionate functioning.

The simplest thermal diode is the thermosyphon in which gravity provides the asymmetry but of course with the restriction on positioning. Gravity will also give a diode effect in the wicked heat pipe since

$$\Delta P_c = \Delta P_l + \Delta P_v \pm \Delta P_g$$

Reversal of direction of flow will reserve the sign of ΔP_g and provided that $|\Delta P_g| > \Delta P_c$, the pipe will behave as a diode.

Kirkpatrick [23] describes two types of thermal diode, one employing liquid trapping and the second, liquid blockage. Referring to Fig. 6.8 with the heat flow shown in Fig. 6.8(a), the heat pipe will behave normally. If the relative positions of the evaporator and condenser are reversed then the condensing liquid is trapped in the reservoir whose wick is not connected to the pipe wick on the left-hand side of the diagram and hence cannot return. The pipe will then not operate and no heat transfer will occur.

Figure 6.9 shows a similar arrangement but in this case excess liquid is placed in the pipe. In Fig. 6.9(a), this liquid will accumulate in the reservoir at the condensing end and the pipe will operate normally. In Fig. 6.9(b) the positions of the evaporator and condenser are reversed and the excess liquid blocks off the evaporator and the pipe ceases to operate.

Spacecrafts have employed thermal diodes for several decades. In Russia, a comparison of cryogenic heat pipe thermal diodes with active systems, such as

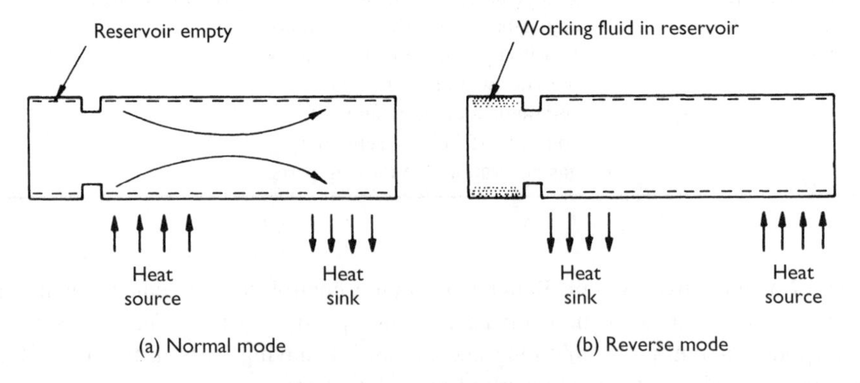

Fig. 6.8 Liquid trap diode.

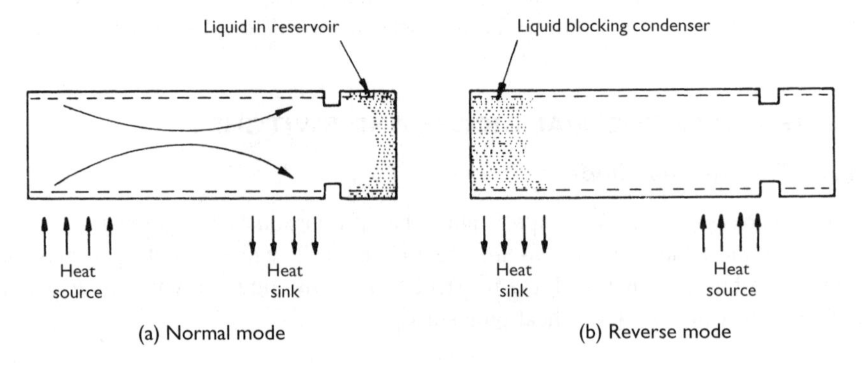

Fig. 6.9 Liquid blockage diode.

Stirling coolers, suggested that high-purity Germanium (HPGe) detectors, used in gamma-ray spectroscopy, could reap several benefits using the thermal diode. The passive nature, of course, is attractive, but in addition weight reduction and reliability featured. Where Stirling coolers are necessary to provide the active cooling, the heat pipe thermal diode is proposed, as shown in Fig. 6.10 to link the HPGe to a space radiator [19]. This particular model is a liquid trap diode, as additionally described for the Russian project MARS-96 in ref. [14].

Thermal diodes in terrestrial applications have been important components of renewable energy systems, where heat/coolth transport in one direction is essential under some circumstances, such as when collecting solar gain in the Winter for space heating. In hot climates, the use of the clear night sky as a radiative heat sink can provide useful cooling duties via thermal diodes [18]. The unit developed in Nigeria employed four methanol working fluid thermal diodes, with their condensers linked to an external radiator and their evaporators inserted into the walls of a cooler box. The radiator had a cooling capacity of over 600 kJ/m^2 per night and was able to cool the chamber to about 13 °C when the ambient (at night) was 20 °C.

Although commercialisation of many of the systems proposed for building heating/cooling that rely on heat pipe thermal diodes is difficult because of the perceived high cost and difficulties in retrofitting such systems, interest does continue. Work involving a number of universities and companies in the United Kingdom and continental Europe on such a system for room heating in winter or cooling in summer was recently carried out under a European Commission-funded study [15].

The thermal diode panels, illustrated in Figs 6.11a and b were manufactured by Thermacore Europe and each panel comprised nine copper–water heat pipes, with

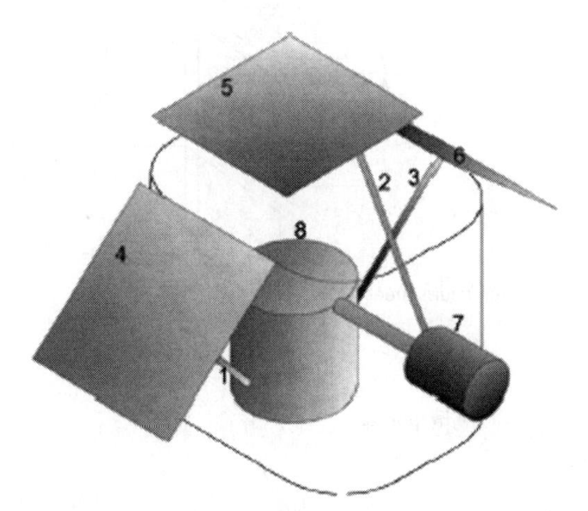

Fig. 6.10 The three-way cooler for an HPGe spectrometer proposed for a lunar lander. The HPGe detector (7) is connected to a Stirling micro-cooler (8) and radiator (5) using a diode heat pipe (2). The micro-cooler compressor (8) is connected to radiators (4 & 6) by two other thermal diodes (1 & 3). Ref. [19].

(a)

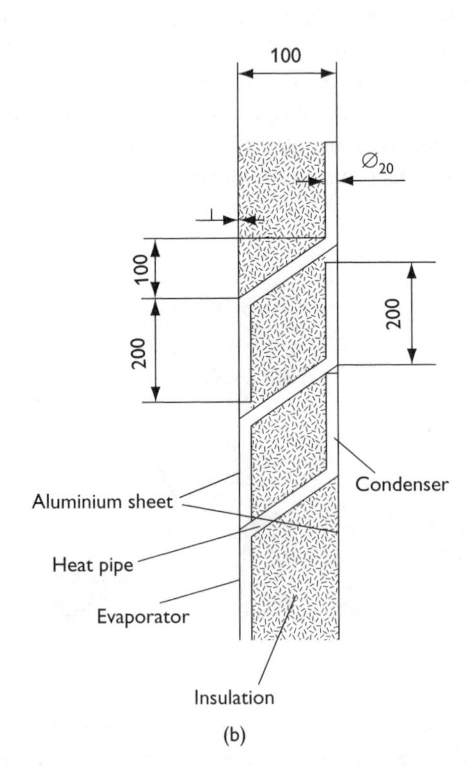

(b)

Fig. 6.11 Thermal diode wall panels developed as a passive thermal control feature for buildings [15].

diameters of 12.7 mm, welded to two aluminium sheets spaced in the configuration shown in Fig. 6.11b. Wicks were put in the lower sections of the heat pipes in order to assist liquid distribution over what were the evaporator sections.

Using the left-hand side of the panel (where the heat pipe evaporators are fixed), as the inside wall of the building, and the outer condenser sections adjacent to the outside wall of the building, in summer the unit provides a nighttime cooling role. The diode effect prevents the ingress of heat during the hot day in Southern Europe, where the system was tested. It was found that the panels with a thickness of 10 cm were able to perform as well as those with a thickness of 4 cm of thermal insulation material, with an apparent thermal conductivity of 0.07 W/mK. When heat transfer was required the apparent thermal conductivity was three to five times this value, depending upon the temperature difference.

For readers interested in other similar uses of heat pipes as thermal diodes in buildings, for heating or cooling, see references [16, 17].

6.2.2 The heat pipe switch

A number of methods for switching off the heat pipe have been referred to. Some examples are discussed by Brost and Schubert [25] and Eddleston and Hecks [26]. Figure 6.12a shows a simple displacement method in which the liquid working fluid can be displaced from an unwicked reservoir by a solid displacer body [25]. Figure 6.12b shows interruption of the vapour flow by means of a magnetically operated vane. The working fluid may be frozen by means of a thermoelectric cooler.

A thermal switch employing bellows was developed in the United States in the early 1980s [27]. Illustrated in Fig. 6.13, this device was capable of transporting about 100 W horizontally and the distance between the source and the sink was approximately 10 cm. Later, Peterson [28] modelled such devices for use with electronic components or multichip modules. One advantage of such a design was the use of the internal pressure in the bellows to ensure good thermal contact

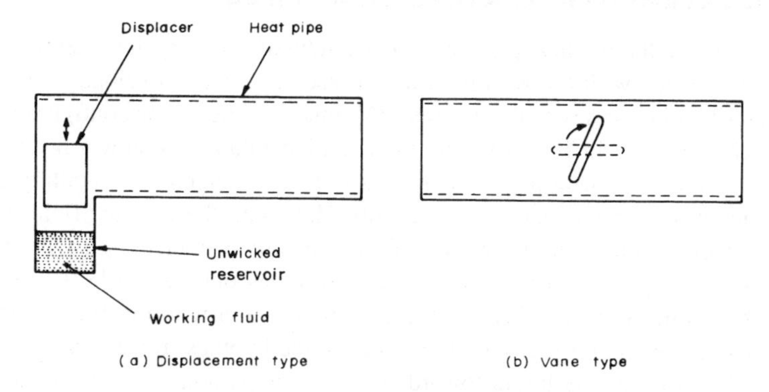

Fig. 6.12 Thermal switches.

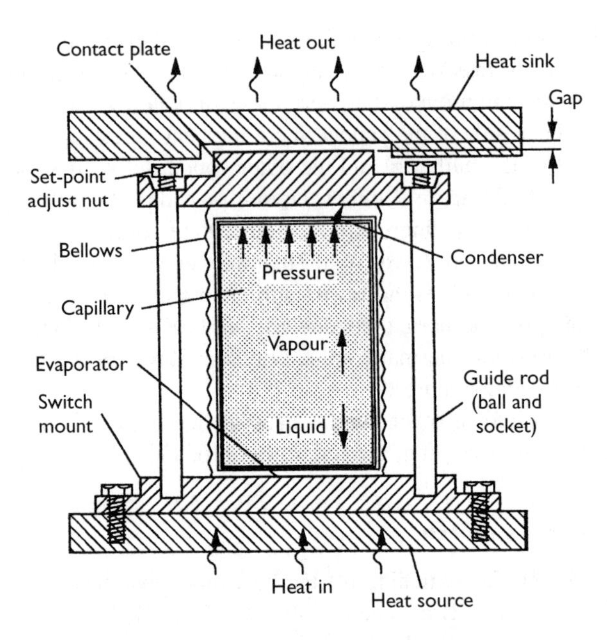

Fig. 6.13 Heat pipe thermal switch.

between the heat pipe and the heat sink when the switch was 'on'. Similar bellows systems were developed in the United Kingdom in the 1970s, the bellows being used to undertake a mechanical function in addition to the basic switching mode created by the presence or absence of a thermal contact.

The heat pipe switch can be particularly useful in cryogenic applications, where heat leakage can affect sensitive instruments in avionics and in spacecraft. This was the reasoning behind the development by Los Alamos National Laboratory [20].

6.3 PULSATING (OSCILLATING) HEAT PIPES

Pulsating, or oscillating, heat pipes comprise a tube of capillary diameter, evacuated and partially filled with the working fluid. These have been developed since originally described in a series of Patents [84–86]. Pulsating heat pipe configurations are shown schematically in Fig. 6.14 and their implementation is shown in Fig. 6.15. Typically, a pulsating heat pipe comprises a serpentine channel of capillary dimension, which has been evacuated, and partially filled with the working fluid. Surface tension effects result in the formation of slugs of liquid interspersed with bubbles of vapour. The operation of pulsating heat pipes was outlined in [87]. When one end of the capillary tube is heated (the evaporator), the working fluid evaporates and increases the vapour pressure, thus causing the bubbles in the evaporator zone to grow. This pushes the liquid towards the low-temperature end (the condenser). Cooling of the condenser results in a reduction of vapour pressure and condensation

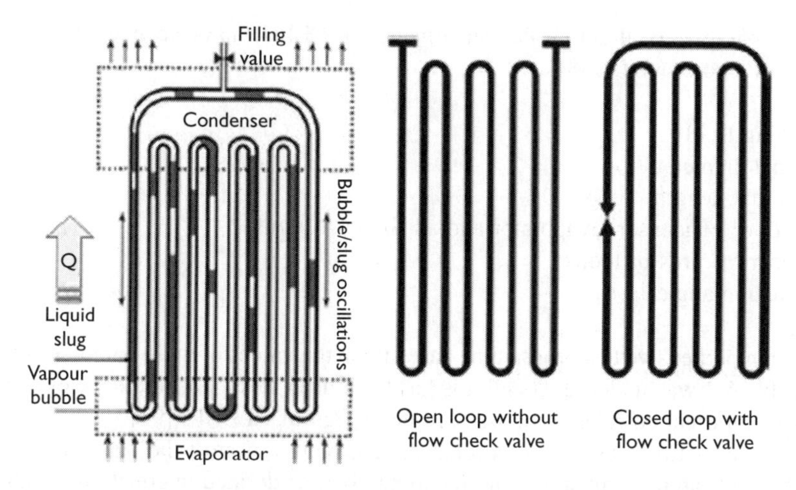

Fig. 6.14 Schematic representation of pulsating heat pipe [88].

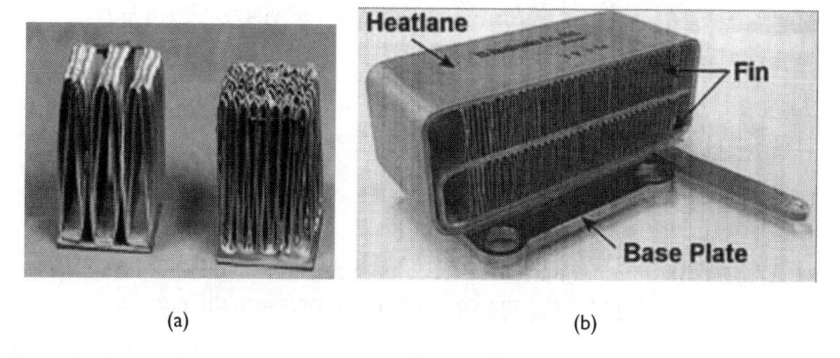

Fig. 6.15 Practical implementation of pulsating heat pipe (a) ref. [88], (b) ref. [89].

of bubbles in that section of the heat pipe. The growth and collapse of bubbles in the evaporator and condenser sections, respectively, results in an oscillating motion within the tube. Heat is transferred through latent heat in the vapour and through sensible heat transported by the liquid slugs.

Closed loop pulsating heat pipes (CLPHPs) perform better than open loop devices because of the fluid circulation that is superposed upon the oscillations within the loop. It has been suggested that further performance improvements may result from the use of check valves within the loop; however, due to the inherently small nature of the device it is difficult and costly to install such valves [86, 90]. Therefore, a closed loop device without a check valve is the most practicable implementation of the pulsating heat pipe.

The parameters affecting the performance of CLPHP have been summarised by Groll and co-workers [88] as

Working fluid
Internal diameter
Total tube length
Length of condenser, evaporator and adiabatic sections
Number of turns or loops
Inclination angle.

A comprehensive test programme based upon the values of these parameters listed in Table 6.2 was undertaken and this has been used to support the development of a semiempirical correlation to assist in the design of CLPHPs [91].

The correlation was based on three dimensionless groups, the Karman number, the liquid Prandtl number and the Jacob number, as defined in equation 6.1a,b and c, respectively, combined with the inclination angle (β, measured in radian) and number of turns, N.

$$\mathrm{Ka}_{\mathrm{liq}} = f \cdot \mathrm{Re}_{\mathrm{liq}}^2 = \frac{\rho_{\mathrm{liq}} \cdot (\Delta P)_{\mathrm{liq}} \cdot D_{\mathrm{i}}^2}{\mu_{\mathrm{liq}}^2 \cdot L_{\mathrm{eff}}} \quad \text{where } L_{\mathrm{eff}} = 0.5(L_{\mathrm{e}} + L_{\mathrm{c}}) + L_{\mathrm{a}} \tag{6.1a}$$

$$\mathrm{Pr}_{\mathrm{liq}} = \left(\frac{C_{\mathrm{p,liq}} \cdot \mu_{\mathrm{liq}}}{k_{\mathrm{liq}}} \right) \tag{6.1b}$$

$$\mathrm{Ja} = \left(\frac{h_{\mathrm{fg}}}{C_{\mathrm{p,liq}} \cdot (\Delta T)_{\mathrm{sat}}^{\mathrm{e-c}}} \right) \tag{6.1c}$$

where $\Delta T_{\mathrm{sat}}^{e-c}$ is the temperature difference between the working fluid in the evaporator and condenser and ΔP_{liq} the corresponding pressure difference.

$$\dot{q} = \left(\frac{\dot{Q}}{\pi D_{\mathrm{i}} \cdot N \cdot 2L_{\mathrm{e}}} \right) = 0.54(\exp(\beta))^{0.48} \mathrm{Ka}^{0.47} \mathrm{Pr}_{\mathrm{liq}}^{0.27} \mathrm{Ja}^{1.43} N^{-0.27} \tag{6.2}$$

Table 6.2 Matrix of parameters tested [88]

Working fluids	D_{i} (mm)	L_{total} (m)	$L_{\mathrm{e}} = L_{\mathrm{a}} = L_{\mathrm{c}}$ (m)	N (number of turns)
Water–ethanol–R-123	2.0	≈5	0.15	5
	2.0, 1.0	≈5	0.10	7
		≈10	0.10	16
			0.15	11
		≈15	0.10	23
			0.15	16

Note: Fill ratio always maintained at 50 per cent in all configurations. All configurations tested at inclinations of 0° (horizontal) to +90° (vertical, evaporator down).

This correlation holds provided the surface tension forces are large relative to gravitational forces. For this to be true, the Bond number, defined by equation 6.3 must be less than ~2 [92].

$$\mathrm{Bo} = \frac{D_\mathrm{i}}{\sqrt{\frac{\sigma}{g}(\rho_\mathrm{l} - \rho_\mathrm{g})}} \tag{6.3}$$

This has been shown to be the condition for discrete bubbles and liquid slugs sustained in a tube when the fluid is stationary and thus determines whether confined bubbles exist during the two-phase flow in a channel [93]. At higher heat fluxes with inclination angles greater than 30° from the horizontal, annular flow predominates. The transitions are shown in Fig. 6.16.

Equation 6.2 fits the data of [88], as shown in Fig. 6.17 [91].

As indicated in Table 6.2, these data were from tests with 50 per cent filling ratio. The studies carried out by Groll et al. indicated that the CLPHP operated satisfactorily with volumetric filling ratios in the range 25–65 per cent as indicated in Fig. 6.18. With greater filling ratios, the pumping action of bubbles was insufficient for good performance. At very low filling ratios, partial dryout of the evaporator was detected. It was also noted that at zero fill, the thermal resistance was effectively

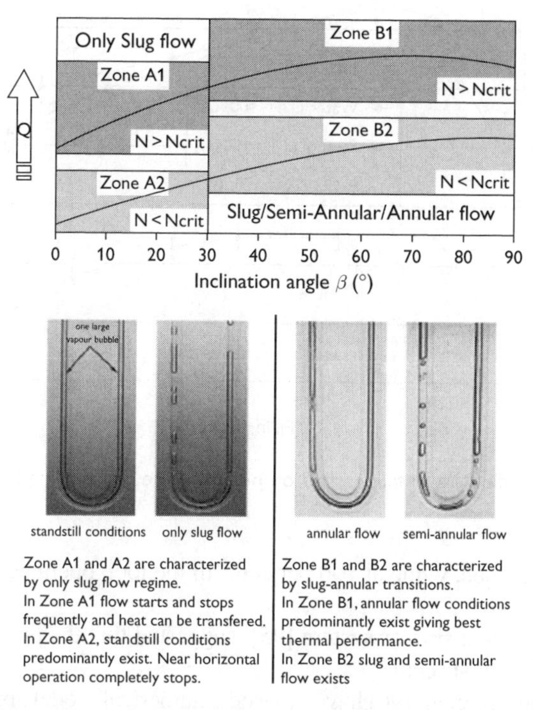

Fig. 6.16 Operating zones in pulsating heat pipes [91].

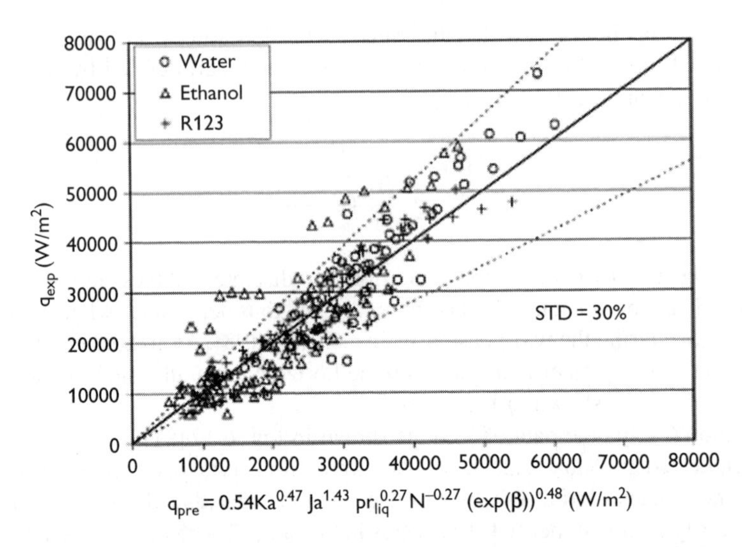

Fig. 6.17 Comparison of equation 6.2 with experimental results [91].

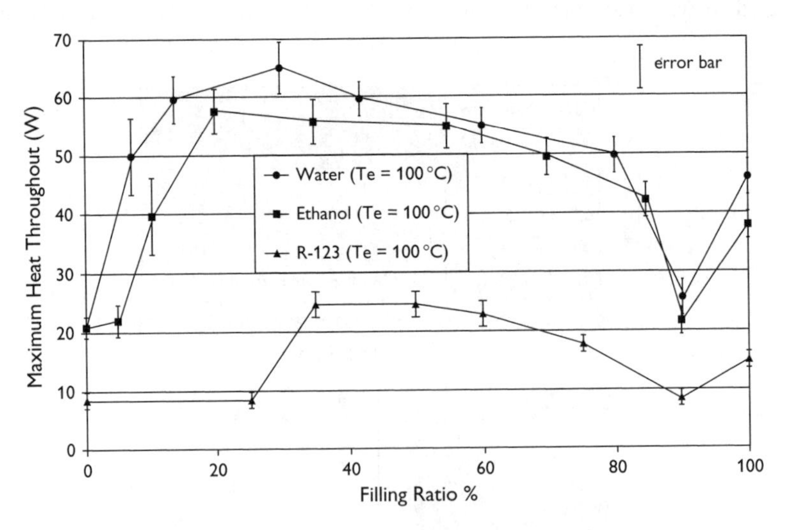

Fig. 6.18 Influence of filling ratio on performance of pulsating heat pipe [94].

that of the copper tube, while at 100 per cent fill the loop operated as a single-phase thermosyphon loop.

Another study, on a similar system [95], suggested that the optimum filling ratio is approximately 70 per cent.

Sakulchangsatjatai et al. [96] have devised a numerical model applicable to closed and open loop pulsating heat pipes. This underpredicted the heat transferred, but

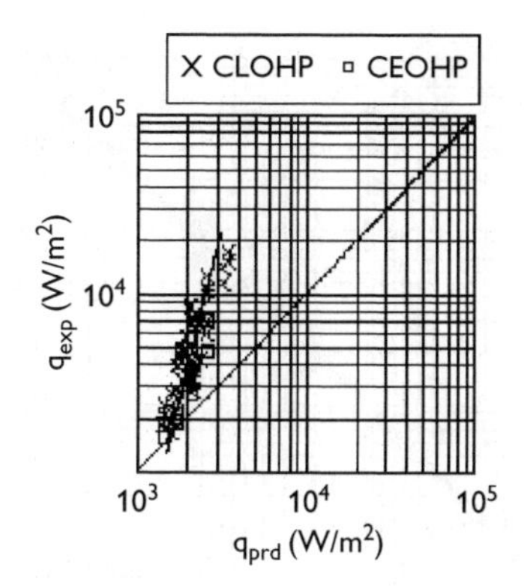

Fig. 6.19 Comparison of experimental and predicted heat transfer [96].

followed the correct trends, as shown in Fig. 6.19, indicating that the following parameters have the greatest influence on the heat pipe performance.

- Evaporator length – as the evaporator length is increased, the amplitude of pulsation increases and the frequency decreases. The net result is a reduction in heat transfer capacity with increasing evaporator length.
- Internal diameter – the heat transfer capacity increases with pipe diameter, provided the diameter is less than the value specified by equation (6.3).
- Working fluid – the analysis of Sakulchangsatjatai et al. indicates that latent heat is the most important fluid property and that low latent heat is desirable since this results in greater vapour generation, and hence fluid movement, for a given heat flux. This conclusion conflicts with the trends predicted by equation (6.2) and the results of other workers [95, 97].

A correlation has been formulated for the maximum heat flux in closed-end oscillating heat pipes (CEOHP) by Akbarzadeh and co-workers [97]. This may be stated as follows:

$$\mathrm{Ku}_0 = 53680 \times \left[\frac{Di}{L_\mathrm{e}}\right]^{1.127} \times \left[\frac{C_\mathrm{p}\Delta T}{h_\mathrm{fg}}\right]^{1.417} \times \left[Di\left[\frac{g(\rho_1 - \rho_\mathrm{v})}{\sigma}\right]^{0.5}\right]^{-1.32} \tag{6.4a}$$

$$\mathrm{Ku}_{90} = 0.0002 \times \left[\frac{Di}{L_\mathrm{e}}\right]^{0.92} \times \left[\frac{C_\mathrm{p}\Delta T}{h_\mathrm{fg}}\right]^{-0.212} \times \left[Di\left[\frac{g(\rho_1 - \rho_\mathrm{v})}{\sigma}\right]^{0.5}\right]^{-0.59}$$
$$\times \left[1 + \left(\frac{\rho_\mathrm{v}}{\rho_1}\right)^{0.25}\right]^{13.06} \tag{6.4b}$$

where Ku is the Kutateladze Number $\{= q/h_{\mathrm{fg}}\rho_\mathrm{v}^{0.5}[\sigma g(\rho_1 - \rho_\mathrm{v})]^{0.25}\}$, h_{fg} the latent heat, and the subscripts 0 and 90 refer to horizontal and vertical orientation, respectively (Fig. 6.20).

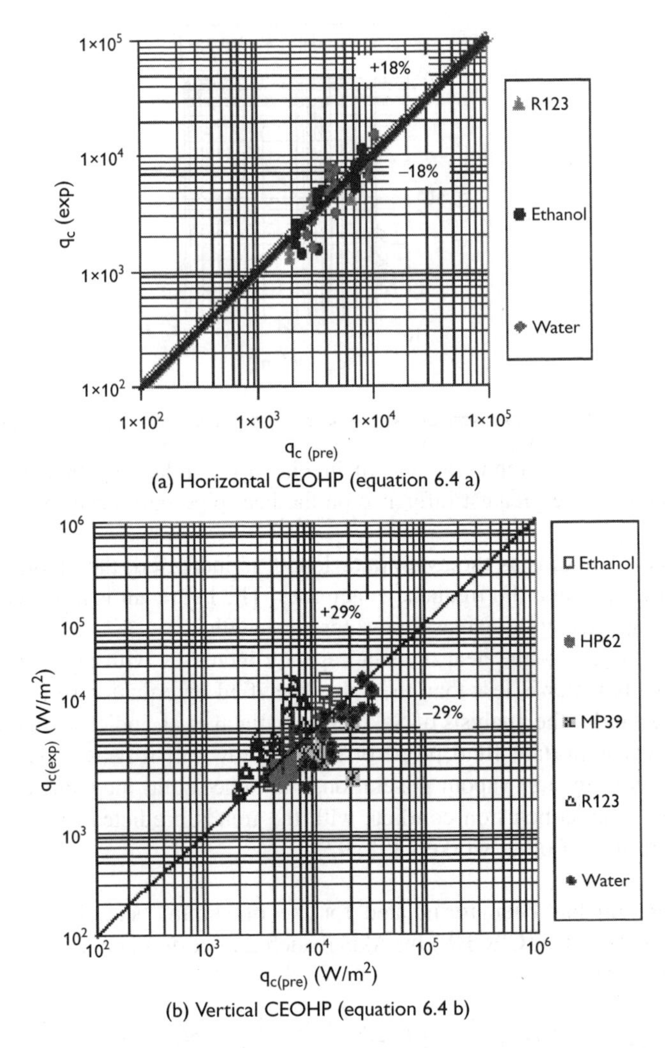

Fig. 6.20 Comparison of maximum heat flux with prediction [97].

The pulsating heat pipe clearly has potential applications, particularly in electronic cooling systems, where it is suggested [98] that the spreading resistance may be 50 per cent of that of a conventional heat sink and the effective conductivity 24 kW/mK, compared to 0.4 kW/mK for metallic copper.

6.4 LOOP HEAT PIPES AND CAPILLARY PUMPED LOOPS

As discussed in Chapter 2, the operation of a heat pipe relies upon the capillary head within the wick which is sufficient to overcome the pressure drops associated with the liquid and vapour flow and the gravitational head. In order to operate with the evaporator above the condenser in a gravitational field, it is necessary for the wick to extend for the entire length of a conventional heat pipe. The capillary head ΔP_c is inversely proportional to the effective pore radius of the wick but independent of length, while the hydraulic resistance is proportional to the length of the wick and inversely proportional to the square of the pore radius. Thus, if the length of a heat pipe operating against gravity is to be increased, a reduction in pore radius is required to provide the necessary capillary head, but this results in an increase in the liquid pressure drop. The conflicting effects of decreasing the pore size of the wick inherently limit the length at which heat pipes operating against gravity can be successfully designed. In the same way, the necessity for the liquid to flow through the wick limits the total length of the classic wicked heat pipe.

Loop heat pipes (LHP) and Capillary pumped loops (CPL) were developed to overcome the inherent problem of incorporating a long wick with small pore radius in a conventional heat pipe. The CPL was first described by Stenger [99], working at the NASA Lewis Research Centre in 1966 and the first LHP was developed independently in 1972 by Gerasimov and Maydanik [100] of the Ural Polytechnic Institute. Development of these heat pipe types has been largely driven by their potential for use in space where initially they were considered for applications requiring high (0.5–24 kW) transport capabilities but the advantages of the two-phase loops with small diameter piping systems and no distributed wicks have led to their exploitation at lower powers [101].

A simple LHP is shown schematically in Fig. 6.21.

The operation of a LHP may be summarised as follows. At start-up, the liquid load is sufficient to fill the condenser and the liquid and vapour lines and there is sufficient liquid in the evaporator and compensation chamber to saturate the

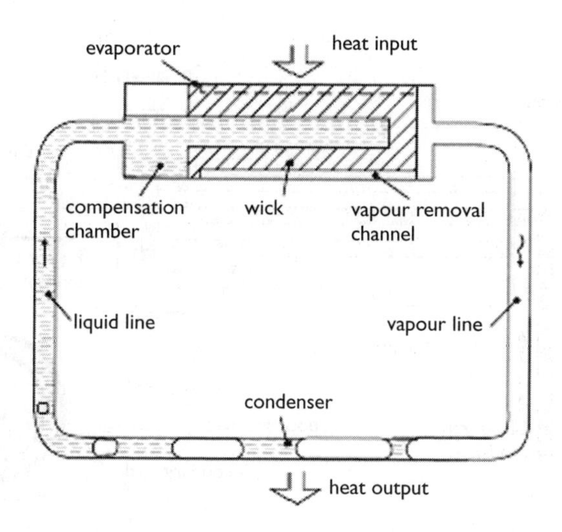

Fig. 6.21 Operating principle of loop heat pipe. [102].

wick (Level A-A, Fig. 6.23). When a heat load is applied to the evaporator fluid evaporates from the surface of the wick, and to a lesser extent in the compensation chamber, but as the wick has an appreciable thermal resistance the temperature and pressure within the compensation chamber is less than that in the evaporator. The capillary forces in the wick prevent the flow of the vapour from evaporator to compensation chamber. As the pressure difference between evaporator and compensation chamber increases, the liquid is displaced from the vapour line and the condenser and returned to the compensation chamber.

A typical LHP evaporator is shown schematically in Fig. 6.22. The evaporation in LHPs takes place on the surface of the wick adjacent to the evaporator wall. Vapour removal channels must be incorporated in the wick or evaporator wall to ensure that the vapour can flow from the wick to the vapour line with an acceptable pressure drop. A secondary wick is used to ensure uniform liquid supply to the primary wick and to provide liquid to the wick in the event of transient dryout.

The LHP has the features of the classical heat pipe with the following additional advantages:

- High heat flux capability
- Capability to transport energy over long distances without restriction on the routing of the liquid and vapour lines
- Ability to operate over a range of 'g' environments
- No wick within the transport lines
- Vapour and liquid flows separated, therefore no entrainment
- May be adapted to allow temperature control, typically in the form of a CPL, as discussed below.

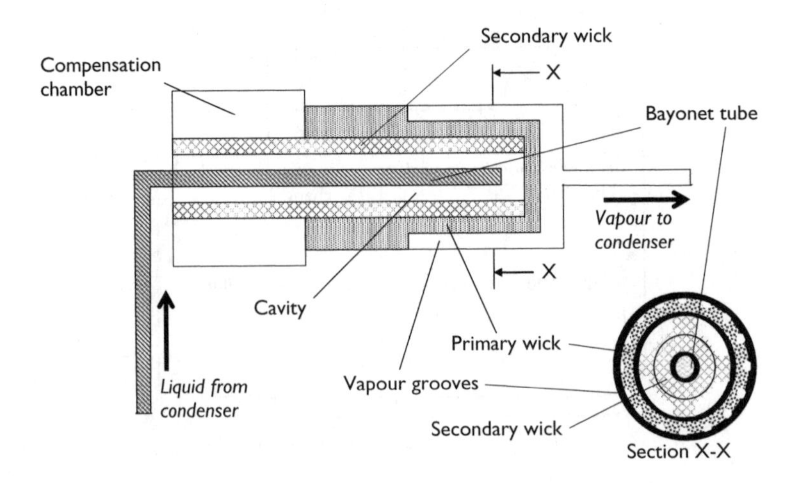

Fig. 6.22 Typical evaporator of loop heat pipe.

Mechanically pumped loops, or run-around-coils share the features listed above but have the added complexity of the pump and require an auxiliary power supply.

When operating at steady state, the thermodynamic cycle of the LHP may be described with reference to Fig. 6.23. The point 1 represents the vapour immediately above the meniscus on the evaporator wick, this vapour is saturated. As the vapour passes through the vapour channels in the evaporator, it is superheated by contact with the hot wall of the evaporator and enters the vapour line at state 2. Flow in the vapour line results in an approximately isothermal pressure drop to state 3 at entry to the condenser. The pressure drop in the condenser is generally negligible and heat transfer results in condensation to saturated liquid (state 4) and subcooling to state 5. This subcooling is necessary to ensure that the vapour is not generated in the liquid return line, either as the pressure drops in the liquid or due to heat transfer from the surroundings. The liquid enters the compensation chamber at state 6 and is heated to the saturation temperature within the compensation chamber that contains saturated liquid at state 7 in equilibrium with the saturated vapour filling the remaining space. Flow through the wick takes the liquid to state 8. The liquid within the wick is superheated, but evaporation does not occur because of the very small pore size and absence of nucleation sites. At the surface of the wick the capillary effect results in the formation of menisci at each pore, the pressure difference across these menisci being ΔP_c.

There are three conditions for a LHP to function. The first is identical to that of a classical heat pipe as discussed in Chapter 2,

$$\Delta P_{c,max} \geq \Delta P_l + \Delta P_v + \Delta P_g \tag{2.2}$$

and the pressure drops are evaluated using similar techniques but note that the liquid pressure drop is due to flow through the liquid line, in addition to through the wick.

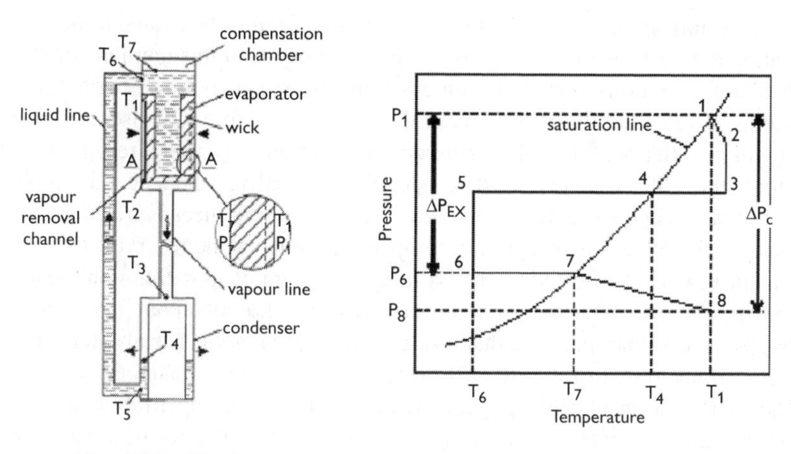

Fig. 6.23 The Loop heat pipe and its thermodynamic cycle [102].

The second condition applicable to LHPs is that the pressure drop between the evaporating surface of the wick and the vapour space in the reservoir, ΔP_{EX}, corresponds to the change in saturation temperature between states 1 and 7. This is the condition for liquid to be displaced from the evaporator to the compensation chamber at start-up. For the relatively small temperature changes in a heat pipe, the slope of the $(dP/dT)_{sat}$ may be regarded as constant and the characteristics of the LHP must be such that

$$\left(\frac{dP}{dT}\right)_{sat} \times \Delta T_{1-7} = \Delta P_{EX} \tag{6.5}$$

The third requirement for correct operation is that the liquid subcooling at exit from the condenser (state 5) must, as stated above, be sufficient to prevent excessive flashing of liquid to vapour in the vapour line. This implies

$$\left(\frac{dP}{dT}\right)_{sat} \times \Delta T_{4-6} = \Delta P_{5-6} \tag{6.6}$$

The claimed advantages of LHPs emphasise their suitability for transmitting heat over significant distances, however, their ability to operate with high heat fluxes and tortuous flow paths makes the use of miniature LHPs attractive in electronic cooling applications. Miniaturisation presents the problem of meeting the condition given in equation 6.6., the necessary temperature difference between the evaporating meniscus and the compensation chamber is difficult to maintain with the thin wick inherent in a miniature system. One solution to this problem is to use wick structures having low thermal conductivity, but this option makes design of the evaporator to minimise the temperature drop between the evaporator case and the wick surface more problematical.

The CPL differs from the LHP in that the compensation chamber is separate from the evaporator as illustrated in Fig. 6.24.

Prior to start-up of a CPL, the two-phase reservoir must be heated to the required operating temperature. The entire loop will then operate at this nominal temperature with slight variations owing to some superheat or subcooling. Upon setting an operating temperature in the reservoir, the internal pressure will rise and the entire loop will be filled with liquid, this process is referred to as pressure priming. After priming, the CPL is ready to start operating. Initially, when heat is applied to the capillary evaporator, its temperature rises as sensible heat is transferred to the working fluid. When the evaporator temperature reaches the reservoir temperature, the latent heat is transferred to the working fluid and evaporation commences. As evaporation proceeds, a meniscus is formed at the liquid/vapour interface, hence develops the capillary pressure that results in fluid circulation. Steady state operation is then similar to that of a LHP with integral compensation chamber.

The LHP benefits at start-up since the evaporator and compensation chamber are integral, while separate compensation chamber of the CPL must be maintained at the appropriate temperature. This can be advantageous if control of the operating

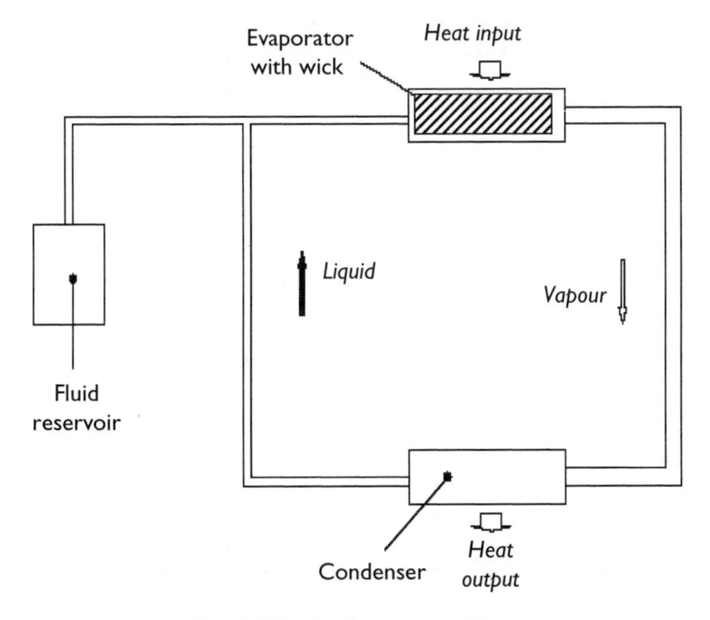

Fig. 6.24 Capillary pumped loop.

conditions of the device is required. Packaging of the evaporator and compensation chamber in a single envelope can be problematic in some applications, in which case the CPL would be preferable. A comprehensive discussion of the relative merits of the two technologies is given in [103].

Correct sizing of the compensation chamber is an essential part of the design of a LHP or CPL. This is relatively simple in principle. Since mass is conserved, the mass, M_{charge}, must be the same at start-up and during operation. Therefore,

$$M_{\text{charge}} = \rho_{\text{l,cold}} \left(\sum V_{\text{l,cold}} + aV_{\text{cc}}\right) + \rho_{\text{v,cold}} \left(\sum V_{\text{v,cold}} + (1-a)\ V_{\text{cc}}\right) \quad (6.7a)$$

$$M_{\text{charge}} = \rho_{\text{l,hot}} \left(\sum V_{\text{l,hot}} + bV_{\text{cc}}\right) + \rho_{\text{v,hot}} \left(\sum V_{\text{v,hot}} + (1-b)\ V_{\text{cc}}\right) \quad (6.7b)$$

where $\sum V$ is the total volume occupied by the liquid or vapour excluding the compensation chamber, and V_{cc} the volume of the compensation chamber.

The compensation chamber has a filling ratio a when cold and b when hot. The liquid and vapour distribution can be determined before and during operation, and appropriate values of a and b are chosen, thus permitting solution of equations (6.7a) and (6.7b) to yield V_{cc}.

Maydanik [102] has classified LHPs in terms of their design configuration, the design of the individual components, their temperature range and control strategy. This results in a range of potential LHP systems as summarised below:

LHP design	LHP dimensions	Evaporator shape	Evaporator design
• Conventional (diode)	• Miniature	• Cylindrical	• One butt-end compensation chamber
• Reversible	• All the rest	• Flat disk-shaped	• Two butt-end compensation chambers
• Flexible		• Flat rectangular	• Coaxial
• Ramified			
1-4 Condenser design	**Number of evaporators and condensers**	**Temperature range**	**Operating-temperature control**
• Pipe-in-pipe	• One	• Cryogenic	• Without active control
• Flat coil	• Two and more	• Low-temperature	• With active control
• Collector		• High-temperature	

Many of the papers published in the area of LHPs describe the design, operation and performance of a particular unit, some, e.g. [104], include additional features, such as a wick structure within the condenser.

Typically, the structural elements of LHPs are stainless steel, but aluminium and copper are also used. Sintered nickel and titanium are commonly used in the manufacture of the wick; these materials are attractive because of their high strength and wide compatibility with working fluids as well as their wicking performance.

The properties of sintered metal wicks are summarised in Table 6.3.

Working fluids are selected using broadly the same criteria as for the classical wicked heat pipe. However, LHPs are more tolerant to the presence of noncondensable gases than the classic heat pipe [101]. A high value of $(\mathrm{d}P/\mathrm{d}T)_{\mathrm{sat}}$ is also desirable, in order to minimise the temperature difference required at start-up. Maydanik suggests that ammonia is the optimum fluid for LHPs operating in the

Table 6.3 Properties of typical sintered wicks [102]

Material	Porosity, %	Effective pore radius, μm	Permeability, $\times 10^{13}$, m^2	Thermal conductivity, W/mK
Nickel	60–75	0.7–10	0.2–20	5–10
Titanium	55–70	3–10	4–18	0.6–1.5
Copper	55–75	3–15	–	–

temperature range −20 °C to 80 °C, while water is best suited to temperatures in the range 100–150 °C.

Typical applications of LHPs in aerospace and electronics cooling are discussed in Chapters 7 and 8.

A representative small CPL is described in Fig. 6.25. This was designed on the basis of the pressure drops in each component, using equations of the form developed in Section 2.3. The calculated maximum pressure drop obtained for either working fluid was less than 150 Pa, which was well below the capacity of the capillary evaporator, which was demonstrated to produce 2250 or 1950 Pa of capillary pressure with acetone and ammonia, respectively, at 30 °C.

Examples of the performance of this unit are given in Fig. 6.26a and b. It can be seen that the temperature difference when operating at equal load is higher for acetone than for ammonia. The loop was demonstrated under start-up conditions and with intermittent load. In the latter case, some temperature spikes were observed, with acetone as the working fluid as indicated in Fig. 6.27. However, the unit rapidly returned to correct operation.

While analysis of the steady state performance of LHPs and CPLs is relatively straightforward and several examples of models are reported in the literature,

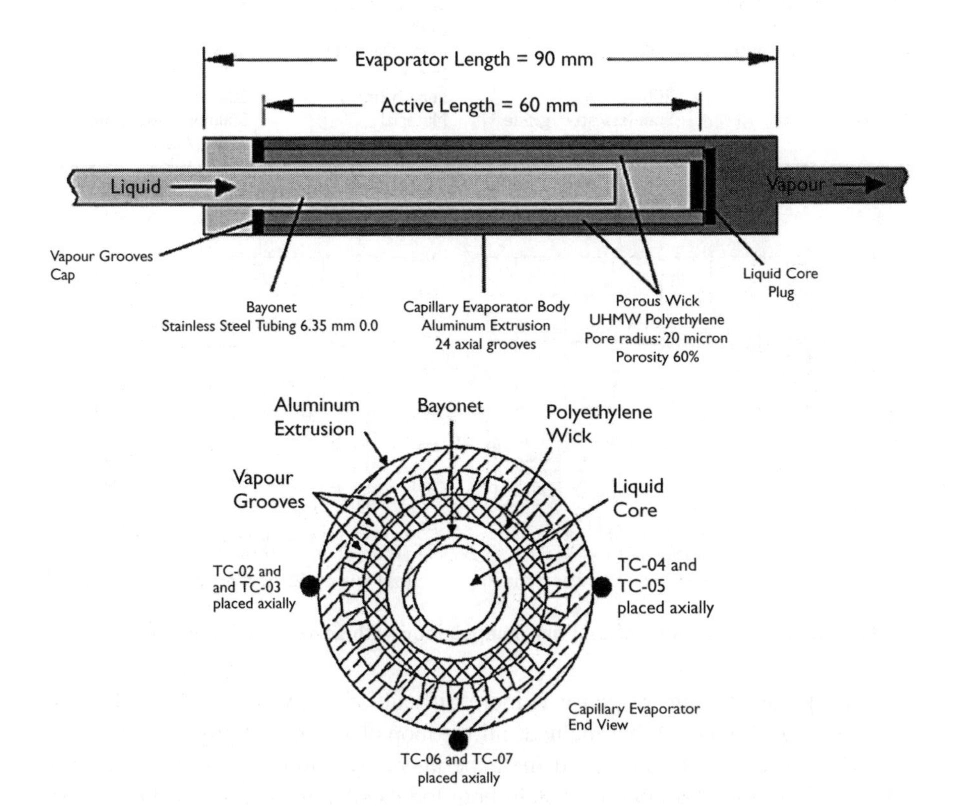

Fig. 6.25 Dimensions of a capillary pumped loop [105].

Geometric characteristics of the CPL

Evaporator		*Two-phase reservoir*	
Total length (mm)	90	Outer diameter (mm)	25.4
Active length (mm)	60	Length	200
Diameter (outer/inner) (mm)	19.0/12.7	Volume (cm^3)	85
Material	Aluminum alloy 6063 (ASTM)	Screen mesh	#200 Stainless steel grade 304L (ASTM)
Number of grooves	24	Material	Stainless steel grade 316L (ASTM)
Grooves height/width (top and base)/angle (mm)	1.5/1.0/0.66/29°		
Polyethylene		*Condenser*	
Mean pore radius (μm)	20	Outer diameter (mm)	6.35
Permeability (m^2)	10^{-12}	Inner diameter (mm)	4.35
Porosity (%)	60	Length (mm)	800
Diameter (outer/inner) (mm)	12.7/7.0	Material	Stainless steel grade 316L (ASTM)
Vapour line		*Liquid line*	
Outer diameter (mm)	6.35	Outer diameter (mm)	6.35
Inner diameter (mm)	4.35	Inner diameter (mm)	4.35
Length (mm)	200	Length (mm)	200
Material	Stainless steel grade 316L (ASTM)	Material	Stainless steel grade 316L (ASTM)

Fig. 6.25 *(Continued)*

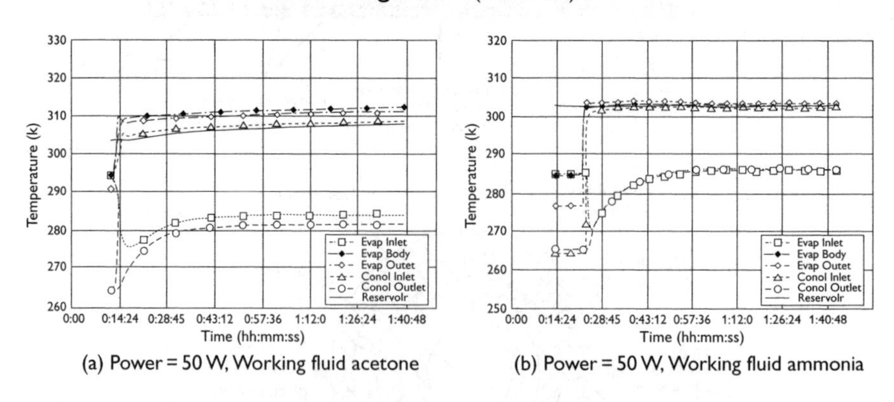

(a) Power = 50 W, Working fluid acetone (b) Power = 50 W, Working fluid ammonia

Fig. 6.26 Performance of capillary pumped loop with acetone and ammonia [105].

e.g. [106], the transient response is more difficult to analyse. Pouzet et al. [107] have developed a model that includes all the loop elements and physical processes in a CPL. The model reproduced the experimentally observed transient phenomena in a CPL and was able to explain both the oscillations observed during rapid

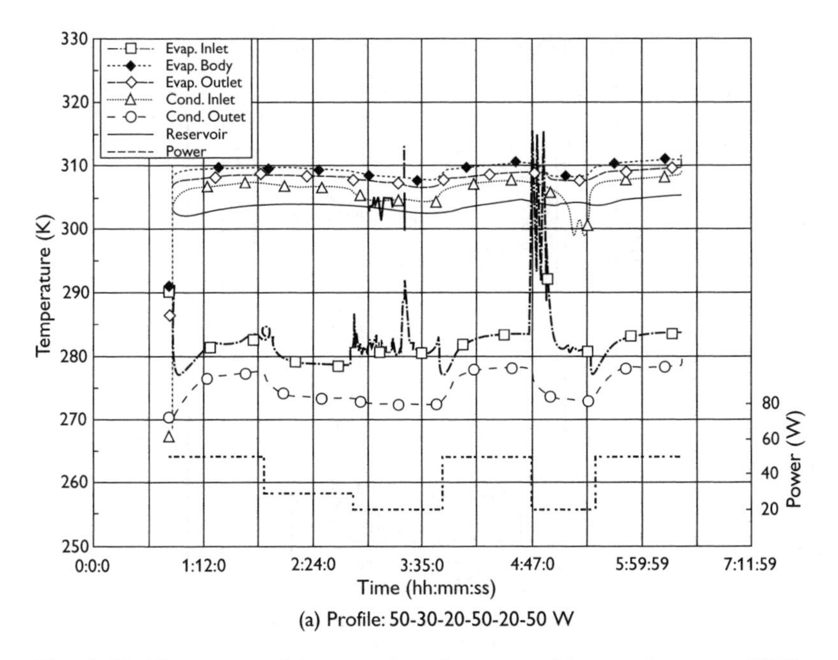

(a) Profile: 50-30-20-50-20-50 W

Fig. 6.27 Transient performance of capillary pumped loop with acetone [105].

changes of load. The model also explained the poor performance of CPLs after an abrupt decrease in load; a common phenomenon that is shown clearly in Fig. 6.27.

6.4.1 Thermosyphon Loops

When working with the evaporator below the condenser in a gravitational field it has been shown that the wick may be omitted from the classical heat pipe to give the two-phase thermosyphon. Similarly, if the condenser is above the evaporator in a loop, then this can operate satisfactorily with no wick. Thermosyphon loops have been investigated by several authors. Khodabandeh [108] has undertaken a detailed study of a loop suitable for cooling a radio base station and recommended suitable correlations for each component of heat transfer and pressure drop in the system (Fig. 6.28).

At steady state, liquid head in the downcomer must equal the sum of the pressure drops around the loop.

$$\rho_1 g h_L = \Delta P_{riser} + \Delta P_{condenser} + \Delta P_{downcomer} + \Delta P_{evaporator} \tag{6.8}$$

For a single-phase pressure drop in the downcomer and condenser, the equations given in Section 2.3.3 were employed. An additional term, ΔP_{LB}, was added for each bend and change in section.

$$\Delta P_{LB} = \xi \frac{G^2}{2\rho_1} \tag{6.9}$$

where ξ is an empirical constant for the geometry.

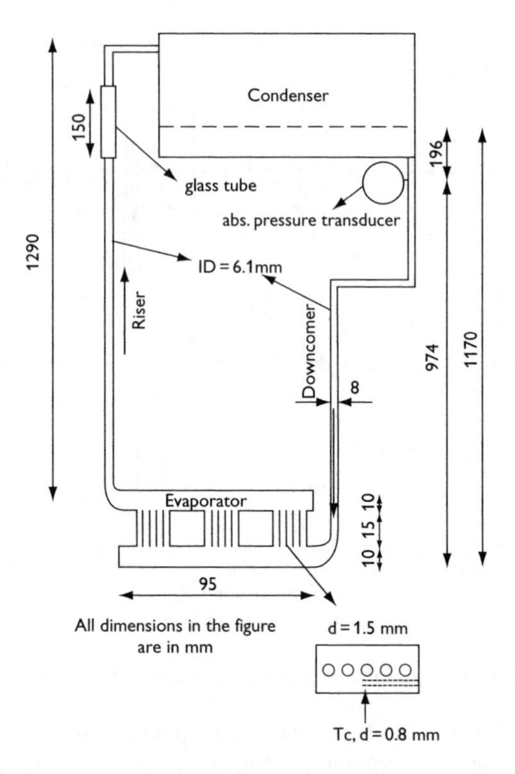

Fig. 6.28 Schematic of thermosyphon loop [108].

The flow in the evaporator and riser is two-phase, hence both void fraction and frictional pressure drop must be determined [109]. A range of void fraction correlations and frictional pressure drop calculations were tested, and it was found that the gravitational pressure drop was best correlated by assuming homogeneous flow, while the Lockhart–Martinelli correlation [110] best fitted the frictional component of pressure drop.

The pressure drop in the short evaporator was principally due to the inertia term and was evaluated assuming homogeneous flow, as recommended by ref. [111],

$$\Delta P_{\text{evaporator}} = G^2 x (v_\text{v} - v_\text{l}) \tag{6.10}$$

The heat transfer in the evaporator [112] was well represented by an expression based on Coopers correlation, modified to account for the vapour mass fraction, x,

$$\alpha = C P_\text{r}^{0.394} \dot{q}^{0.54} \frac{1}{(1-x)^{0.65}} M^{-0.5} \tag{6.11}$$

with $C = 540$.

The thermal resistance at the condenser was dominated by the airside because of the high value of condensing heat transfer coefficient, thus the condensing coefficient was not evaluated.

The effect of noncondensable gas in an industrial scale (48 kW) loop thermosyphon heat exchanger has been investigated by Dube et al. [113]. As with all heat pipes, noncondensable gases may be present due to the release of gases dissolved in the working liquid when it boils, produced by reactions between the working fluid and the material of the container walls and leakage of atmospheric gases inwards through the casing. Additionally, noncondensable gases may be introduced into the system as part of a control strategy [114]. These studies indicated that with an appropriately sized reservoir situated at the outlet of the condenser, the adverse effects of noncondensable gases could be eliminated.

6.5 MICRO-HEAT PIPES

Many microsystems proposed (and in some cases constructed) are directed at mimicking, or at least complementing one or more features of the living tissue, in particular, the human body. The manufacture of replacement parts for our bodies, whether done with biological or nonbiological materials and structures, will necessitate excursions into fluid flow at the bottom end of the microfluidics scale (microfluidics being fluid flow at the microscale), encroaching on nanotechnology. It is therefore appropriate (and hopefully interesting for some readers) to 'set the scene' for micro-heat pipes – which operate mainly at the 'upper' end of the microfluidic size range, by briefly examining a feature of the human body, which the vast majority of us have in abundance – the eccrine sweat gland [21, 22].

An interesting aspect of sweat glands, and their behaviour during thermogenic, or heat-induced, sweating is that they can be both active and passive. The secretion of liquid during active sweating involves the single-phase liquid flow along the sweat duct, with expulsion and evaporation. The resting sweat gland, it is hypothesised, functions in thermoregulation by the two-phase heat transfer, rather like a micro-heat pipe with a duct of typically 14 μ in diameter – perhaps in this decade we should refer to it as a 'nano-heat pipe'! (An average young man has 3 million sweat glands, of which more than 10 per cent are on the head and the highest density is on the soles of the feet – 620/cm^2). Vapour exudation is illustrated in Fig. 6.29.

One may approach the fluid dynamics of these microchannels from a number of directions – conventional single- or two-phase flow theory, or modified correlations from researchers such as Choi et al. [24] (sometimes limited to a single-phase flow), before considering mechanisms such as diffusion, transpiration streaming and osmosis – more commonly associated with water flows in plants and trees, and only the latter having been seriously considered in heat pipes.

Table 6.4 gives some of the characteristics of the porous structures which might be said to represent microfluidics in the 1960s – heat pipe wick structures. Interestingly, while capillary forces perhaps cannot dominate larger scale systems, they can have a positive or negative impact on microfluidic devices, and interface phenomena in

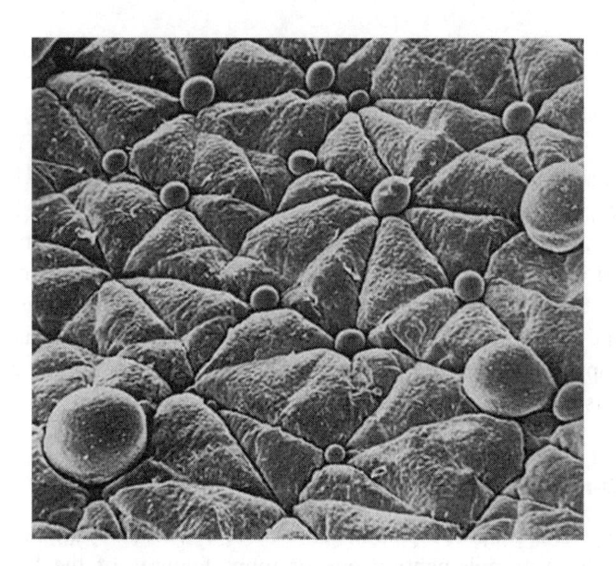

Fig. 6.29 Vapour bubbles shown at the exit of sweat glands by a scanning electron micrograph of forearm skin [22].

Table 6.4 Characteristics of capillary structures

Material	Porosity, %	Effective pore radius, μm	Permeability $\times 10^{13}$, m^2	Thermal conductivity, W/mK
Nickel	60–75	0.7–10	0.2–20	5–10
Titanium	55–70	3–10	4–18	0.6–1.5
Copper	55–75	3–15	–	–

general can become important players in microsystems where fluid flow is a feature. (The heat pipe specialist may well see opportunities for his or her expertise in the wider area of microfluidics – both in extending heat pipe models as dimensions reduce or in modelling other systems.)

Although reference was made in the 1970s, as mentioned above, to very small 'biological heat pipes' [21], the first published data on the fabrication of micro-heat pipes did not appear until 1984 [54]. In this paper, Cotter also proposed a definition of a micro-heat pipe as being one in which the mean curvature of the vapour–liquid interface is comparable in magnitude to the reciprocal of the hydraulic radius of the total flow channel.

This definition was later simplified by Peterson [51] so that a micro-heat pipe could be defined as one that satisfies the condition that

$$\frac{r_c}{r_h} \geq 1$$

Vasiliev et al [52] define micro- and mini-heat pipes (HP) as follows:

micro HP	mini HP	HP of 'usual' dimension
$r_c \leq d_v < l_c$	$r_c < d_v \leq l_c$	$r_c < l_c < d_v$

where r_c is the effective liquid meniscus radius, d_v the lesser dimension of vapour channel cross section, l_c the capillary constant of fluid:

$$l_c = \sqrt{\frac{\sigma}{g(\rho_l - \rho_v)}}$$

Conventional small-diameter wicked heat pipes of circular (and noncircular) cross section have been routinely manufactured for many years. Diameters of 2–3 mm still permit wicks of various types (e.g. mesh, sinter, grooves) to be employed. However, when internal diameters become of the order of 1 mm or less, the above condition applies.

The early micro-heat pipe configurations studied by Cotter include one having a cross section in the form of an equilateral triangle, as illustrated in Fig. 6.30.

The length of each side of the triangle was only 0.2 mm. More recently, Itoh in Japan has produced a range of micro-heat pipes with different cross-sectional forms. Some of these are illustrated in Fig. 6.31. It can be seen that the function of the conventional wick in the heat pipe is replaced by profiled corners that trap and carry the condensate to the evaporator [53].

Heat transport capabilities of micro-heat pipes are, of course, quite low. The data below, taken from ref. [53], are maximum thermal performances at different

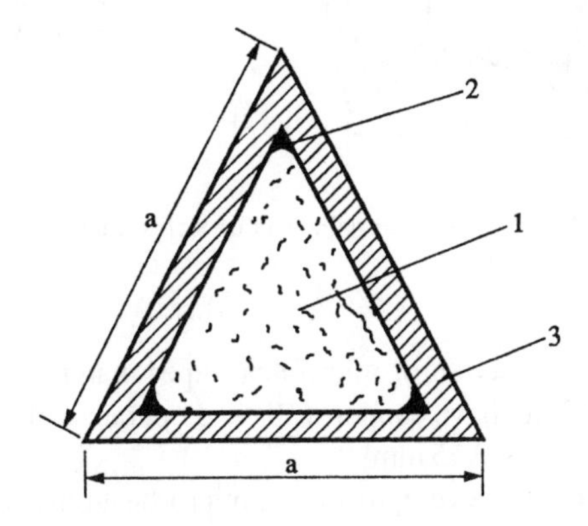

Fig. 6.30 Schematic diagram of Cotter's micro-heat pipe. 1, vapour; 2, liquid; 3, wall; $a = 0.2$ mm.

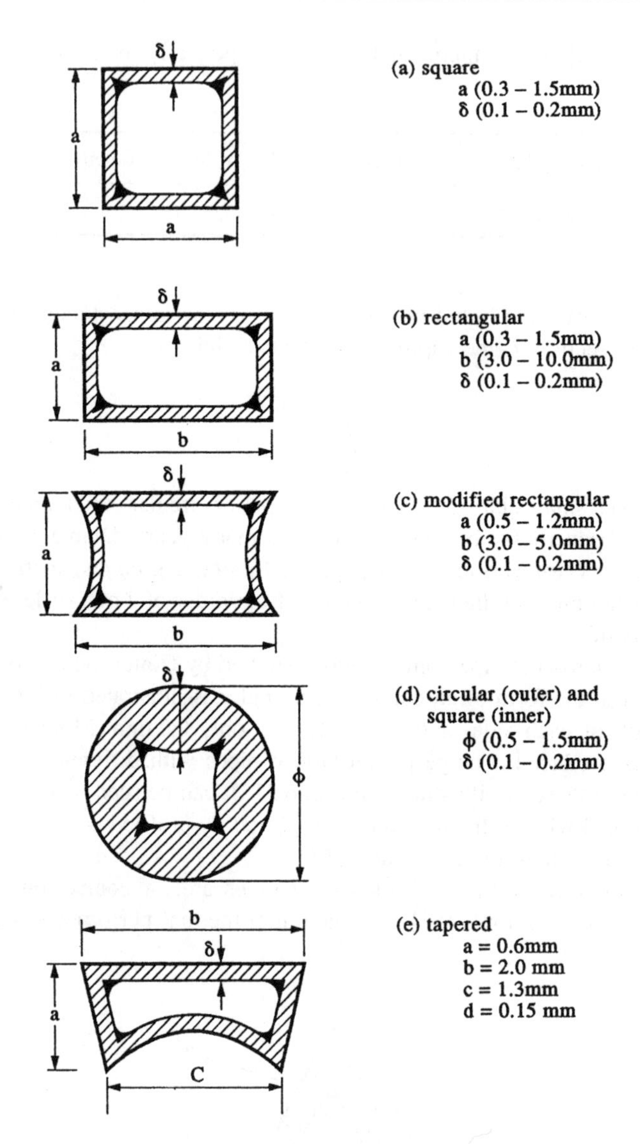

Fig. 6.31 Cross-sectional areas of Itoh's micro-heat pipes.

inclination angles of a selection of micro-heat pipes manufactured by Itoh. All use water as the working fluid, have copper walls and operate with a vapour temperature of 50 °C. Wall thickness is 0.15 mm.

Itoh later reported on the development of a micro-heat pipe having the dimensions shown in Fig. 6.32 with a thickness of only 0.1 mm, and a length of 25 mm, groove depth is 0.01 mm and minimum wall thickness 0.025 mm.

External cross section (mm)	Length (mm)	Heat Transfer (W)		
		−90 °C	0°	90 °C
0.8 × 1.0	30	0.25	0.3	0.32
2.0 × 2.0	65	1.2	1.5	1.7
1.2 × 3.0	150	0.6	2.2	4.0
0.5 × 5.0	50	0.4	1.1	2.1

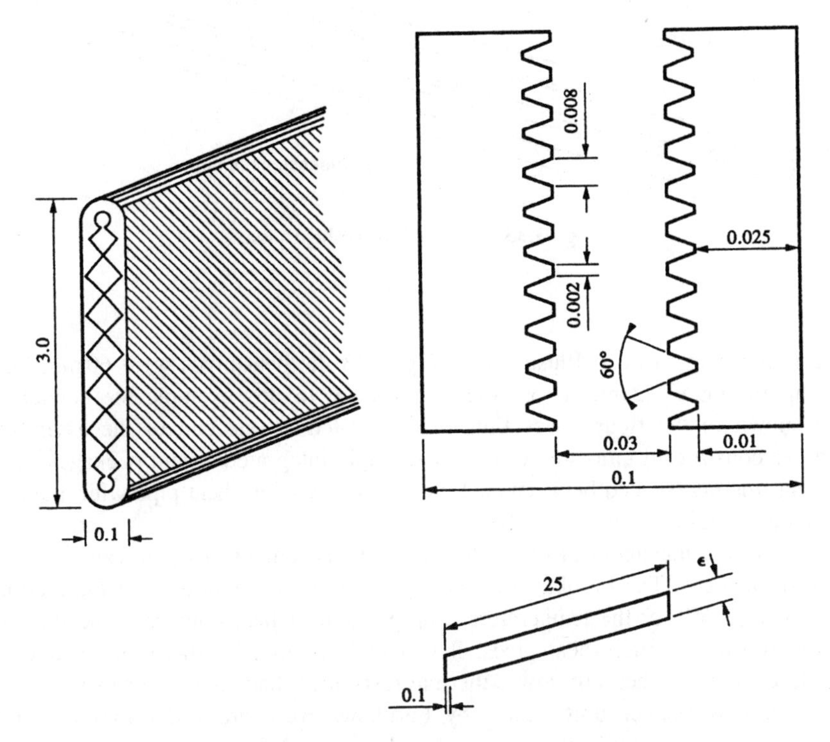

Fig. 6.32 Itoh's latest micro-heat pipe design.

It is pointed out by Cotter that the cooling rates achievable using micro-heat pipes in, for example, electronics cooling is considerably inferior to that offered by forced convection systems. However, the other benefits attributed to heat pipes such as isothermalisation, passive operation and the ability to control temperatures in environments where heat loads vary, do suggest that micro-heat pipes have a potential role to play in a growing number of applications.

The field of electronics cooling was the first application area investigated. Peterson [51] showed a trapezoidal micro-heat pipe arrangement for the thermal control

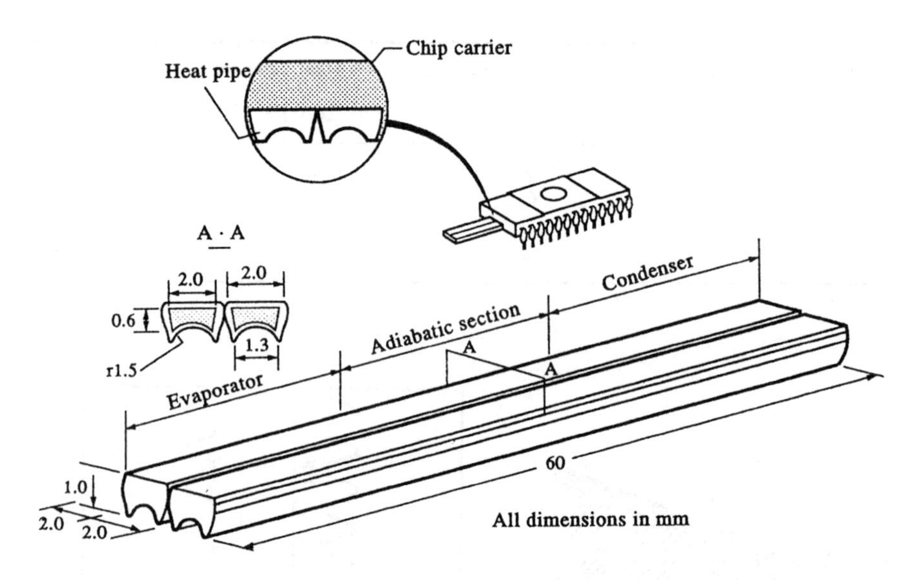

Fig. 6.33 Tapered micro-heat pipe.

of ceramic chip carriers. Illustrated in Fig. 6.33 the heat pipes were 60 mm long and approximately 2 mm across the top surface. In this example the condenser (not shown in the figure) was finned and located at one end of the assembly. Itoh has constructed units that can carry multiple integrated circuit packages, with cooling fins interspaced between each i.c.p. Thus, a micro-heat pipe with multiple evaporators and condensers can be formed.

The thermal management of electronics systems remains a prime candidate for micro-heat pipes. The Furukawa Company stressed the importance of the thermal resistance (K/W), or the temperature drop per unit of heat transfer, in developing high-performance micro-heat pipes. It was claimed that internally grooved wick structures had an inherently lower thermal resistance than sintered or mesh wicks. The high-performance units tested by Furukawa were grooved (pipes of 4 mm diameter – perhaps mini rather than micro by some definitions) and were able to transport up to about 25 W while having a thermal resistance of about 0.1 K/W. At 10 W heat transport, the resistance dropped to <0.05 K/W, while the conventional micro-heat pipe dried out at 15 W and had a consistently greater thermal resistance. In all cases, water was the working fluid, and the high-performance grooved unit was operated successfully at a tilt of 30° against gravity.

By bonding an array of wires to a plate, so that the line contact between the wire and the plate allows a liquid flow path to be formed, generating capillary forces, a new form of micro-heat pipe can be made [55].

The analysis of microgrooved heat pipe structures, where grooves are of a polygonal form, has been detailed recently by Suman et al. [56], working in the United

States. Structures of a similar form to those illustrated in Figs 6.30 and 6.31 were rigorously analysed and a performance factor was obtained for heat pipes of triangular or rectangular shape. Here the microchannel dimensions were measured in the tens and hundreds of microns.

In Belarus, Aliakhnovich et al. [57] have recently reported on a pulsating micro-heat pipe, where the wick can be discarded completely (see also Section 6.3). Capillary channels of 0.5 mm and greater were investigated. The minimum thermal resistance achieved (0.02 K/W) corresponded to the maximum performance of the micro-heat pipe array as determined by the capillary limit of 1.5 W. Water was the working fluid and the overall dimensions of the array were $78 \times 10 \times 2\,\text{mm}^3$. Maximum heat fluxes were rather low, however.

Micro-heat pipes have also been used for cooling laser diodes, bringing about a temperature reduction of approximately 7 °C in the chip, which had been at 47 °C without the heat pipe. The maximum power input to the diodes was of the order of 250 mW. In a number of similar applications, the micro-heat pipe, it is claimed, is preferable to thermoelectric cooling elements, in spite of the greater cooling capacity of the latter. The heat pipe is smaller, lighter and of lower cost.

Applications have also been briefly reported in the medical field, where local heating and/or cooling may be required as a treatment.

6.6 USE OF ELECTROKINETIC FORCES

Earlier editions of *Heat Pipes* dealt in some detail with a variety of electric field effects that could be used to assist heat pipe performance. These included electro-osmosis, electrohydrodynamics (EHD) and ultrasonics.

Within the heat-pipe-related two-phase systems (liquid–vapour) the influence of capillary action can be high, from a negative or positive viewpoint. Where the force is used to drive liquid, as in the micro-heat pipe discussed in Section 6.5, there is a wealth of literature examining optimum configurations of channel cross section, pore sizes, etc. However, the impression is that in many microfluidic devices[1] capillary action is a disadvantage that needs to be overcome by external ‘active’ liquid transport mechanisms. Perhaps there is room for a more positive examination of passive liquid transport designs, but the use of electric fields and other more complicated procedures can provide the necessary control that may not always be available using passive methods alone. Some of these ‘active’ methods are described below.

6.6.1 Electrokinetics

Electrokinetics is the name given to the electrical phenomena that accompany the relative movement of a liquid and a solid. These phenomena are ascribed to the

[1] Microfluidics, or the flow of fluids at the microscale, is discussed in the context of micro-heat pipes in Section 6.5.

presence of a potential difference at the interface between any two phases at which movements occur. Thus, if the potential is supposed to result from the existence of electrically charged layers of opposite sign at the interface, then the application of an electric field must result in the displacement of one layer with respect to the other. If the solid phase is fixed while the liquid is free to move, as in a porous material, the liquid will tend to flow through the pores as a consequence of the applied field. This movement is known as electro-osmosis.

Electro-osmosis using an alternating current has been used as the basis of microfluidic pumps, allowing normal batteries to be used at the modest applied voltages. This helps to integrate such a pumping method with portable lab-on-a-chip devices. Cahill et al. in Zurich [58] have shown that a velocity of up to $100\,\mu m/s$ could be achieved for applied potentials of less than $1\,V_{rms}$. The impact of the chemical state of the surfaces of the channel on electro-osmotic flow has also been investigated. However, of significance to reactions and other unit operations carried out in microfluidic devices, although not so important for heat pipes, is the ability to enhance mixing [70]. Heat pipe work is reported in [61–64]. For the reader interested in early work on osmotic heat pipes (without an applied electric field), see [59, 60].

6.6.2 Electrohydrodynamics (EHD)

For those willing to contemplate more complex systems, the EHD (or possibly MHD – magnetohydrodynamics; see [65, 66]) can be used for fluid propulsion, mixing and separations as well as heat pipe flow enhancement. Qian and Bau [68] have tested an MHD stirrer, illustrated in Fig. 6.34. When a potential difference (PD) is applied across one or more pairs of electrodes, the current that results interacts with the magnetic field to induce Lorentz forces and fluid motion. The alternating application of the PD results in chaotic advection and mixing, but the authors point out that the system needs perfecting.

The work of Jones [69] involved replacing the conventional capillary structure in a heat pipe with electrodes that generated an EHD force.

The heat pipe is restricted to the use of insulating dielectric liquids as the working fluid, but as these tend to have a poor wicking capability and can be used in the vapour temperature range betwen 150 °C and 350 °C, where suitable working fluids are difficult to find performance enhancement is useful. This particular type of unit is still attracting considerable interest.

The EHD heat pipe proposed by Jones consisted of a thin-walled tube of aluminium or some other good electrical conductor, with end caps made of an insulating material such as plexiglass. A thin ribbon electrode is stretched and fixed to the end caps in such a way that a small annulus is formed between it and the heat pipe wall over the complete length of the heat pipe. (This annulus is only confined to about 20 per cent of the heat pipe circumference, and provision must be made for distributing the liquid around the evaporator by conventional means.)

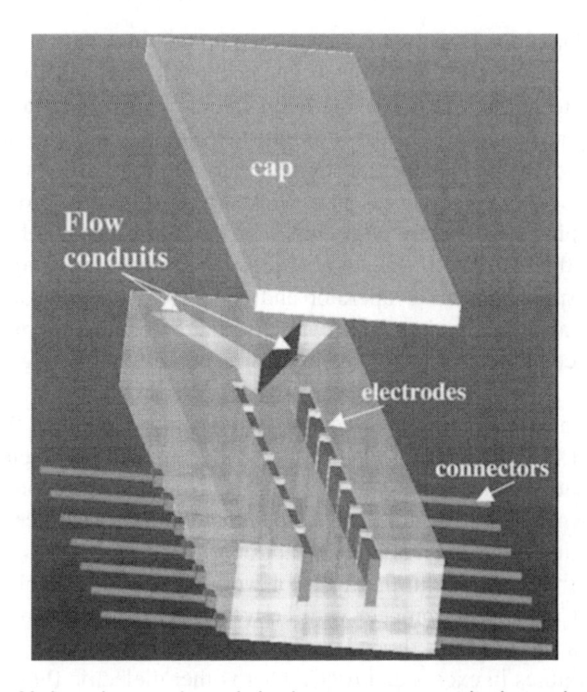

Fig. 6.34 The Y-shaped microchannels lead into a section with electrodes on both walls. The gap between adjacent electrodes is 0.5 mm [68]

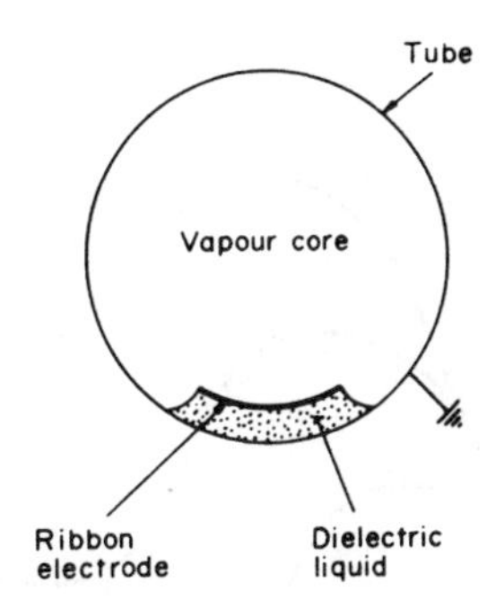

Fig. 6.35 Electrohydrodynamic liquid pump in a heat pipe.

When a sufficiently high voltage is applied, the working fluid collects in the high electric field region between the electrode and the heat pipe wall, forming a type of artery as shown in Fig. 6.35.

Evaporation of the liquid causes an outward bulging of the liquid interface. This creates an inequality in the electromechanical surface forces acting normal to the liquid surface, causing a negative pressure gradient between condenser and evaporator. Thus, a liquid flow is established between the two ends of the heat pipe.

Jones calculated that Dowtherm A could be pumped over a distance approaching 50 cm against gravity, much greater than achievable with conventional wicks. Applications of this technique could include temperature control and arterial priming.

The use of an insulating dielectric liquid somewhat limits the applicability of such heat pipes; but within this constraint, performances have shown 'significant improvement' over comparable capillary wicked heat pipes. Loehrke and Debs [29] tested a heat pipe with Freon 11 as the working fluid, and Loehrke and Sebits [30] extended this work to flat plate heat pipes, both systems using open grooves for liquid transport between evaporator and condenser. It is of particular interest to note that evaporator liquid supply was maintained even when nucleate boiling was taking place in the evaporator section, corresponding to the highest heat fluxes recorded.

Jones [69], in a study of EHD effects on film boiling, found that the pool boiling curve revealed the influence of electrostatic fields of varying intensity, and both the peak nucleate flux and minimum film flux increased as the applied voltage was raised, as shown in Fig. 6.36. Jones concluded that this was a surface hydrodynamic mechanism acting at the liquid–vapour interface.

Because of the limitation in the range of working fluids available, it is unlikely that EHD heat pipes will compete with conventional or other types where water can be used as the working fluid. However Jones and Perry [31] suggest that at vapour temperatures in excess of 170 °C, where other dielectric fluids are available, advantages may result from the lower vapour pressures accrueing to their use. To date, no experiments appear to have been carried out on EHD heat pipes over such a temperature range (e.g. 170–300 °C), however.

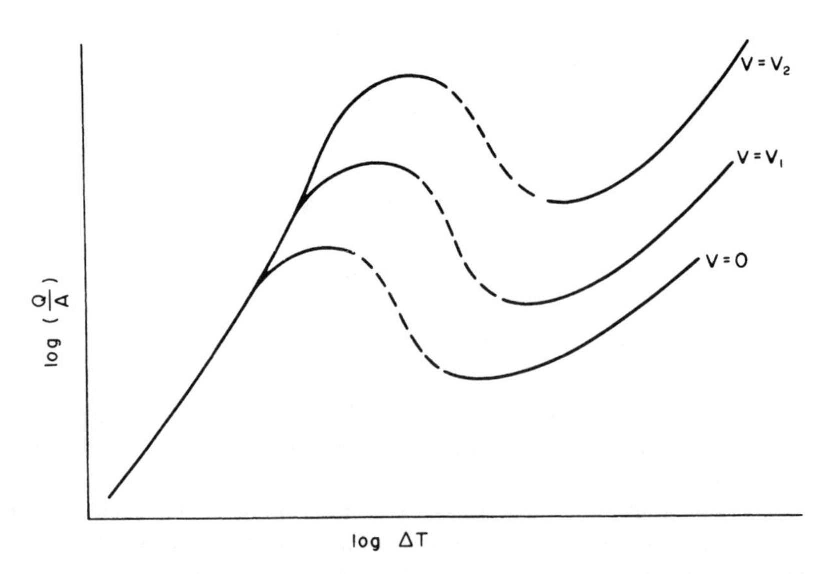

Fig. 6.36 Electrohydrodynamics – the effect of an electrostatic field on the pool boiling curve [69].

Kikuchi [33] has demonstrated that it is possible to plot an envelope showing the operating limits of an EHD heat pipe, in an identical way to that for capillary driven heat pipes. Illustrated in Fig. 6.37 the limits are as follows:

(i) Vapour breakdown limit
(ii) EHD wave speed limit
(iii) Vapour sonic limit
(iv) Voltage supply limit
(v) Size limit

The electrode spacing *S* has a strong influence on the vapour breakdown limit, and the size limit is present because EHD liquid flow structures brought close together can interact with one another.

Kikuchi also constructed a flat plate heat pipe that utilised EHD forces [34].

A useful review of EHD heat pipes, based on the extensive research carried out in the former Soviet Union, was written by Bologa and Savin [35]. The basic use of EHD forces to enhance evaporation and condensation, as evaluated in many instances in shell-and-tube and other heat exchangers, can extend to liquid transport, as illustrated above. A further variation is the introduction of a pulsed electric field to accelerate the flow of liquid in capillary porous structures. Variations in either the steady state or the fluctuating EHD field can, depending upon the porous structure used, be employed to regulate the heat transfer properties of the heat pipe over a wide range.

Figure 6.38 shows the effect of EHD on the temperature difference between the evaporator and the condenser of a heat pipe as the heat load is increased. It can

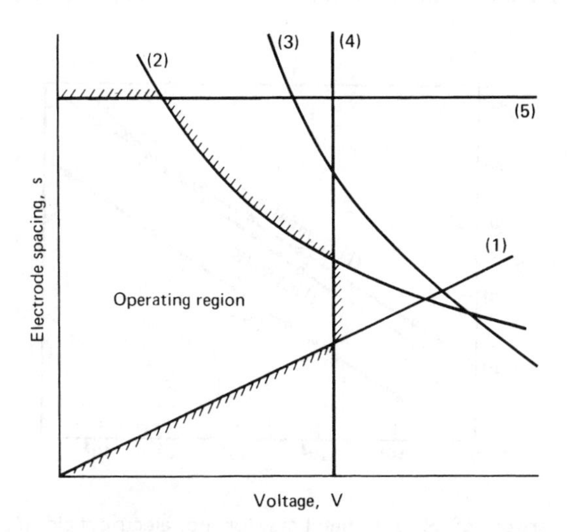

Fig. 6.37 Theoretical operational limits of an electrohydrodynamic heat pipe [65].

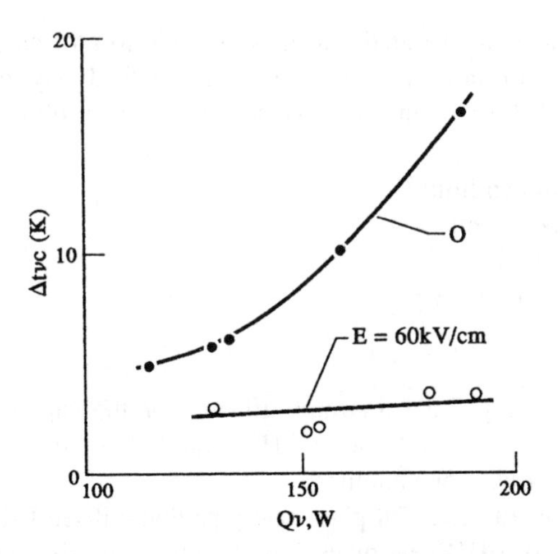

Fig. 6.38 Temperature difference between evaporator and condenser – with and without electrohydrodynamics.

be seen that EHD effects are effective in minimising the ΔT, with, it is claimed, negligible additional power consumption.

Proposals have also been put forward in the former USSR and China [36] for using static or pulsed electric fields to enhance external forced convection heat transfer – 'Corona wind' cooling. The effect of these fields on Nusselt number, as a function of *Re* for applied voltages of up to 30 kV, is shown in Fig. 6.39.

In the field of heat pipes, as in many other heat and mass transfer applications, the use of EHD remains largely confined to the laboratory. Whether potential appli-

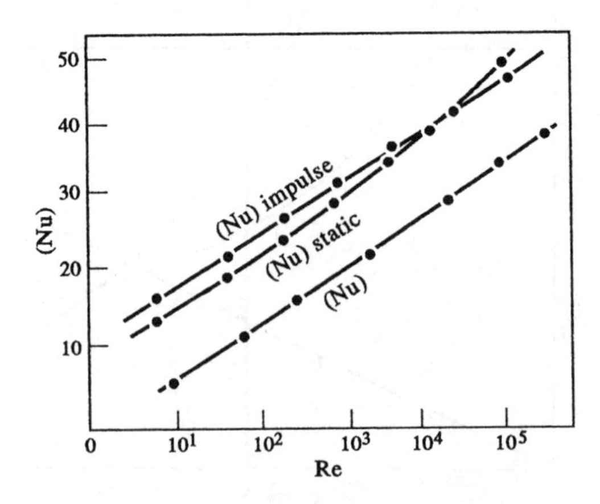

Fig. 6.39 Comparison of Nusselt numbers for no electric field (N_u) and static or impulse field.

cations become realised in practice depends upon trends in the cost and reliability of the power sources and their ease of integration into the heat pipe environment. The area of microfluidics may be the one where practical use is realised.

This may arise from research in the United States [67] where a variant on EHD – the EHD conduction pump – is being investigated for two-phase loops.

As with other fluid transport systems used in heat pipes, the prefix 'micro' may be applied to EHD heat pipes. Yu et al. [71] have reported the active thermal control of an EHD-assisted micro-heat pipe array. The authors had earlier shown that EHD had led to a sixfold increase in the heat dissipation capability of a micro-heat pipe. The extension to active thermal control, as may be achieved with active feedback-controlled VCHPs discussed earlier in this Chapter, may be useful in some applications.

However, the reader may like to consider that the superimposition of electric fields of other types to enhance the flow of liquids through a heat pipe could be used to effect similar active control.

6.6.3 Optomicrofluidics

Research at the Georgia Institute of Technology [72] is directed at using a light beam as an energy source for liquid manipulation. A dye injected into a larger droplet was mixed in the bulk liquid drop by, as with MHD, chaotic advection. In other experiments, the modulated light field can be used to drive liquids and droplets using thermocapillary forces thus generated. (Of course, it may be ultimately shown that thermocapillary driving forces could be dominant in a system where previously conventional capillary forces had, perhaps to the detriment of performance, governed flow characteristics). Even optical tweezers are proposed in the patent literature for micromanipulation.

For those wishing to follow-up the numerous techniques for influencing capillary forces, the paper by Le Berre et al. [73] on electrocapillary forces actuation in microfluidic elements is worthy of study. At the small scale, where particles might be involved in the fluid stream, the impact of electrophoresis and thermophoresis – the motion of a particle induced by a temperature gradient (from a hot to a cold region) should not be neglected.

6.7 ROTATING HEAT PIPES

(In this Section, unless otherwise stated, the rotating heat pipe is of the axial type, where the evaporator and condenser are separated in the direction of the axis of rotation, as in a turbine shaft. A radial rotating heat pipe would, for example, be that in a turbine blade, where the evaporator was radially displaced from the condenser.)

The rotating heat pipe is a two-phase thermosyphon in which the condensate is returned to the evaporator by means of centrifugal force. The rotating heat pipe consists of a sealed hollow shaft, commonly having a slight internal taper along its axial length (although the taper is not strictly necessary but, which will be described

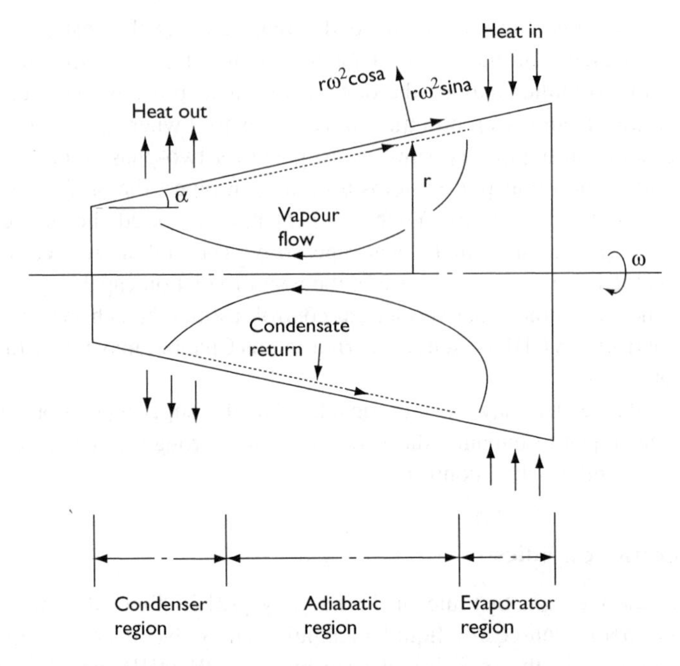

Fig. 6.40 Rotating heat pipe.

later, can benefit performance) and containing a fixed amount of working fluid, Fig. 6.40.

The rotating heat pipe, like the conventional capillary heat pipe, is divided into three sections, the evaporator region, the adiabatic region and the condenser region. However, the rotation about the axis will cause a centrifugal acceleration $\omega^2 r$ with a component $\omega^2 r \sin\alpha$ along the wall of the pipe. The corresponding force will cause the condensed working fluid to flow along the wall back to the evaporator region.

The first reference to the rotating heat pipe was given in an article by Gray [37].

Centrifugal forces will significantly affect the heat and mass transfer effects in the rotating heat pipe and the three regions will be considered in turn.

Published work on evaporation from rotating boilers, Gray et al. [38], suggests that high centrifugal accelerations have smooth, stable interfaces between the liquid and the vapour phases. Using water at a pressure of 1 atm and centrifugal accelerations up to 400 g heat fluxes of up to $257\,\mathrm{W/cm^2}$ were obtained. The boiling heat transfer coefficient was similar to that at 1 g, however the peak, or critical flux, increased with acceleration, Costello and Adams [39] derived a theoretical relationship which predicts that the peak heat flux increases as the one-fourth power of acceleration.

In the rotating condenser region a high condensing coefficient is maintained due to the efficient removal of the condensate from the cooled liquid surface by centrifugal action. Ballbach [40] carried out a Nusselt-type analysis but neglected vapour drag effects. Daniels and Jumaily [41] carried out a similar analysis but have taken account of the drag force between the axial vapour flow and the rotating liquid surface. They concluded that the vapour drag effect was small and could be neglected except at high heat fluxes. These workers also compared their theoretical predictions with measurements made on rotating heat pipes using Arcton 113, Arcton 21 and water as the working fluid. They stated that there is an optimum working fluid loading for a given heat pipe geometry, speed and heat flux. The experimental results appear to verify the theory over a range of heat flow and discrepancies can be explained by experimental factors. An interesting result [42] from this work is the establishment of a figure of merit M for the working fluid where

$$M' = \frac{\rho_l^2 L k_l^3}{\mu_l}$$

ρ_l the liquid density, L the latent heat of vapourisation, k_l the liquid thermal conductivity, and μ_l the liquid viscosity.

M' is plotted against temperature for a number of working fluids in Fig. 6.41.

Normally the rotating heat pipe should have a thermal conductance comparable to or higher than that of the capillary heat pipe. The low equivalent conductance quoted by Daniels and Jumaily [41] may have been due to a combination of very low thermal conductivity of the liquid Arcton[2] and a relatively thick layer in the condenser.

In the adiabatic region, as in the same region of the capillary heat pipe, the vapour and liquid flows will be in the opposite directions, with the vapour velocity much higher than the liquid.

6.7.1 Factors limiting the heat transfer capacity of the rotating heat pipe

The factors which will set a limit to the heat capacity of the rotating heat pipe will be sonic, entrainment, boiling and condensing limits (and noncondensable gases). The sonic limit and noncondensable gas effects are the same as for the capillary heat pipe. Entrainment will occur if the shear forces due to the counter flow vapour are sufficient to remove droplets and carry them to the condenser region. The radial centrifugal forces are important in inhibiting the formation of the ripples on the liquid condensate surface which precede droplet formation.

The effect of rotation on the boiling limit has already been referred to in ref. [38, 39], as has the condensing limit [41].

The Second International Heat Pipe Conference provided a forum for reports of several more theoretical and experimental studies on rotating heat pipes.

---

[2] Arcton 21 and Arcton 113 are chlorofluorocarbons and would no longer be the viable working fluids as they have high ozone depletion potentials.

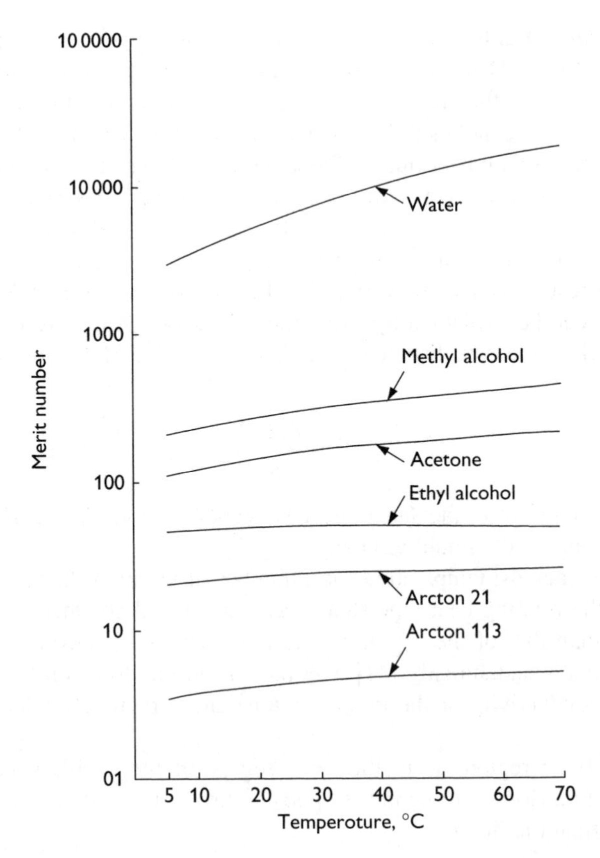

Fig. 6.41 Property group (figure of merit) versus temperature for various working fluids in a rotating heat pipe.

Marto [43] reported on recent work at the Naval Post-graduate School, Monterey, where studies on this type of heat pipe had been continuing since 1969. The developments at Monterey concern the derivation of a theoretical model for laminar film condensation in the rotating heat pipe, taking into account vapour shear and vapour pressure drop. Marto found that these effects were small, but he recommended that the internal condensation resistance be of the same order as the condenser wall resistance and outside convection resistance. Thus, it is desirable in rotating heat pipes to make the wall as thin as possible, although in rotating electrical machines this is often inconsistent with structural requirements.

Vasiliev and Khrolenok [44] extended the analysis to the evaporator section, and recommended that

(i) The most favourable mode of evaporator operation is fully developed nucleate boiling with minimal free convection effects because high heat transfer coefficients are obtained, independent of liquid film thickness and rotational speed.

(ii) For effective operation when boiling is not fully developed, the liquid film, preferably as thin as possible, should cover the whole evaporator surface.

(iii) Condenser heat flux is improved with increasing rotational speed. (For a 400 mm long × 70 mminternal diameter unit, with water as the working fluid, Vasiliev achieved condenser coefficients of 5000 W/m^2 °C at 1500 rpm with an axial heat transport of 1600 W.)

In a useful review of the performance of rotating heat pipes, Marto and Weigel [45] reported their experiments on rotating heat pipes employing several forms of cylindrical condenser sections, including the use of cylinders having smooth walls, fins and helical corrugations. Water, ethanol and R113 were studied as working fluids, and heat transfer coefficients in the condenser section were measured as a function of rotational speed (varied between 700 and 2800 rpm).

The results showed that in a rotating heat pipe it was not necessary for the condenser section to be tapered, and cheap smooth wall condensers, which are slightly inferior in heat transport capability, could be used. However, much improved performance could be achieved with finned condensers, one particular unit produced by Noranda and employing spiralled fins, giving enhancements of 200–450 per cent. The authors proposed that the surface acted as pump impellers to force the condensate back to the evaporator section.

Recent work on rotating heat pipes at speeds of up to 4000 rpm suggests that the effect of a taper in the condenser section can be positive [46]. In a heat pipe where tests were carried out with the fluid inventory (water in a copper pipe) varying between 5 per cent and 30 per cent of the internal volume, one set of experimental data showed that for a given temperature difference and fluid inventory, the heat transfer in the tapered heat pipe was slightly over 600 W, while in the cylindrical unit it was just over 200 W. A reduction in fluid inventory aided the cylindrical unit more (rising to 450 W) than the tapered unit, which stayed about the same. The differences are attributed to the reduction in the thermal resistance across the film.

The reader may find the review paper by Peterson and Win [47] useful as a summary of rotating heat pipe theory and performance.

6.7.2 Applications of rotating heat pipes

The rotating heat pipe is obviously applicable to rotating shafts having energy-dissipating loads, for example the rotors of electrical machinery, rotary cutting tools, heavily loaded bearings and the rollers for presses. Polasek [48] reported experiments on cooling on a.c. motor incorporating a rotating heat pipe in the hollow shaft, Fig. 6.42. He reported that the power output can be increased by 15 per cent with the heat pipe, without any rise in winding temperature. Gray [37] suggested that the use of a rotating heat pipe in an air-conditioning system, Groll et al. [49] reported the use of a rotating heat pipe for temperature flattening of a

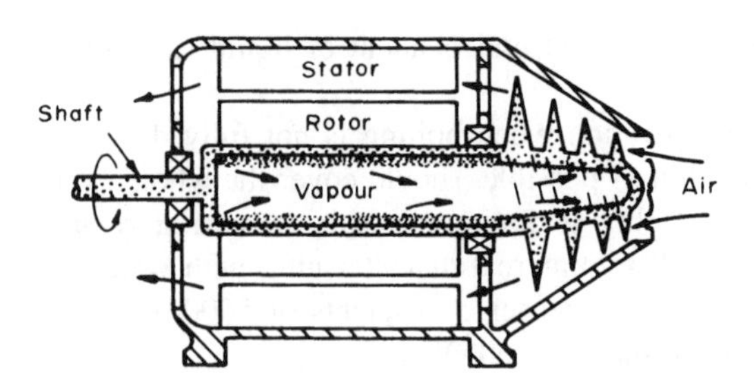

Fig. 6.42 Application of rotating heat pipes to cooling of motor rotors.

rotating drum. The drum was used for stretching plastic fibres and rotated at 4000–6000 rpm, being maintained at a temperature of 250 °C. Groll selected diphenyl as the working fluid.

As well as the use of 'conventional' rotating heat pipes to cool electric motors, as illustrated in Fig. 6.43 where the central shaft is in the form of a heat pipe, other geometries have been adopted. Groll and his colleagues at IKE, Stuttgart [50], arranged a number of copper/water heat pipes, each 700 mm long and 25 mm outside diameter, around the rotor, as shown in the end view in Fig. 6.44. It was found that adequate pumping capability was provided by the axial component of the centrifugal force, but when the rotor was stationary, circumferential grooves were needed in order to ensure that the heat pipe functioned. Each heat pipe was capable of transferring in an excess of 500 W, and rotor speeds were up to 6000 rpm.

The disc-shaped rotating heat pipe proposed by Maezawa et al. [74] also utilises centrifugal forces for condensate return, but in this case the unit resembles a wheel

Fig. 6.43 Electric motors employing rotating heat pipes (Courtesy Furukawa Electric Co. Ltd, Tokyo).

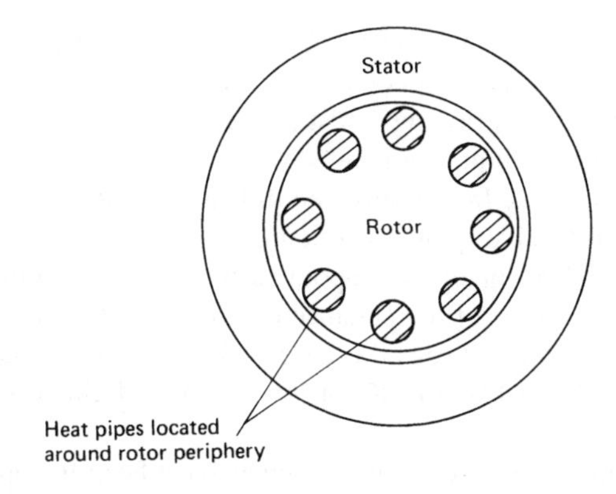

Fig. 6.44 Location of heat pipes outside rotor centreline.

or disc on a shaft, rather than being an integral part of the shaft itself. The disc-shaped rotating heat pipe is illustrated in Fig. 6.45, where it is compared to a 'conventional' rotating heat pipe. Vapour and liquid flow is predominantly radial, as opposed to the axial flow in the more common form. Using this type of heat pipe, work has been carried out on cooling the brakes of heavy road vehicles, and the experiments have successfully demonstrated that brake temperatures can be significantly reduced. Ethanol was used as the working fluid.

The disc heat pipe has its equivalent in an interesting area of chemical engineering – that of intensified process technology. Protensive, a UK-based company has employed enhanced two-phase heat transfer on spinning discs, rotating at

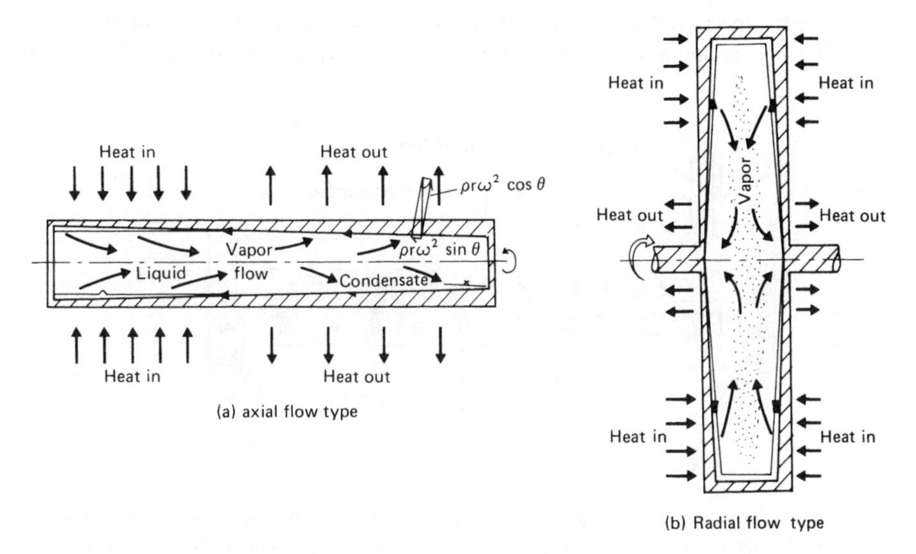

Fig. 6.45 The disc-shaped rotating heat pipe (b) [74].

500–10 000 rpm, for the thermal control of chemical reactions on the 'spinning disc reactor' and for stripping and concentration of chemicals [75]. The equipment for high-capacity evaporative stripping, illustrated in Fig. 6.46, uses the spinning disc to generate a thin film of liquid organic solvent that is readily evaporated via heat input to the spinning disc. In variants, the disc itself may be hollow and act as a vapour chamber. Protensive quoted an example of toluene evaporation on a 1-m-diameter disc where 1.5 tonne/h is evaporated with a temperature driving force of 20 °C. An interesting aspect of the enhancement is the presence of surface waves – perhaps worth examining in disc and other rotating heat pipe concepts – that allow film coefficients of 20–50 $kW/m^2/K$ for low viscosity fluids (of the types used in heat pipes).

Most applications of axial rotating heat pipes have been for operation at modest rotational speeds, typically up to 3000 rpm. However, the demands for high-speed rotating motors, and systems where generators or motors may be directly connected to high-speed prime movers, such as gas turbines, have spurred interest in higher rotational speeds, as high as 60 000 rpm.

Among the first reports on work at very high speeds was that of Ponnappan et al. [76] presented at the 10th International Heat Pipe Conference. The ultimate use was to be in a generator directly coupled to an aircraft gas turbine. With power densities of up to 11 kW/kg being predicted, and rotor tip velocities of over 300 m/s, the machine efficiency was such that about 20 per cent of the rated power capacity was dissipated as a thermal energy in the generator. In the specific unit for which the rotating heat pipe was investigated, the switched reluctance generator heat rate was over 1.9 kW, and the rotor temperature needed to be maintained below 388 °C, when rotating at 40 000 rpm. The shaft size was 12.3 cm in length and 2.54 cm in diameter.

The heat pipe tested had a tapered condenser section, leading into an untapered adiabatic and condenser sections. Water was used as the working fluid in a 316

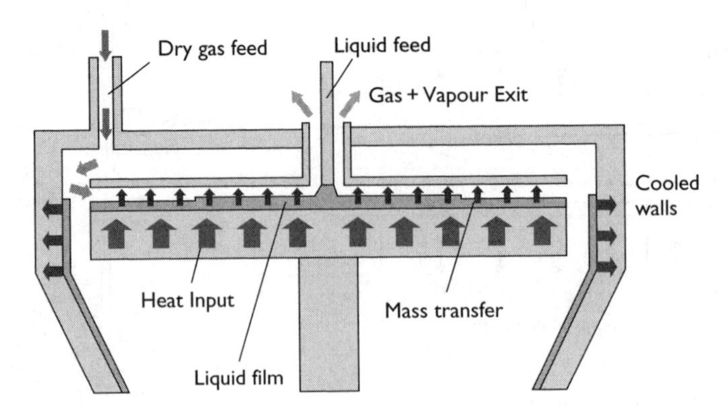

Fig. 6.46 The spinning disc reactor, with liquid introduced to the centre of the disc, and rotation driving it to the periphery, mirrors the fluid movement within disc heat pipes (Courtesy Protensive Ltd.).

stainless steel, and with a total heat pipe internal volume of 57.44 cm^3, the water fill was 0.0105 kg, representing 18.3 per cent of the internal volume at ambient temperature. Design capacity was 2 kW, and the maximum operating temperature was 250 °C.

The paper presents full performance data, and early data on modelling. However, with a heat input of 1.3 kW, at 30 000 rpm,[3] the steady state temperature distribution with jet oil cooling of the condenser, indicated that the temperature difference between extreme ends of the evaporator and condenser was 20 °C at a vapour temperature of about 200 °C. Conventional rotating heat pipe theory gave unsatisfactory agreement, but the authors point out, interestingly, that at centrifugal accelerations of 1000–9000 G at the inner wall, the impact of this on hydrodynamics and heat transfer is not known. Further data from Ponnappan and co-workers [77] detail experiments with water and the application again is in the important and upcoming area of more-electric (or all-electric) aircraft.

It was proposed by Song et al. [78] that the discrepancies between theory and experiment were due to the large fluid loading in the heat pipe, affecting film thickness in the condenser.

The R&D Centre at Hiroshima Machine Tool Works in Japan successfully operated a high-speed rotating heat pipe (24 000 rpm) for cooling the inner-race bearings on a spindle motor. The 44-mm outside diameter heat pipe was able to decrease the bearing temperature from 70 °C to 45 °C, thus avoiding a seizure problem [79].

6.7.3 Microrotating heat pipes

The ability to miniaturise systems, or to construct micro versions of equipment, is increasingly necessary as the demand for increased power out of ever smaller unit operations, be they static or dynamic, has to be met. As seen in Section 6.5 in this chapter, the heat pipe has had to develop in its micro form, in common with many other thermal control technologies.

Modelling of the heat transfer in small rotating heat pipes has been carried out by Lin et al. [80]. In this analysis, the rotating unit had axial internal grooves of triangular cross section, running the length of the internal surface. The unit used ammonia working fluid in the simulation and had an internal diameter of 4 mm and a rotational speed of 3000 rpm. Interestingly, it was found that the influence of rotation on the heat transfer in the micro region of evaporation could be neglected.

Ling et al. [81] carried out experimental work on radially rotating miniature high-temperature heat pipes. The radial rotating form would be suitable for cooling gas turbine blades, for example. In such environments, as turbine inlet temperatures are consistently on an increasing trend, the use of such heat pipes becomes important if the blade life is to be maintained in the increasingly hostile conditions. It is sometimes found that conventional air cooling methods are no longer sufficient to allow the blades to perform adequately.

[3] If directly coupled to the gas turbine, rotational speeds might reach 60 000 rpm.

The heat pipes tested had diameters of 2 and 1.5 mm and lengths of 80 mm. The construction material was s/s 304 W, and the working fluid was sodium. Operating temperature was up to 900 °C. Heat transfer capabilities were up to 280 W in the 2 mm diameter unit, and slightly less in the smaller heat pipe. The test facility was capable of a maximum speed of 3600 rpm, quite modest in gas turbine terms. The work demonstrated the adverse effect of noncondensible gases in high-temperature rotating heat pipes (in common with its effect on other types), and the experimental data emphasised the need for compatibility and high-quality manufacturing procedures, should heat pipes 'fly' in such applications.

Finally, the 'radial rotating heat pipe', albeit with a rather low rotational speed, has been proposed for one application where centripetal forces are unlikely to contribute greatly to performance! Patent GB 2409716 [82] describes heat pipes used in an automotive steering wheel, these connecting the hub to the rim. The effect of these, depending upon the function required, is to heat or cool the rim to maintain a comfortable feel for the driver!

6.8 MISCELLANEOUS TYPES

6.8.1 The sorption heat pipe (SHP)

As mentioned in the Introduction, there are instances where heat pipes alone may not satisfy the needs of high-heat flux control. One approach is to combine the heat pipe with another heat and/or mass transfer phenomenon to improve the overall system capability. A device that combines heat pipes and sorption (in this case adsorption) is the SHP [83].

Proposed for terrestrial and space applications, the basic unit is illustrated in Fig. 6.47, and in more functional form in Fig. 6.48.

Solid adsorption cooling systems are well-known in the air-conditioning field (competing, so far unsuccessfully, with the more familiar absorption cycle units), where ammonia gas and activated carbon are a typical gas–solid pair. In the system proposed here, it is therefore logical to select ammonia as the heat pipe working

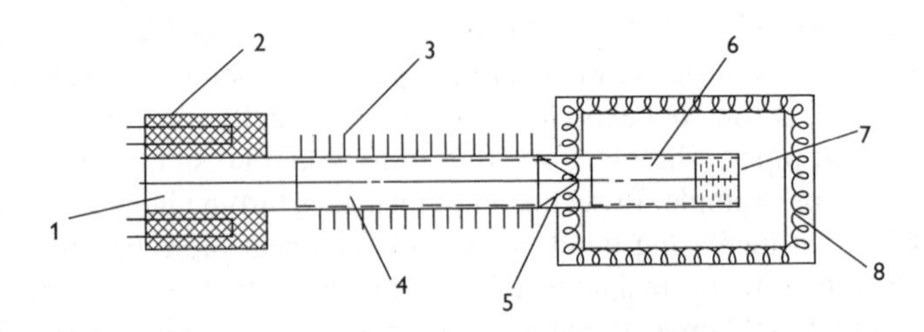

Fig. 6.47 Sorption heat pipe showing (1) vapour channel; (2) porous sorption structure; (3) finned heat pipe evaporator/condenser; (4) heat pipe wick; (5) porous valve; (6) heat pipe low-temperature evaporator (wicked); (7) working fluid accumulated in the evaporator; (8) cold box [83].

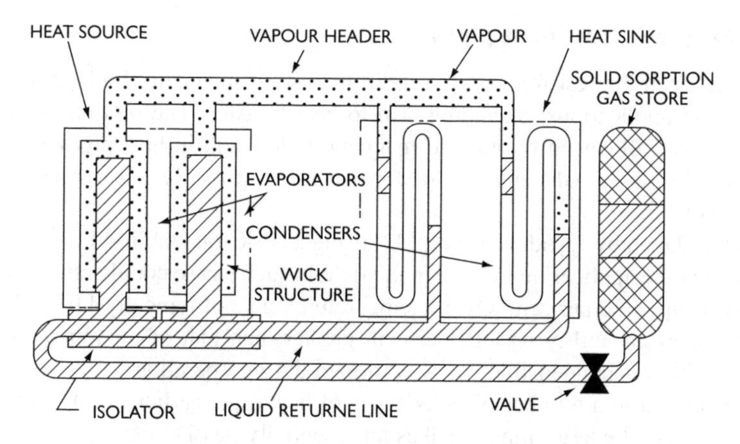

Fig. 6.48 Schematic of the sorption heat pipe [83].

fluid, as it is readily adsorbed and desorbed into and from the sorption structure shown in Fig. 6.48, which is also activated carbon.

The cycle is more complex than for a simple heat pipe and involves the following stages:

(i) desorption by applying heat to the sorption structure (2) that also has heaters embedded in it;
(ii) condensation of the desorbed fluid (ammonia) in (3);
(iii) excess ammonia liquid is filtered through the porous valve (5) into the evaporator (6);
(iv) liquid in (3) that saturates the wick (4) returns via the wick to the bed (2), enhancing heat and mass transfer there; and
(v) when desorption is complete the heaters in (2) are turned off, the working fluid collects in (6) and the pressure in the bed (2) decreases.

In phase 2 (sorption cycles of this type are cyclic), the following processes occur:

(i) Valve (5) is opened allowing the vapour pressure in the heat pipe to equalise.
(ii) During liquid evaporation (a) the air in the cold box (8) is cools down.
(iii) Valve (5) is closed once the sorbent bed is saturated with ammonia.
(iv) The bed (2) starts to cool down, aided by the heat pipe condenser (3).

It is claimed that the system is superior to a LHP with the same working fluid and evaporator dimensions, and that the unit can be considered for sensor cooling for space uses. In particular, the integration of the sorption cooler with a LHP is seen as beneficial, and measurements with heat fluxes of 100–200 W/cm^2 give average evaporator thermal resistances of 0.07–0.08 K/W.

6.8.2 Magnetic fluid heat pipes

There has been interest for many years in the use of magnetic fluids in a range of engineering applications, extending also to heat transfer. The motion of magnetic fluids can be influenced by applied magnetic fields, and in heat pipes it has been proposed that this could enhance performance and possibly allow a new control mechanism.

Work in Japan by Jeyadevan et al. [32] using a citric ion-stabilised magnetic fluid showed that the application of a magnetic field in such a case could enhance the heat transfer by up to 30 per cent. Over a temperature range of typically 110–130 °C, the heat transport capability was 10 per cent greater than that of a comparable water heat pipe. No noncondensible gas was formed during operation.

As described in a number of other types of heat pipe, earlier in this chapter, the need for external energy inputs makes an essentially passive device an active one. Thus, the reasons for employing magnetic fields need, in the opinion of the authors, to be strong before such a step should be investigated.

REFERENCES

[1] Wyatt, T. A controllable heat pipe experiment for the SE-4 satellite JHU Tech. Memo APL-SDO-1134, John Hopkins University. Appl. Phys. Lab. March 1965, AD 695433.

[2] Kosson, R. et al. Development of a high capacity variable conductance heat pipe. AIAA Paper 73–728, 1973.

[3] Marcus, D.B. and Fleischman, G.L. Steady state and transient performance of hot reservoir gas controlled heat pipes. ASME Paper 70-HT/SpT-11, 1970.

[4] Turner, R.C. The constant temperature heat pipe — a unique device for thermal control of spacecraft components. AIAA with Thermophysics Conference, Paper 69–632, June 1969 (RCA).

[5] Rogovin, J. and Swerdling, B. Heat pipe applications to space vehicles. AIAA Paper 71–421, 1971.

[6] Bienert, W. and Brennan, P.J. Transient performance of electrical feedback-controlled variable conductance heat pipes. ASME Paper 71-Av-27, 1971.

[7] Bienert, W., Brennan, P.J. and Kirkpatrick, J.P. Feedback-controlled variable conductance heat pipes. AIAA Paper 71–421, 1971.

[8] Bienert, W. et al. Study to evaluate the feasibility of a feedback-controlled variable conductance heat pipe. Contract NAS2-5772, Tech. Summary Report DTM-70-4. Dynatherm, September 1970.

[9] Depew, C.A., Sauerbrey, W.J. and Benson, B.A. Construction and testing of a gas-loaded passive control variable conductance heat pipe. AIAA Paper 73–727, 1973.

[10] Marcarino, P. and Merlone, A. Gas-controlled heat-pipes for accurate temperature measurements. Paper presented at 12th International Heat Pipe Conference, Moscow. Appl. Therm. Eng., Vol. 23, pp 1145–1152, 2003.

[11] Marcarino, P., Merlone, A. and Dematteis, R. Determination of the mercury vapour P–T relation using a heat pipe, in: Fellmuth, B., Seidel, J. and Scholz, G. (Eds). Proceedings of 8th International Symposium on Temperature and Thermal Measurements in Industry and Science, 2002, VDE Verlag, Berlin, pp 1203–1208.

[12] Ohashi, K., Hayakawa, H., Yamada, M., Hayashi, T. and Ishii, T. Preliminary study on the application of the heat pipe to the passive decay heat removal system of the modular HTR. Prog. Nucl Energy., Vol. 32, No. 3/4, pp 587–594, 1998.

[13] Watanabe, K., Kimura, A., Kawabata, K., Yanagida, T. and Yamauchi, M. Development of a variable-conductance heat-pipe for a sodium–sulphur (NAS) battery. Furukawa Rev., No. 20, pp 71–76, 2001.

[14] Pashkin, A., Prokopenko, I., Rybkin, B., Chernenko, A., Kostenko, V. and Mitrofanov, I. Development of cryogenic heat pipe diodes for Germanium gamma-spectrometer thermal control system. Proceedings of IV Minsk International Seminar: 'Heat Pipes, Heat Pumps, Refrigerators'. Minsk, Belarus, 4–7 September 2000.

[15] Varga, S., Oliveira, A.C. and Afonso, C.F. Characterisation of thermal diode panels for use in the cooling season in buildings. Energy Build., Vol. 34, pp 227–235, 2002.

[16] Bahr, A. and Piwecki, H. Passive solar heating with heat storage in the outside walls. European Commission Report EUR 7077EN. JRC Ispra, Italy, 1981.

[17] Ochi, T., Ogushi, T. and Aoki, R. Development of a heat pipe thermal diode and its heat transport performance. JSME Int. J., Vol. 39, No. 2, pp 419–425, 1996.

[18] Ezekwe, C.I. Performance of a heat pipe assisted night sky radiative cooler. Energy Conversion Manag., Vol. 30, No. 4, pp 403–408, 1990.

[19] Chernenko, A., Kostenko, V., Loznikov, V., Semena, N., Konev, S., Rybkin, B., Paschin, A. and Prokopenko, I. Optimal cooling of HPGe spectrometers for space-born experiments. Nucl. Instrum. Methods in Phys. Res. A., Vol. 442, pp 404–407, 2000.

[20] Prenger, F.C., Stewart, W.F. and Runyan, J.E. Development of a cryogenic heat pipe. Proceedings of International Cryogenic Engineering Conference and International Cryogenic Materials Conference, Albuquerque, New Mexico, 12–16 July 1993.

[21] Thiele, F.A.J., Mier, P.D. and Reay, D.A. Heat transfer across the skin: the role of the resting sweat gland. Proceedings of the Congress on Thermography, Amsterdam, June 1974.

[22] Thiele, F.A.J., Reay, D.A. and Mali, J.W.H. A possible contribution to heat transfer through the skin by the eccrine (atrichial) sweat gland. Chapter II, (Ed.) Jadassohn J., in: Handbuch der Haut- und Geschlechtskrankheiten. Springer-Verlag, Berlin, 1981.

[23] Kirkpatrick, J.P. Variable conductance heat pipes – from the laboratory to space. 1st International Heat Pipe Conference, Stuttgart, 1973, October 15–17.

[24] Choi, S.B. et al. Liquid flow and heat transfer in microtubes. In: Micromechanical Sensors, Actuators and Systems, Choi et al, (Ed.) ASME DSC, Vol. 32, pp 123–134, 1991.

[25] Brost, O. and Schubert, K.P. Development of alkali-metal heat pipes as thermal switches. In: Micromechanical Sensors, Actuators and Systems, Choi et al., (Ed.) ASME DSC, Vol. 32, pp 123–134, 1991.

[26] Eddleston, B.N.F. and Hecks, K. Application of heat pipes to the thermal control of advanced communications spacecraft. In: Choi, (Ed.). Micromechanical Sensors, Actuators and Systems, Vol. 32, ASME DSC, pp 123–134, 1991.

[27] Wolf, D.A. Flexible heat pipe switch. Final Report, NASA Contract NASS-255689, October 1981.

[28] Peterson, G.P. Analytical development and computer modelling of a bellows type heat pipe for the cooling of electronic components. ASME Winter Annual Meeting, Paper 86-WA/HT-89, Ansheim, California, 1986.

[29] Loehrke, R.I. and Debs, R.J. Measurements of the performance of an electrohydrodynamic heat pipe. AIAA Paper 75–659, 10th Thermophysics Conference, 1975.

[30] Loehrke, R.I. and Sebits, D.R. Flat plate electro-hydrodynamic heat pipe experiments. Proceedings of 2nd International Heat Pipe Conference, Bologna. ESA Report SP 112, 1976.

[31] Jones, T.B. and Perry, M.P. Electro-hydrodynamic heat pipe experiments. J. Appl. Phys. Vol. 45, No. 5, 1974.

[32] Jeyadevan, B., Koganezawa, H. and Nakatsuka, K. Performance evaluation of citric ion-stabilised magnetic fluid heat pipe. J. Magnetism Magn. Mater., Vol. 289, pp 253–256, 2005.

[33] Kikuchi, K. Study of EHD heat pipe. Technocrat, Vol. 10, No. 11, pp 38–41, November 1977.

[34] Kikuchi, K. et al. Large scale EHD heat pipe experiments. In: Advanced in Heat Pipe Technology, Reay D.A. (Ed.) Proceedings of IV International Heat Pipe conference, London. Pergamon, Oxford, 1981.

[35] Bologa, M.K. and Savin, I.K. Electro-hydrodynamic heat pipes. Proceedings of 7th International Heat Pipe Conference, Minsk, May 1990. Hemisphere, New York, 1991.

[36] Kui, L. The enhancing heat transfer of heat pipes by the electric field. Proceedings of 7th International Heat Pipe Conference, Minsk, May 1990. Hemisphere, New York, 1991.

[37] Gray, V.H. The rotating heat pipe. A wickless hollow shaft for transferring high heat fluxes. ASME Paper No. 69-HT-19, 1969.

[38] Gray, V.H., Marto, P.J. and Joslyn, A.W. Boiling heat transfer coefficients: interface behaviour and vapour quality in rotating boiler operation to 475 g. NASA TN D-4136, March 1968.

[39] Costello, C.P. and Adams, J.M. Burnout fluxes in pool boiling at high accelerations. Mechanical Engineering Department, University of Washington, Washington DC, 1960.

[40] Ballback, L.J. The operation of a rotating wickless heat pipe. M.Sc. Thesis, United States Naval Postgraduate School, Monterey, CA, 1969.

[41] Daniels, T.C. and Al-Jumaily, F.K. Theoretical and experimental analysis of a rotating wickless heat pipe. Proceedings of 1st International Heat Pipe Conference, Stuttgart, October 1973.

[42] Al-Jumaily, F.K. An investigation of the factors affecting the performance of a rotating heat pipe. Ph.D. Thesis, University of Wales, December 1973.

[43] Marto, P.J. Performance characteristics of rotating wickless heat pipes. Proceedings of 2nd International Heat Pipe Conference, Bologna. ESA Report SP 112, 1976.

[44] Vasiliev, L.L. and Khrolenok, V.V. Centrifugal coaxial heat pipes. Proceedings of 2nd International Heat pipe Conference, Bologna. ESA Report SP 112, 1976.

[45] Marto, R. and Weigel, H. The development of economical rotating heat pipes. Advances in Heat Pipe Technology. In: Reay D.A. (Ed.) Proceedings of IV International Heat Pipe Conference, Pergamon Press, Oxford, 1981.

[46] Song, F., Ewing, D. and Ching, C.Y. Experimental investigation on the heat transfer characteristics of axial rotating heat pipes. Int. J. Heat Mass Transf., Vol. 47, pp 4721–4731, 2004.

[47] Peterson, G.P. and Win, D. A review of rotating and revolving heat pipes. ASME Paper 91-HT-24, ASME, New York, 1991.

[48] Polasck, F. Cooling of a.c. motor by heat pipes. Proceedings of 1st International Heat Pipe Conference, Stuttgart, October 1973.

[49] Groll, M., Kraus, G., Kreel, H. and Zimmerman, P. Industrial applications of low temperature heat pipes. Proceedings of 1st International Heat Pipe Conference, Stuttgart, October 1973.

[50] Groll, M. et al. Heat pipes for cooling of an electric motor. Paper 78–446, Proceedings of III International Heat Pipe Conference, Palo Alto. AIAA Report CP784, New York, 1978.

[51] Peterson, G.P. Investigation of miniature heat pipes. Final Report, Wright Patterson AFB, Contract F33615-86-C-2733, Task 9, 1988.

[52] Vasiliev, L.L. et al. Copper sintered powder wick structures of miniature heat pipes. VI Minsk International Seminar 'Heat Pipes, Heat Pumps, Refrigerators'. Minsk, Belarus, 12–15 September 2005.

[53] Itoh, A. and Polasek, F. Development and application of micro heat pipes. Proceedings of 7th International Heat Pipe Conference, Minsk, May 1990. Hemisphere, New York, 1990.

[54] Cotter, T.P. Principles and prospects for micro heat pipes. Proceedings of 5th International Heat Pipe Conference, Vol. 1, Tsukuba. 1984. pp 328–335.

[55] Launay, S. et al. Investigation of a wire plate micro heat pipe array. Int. J. Therm. Sci., Vol. 43, pp 499–507, 2004.

[56] Suman, B. and Kumar, P. An analytical model for fluid flow and heat transfer in a micro-heat pipe of polygonal shape. Int. J. Heat Mass Transf., Vol. 48, No. 21–22, pp 4498–4509, 2005.

[57] Aliakhnovich, A. et al. Investigation of a heat transfer device for electronics cooling. VI Minsk Int. Seminar 'Heat Pipes, Heat Pumps, Refrigerators'. Minsk, Belarus, 12–15 September 2005.

[58] Cahill, B.P. et al. Electro-osmotic pumping on application of phase-shifter signals to interdigitated electrodes. Sen. Actuators B., Vol. 110, No. 1, pp 157–163, 2005.

[59] Minning, C.P. et al. Development of an osmotic heat pipe. AIAA Paper 78–442, Proceedings of III International Heat Pipe Conference, Palo Alto. AIAA Report CP784, 1978.

[60] Minning, C.P. and Basiulis, A. Application of osmotic heat pipes to thermal-electric power generation systems. Advances in Heat Pipe Technology. In: Reay D.A. (Ed.) Proceedings IV International Heat Pipe Conference, Pergamon Press, Oxford, 1981.

[61] Dresner, L. Electrokinetic phenomena in charged microcapiliaries. J. Phys. Chem., Vol. 67, p 1635, 1963.

[62] Burgeen, D. and Nakache, F.R. Electrokinetic flow in ultrafine capillary slits. J. Phys. Chem, Vol. 68, p 1084, 1964.

[63] Abu-Romia, M.M. Possible application of electro-osmotic flow pumping in heat pipes. AIAA Paper 71–423, 1971.

[64] Cosgrove, J.H. et al. Operating characteristics of capillary-limited heat pipes. J. Nucl. Energy, Vol. 21, pp 547–558, 1967.

[65] Jones, T.B. Electro-hydrodynamic heat pipes. Int. J. Heat Mass Trans., Vol. 16, pp 1045–1048, 1973.

[66] Ma, S. et al. Heat transport enhancement of monogroove heat pipe with electrohydrodynamic pumping. J. Thermophy Heat Transf., Vol. 11, No. 3, pp 454–460, 1997.

[67] Feng, Yinshan et al. Refrigerant flow controlled/driven with electrohydrodynamic conduction pump. Proceedings of VI Minsk International Seminar 'Heat Pipes, Heat Pumps, Refrigerators', Minsk, Belarus, 12–15 September 2005.

[68] Shizhi Qian and Haim H. Bau. Magneto-hydrodynamic stirrer for stationary and moving fluids. Sens Actuators B, Vol. 106, No. 2, pp 859–870, 2005.

[69] Jones, T.B. Electro-hydrodynamic effects on minimum film boiling. Report PB-252 320, Colorado State University, 1976.

[70] Hsin-Yu Wu and Cheng-Hsiennsk, Liu. A novel electrokinetic micromixer. Sens. Actuators A, Vol., 118, pp 107–115, 2005.
[71] Yu, Zhiquan et al. Temperature control of electrohydrodynamic micro heat pipes. Exp. Therm. Fluid Sci., Vol. 27, pp 867–875, 2003.
[72] Garnier, N., Grigoriev, R.O. and Schatz, M.F. Optical manipulation of microscale fluid flow. Phys. Rev. Lett, Vol. 91, Paper 054501, 2003. http://cns.physics.gatech.edu/~roman/muflu.html
[73] Le Berre, M. et al. Electrocapillary force actuation of microfluidic elements. Microelectronic Eng., Vol. 78–79, pp 93–99, 2005.
[74] Maezawa, S. et al. Heat transfer characteristics of disc-shaped rotating, wickless heat pipes. Advances in heat pipe technology; in: Reay D.A. (Ed.) Proceedings of IV International Heat Pipe Conference, Pergamon Press, Oxford. 1981.
[75] Protensive Ltd. Web Site. http://www.protensive.co.uk/pages/technologies/category/categoryid=heat
[76] Ponnappan, R., He, Q., Baker, J., Myers, J.G. and Leland, J.E. High speed rotating heat pipe: analysis and test results. Paper H3-10, Proceedings of 10th International Heat Pipe Conference, Stuttgart, 21–25 September 1997.
[77] Ponnappan, R., He, Q. and Leland, T.E. Test results of water and methanol high-speed rotating heat pipes. J. Thermophys. Heat Trans., Vol. 12, No. 3, pp 391–397, 1998.
[78] Song, F., Ewing, D. and Ching, C.Y. Fluid flow and heat transfer model for high speed rotating heat pipes. Int. J. Heat Mass Transf., Vol. 46, pp 4393–4401, 2003.
[79] Hashimoto, R., Itani, H., Mizuta, K., Kura, K. and Takahashi, Y. Heat transport performance of rotating heat pipes installed in a high-speed spindle. Mitsubishi Heavy Industries Tech. Rev., Vol. 33, No. 2, 1996.
[80] Lin, L. and Faghri, A. Heat transfer in micro region of a rotating miniature heat pipe. Int. J. Heat Mass Transf., Vol. 42, pp 1363–1369, 1999.
[81] Ling, J. and Cao, Yiding. Closed-form analytical solutions for radially rotating miniature high-temperature heat pipes including non-condensible gas effects. Int. J. Heat Mass Transf., Vol. 43, pp 3661–3671, 2000.
[82] GB Patent 2409716. Assigned to autoliv development (Sweden). A steering wheel with heat pipes and a plastic thermally insulating hub. Publication date: 6 July 2005.
[83] Vasiliev, L. and Vasiliev, L. Jr. Sorption heat pipe – a new thermal control device for space and ground application. Int. J. Heat Mass Transf., Vol. 48, pp 2464–2472, 2005.
[84] Akachi, H. US Patent No. 4921041, 1990.
[85] Akachi, H. US Patent No. 5219020, 1993.
[86] Akachi H. US Patent No. 5490558, 1996.
[87] Polasek, F. and Rossi, L. Thermal control of electronic equipment and two-phase thermosyphons, 11th IHPC, 1999.
[88] Charoensawan, P., Khandekar, S., Groll, M. and Terdtoon, P. Closed loop pulsating heat pipes, Part A: parametric experimental investigations, Appl. Therm. Eng., Vol. 23, 2009–2020, 2003.
[89] Vogel, M. and Xu, G. Low profile heat sink cooling technologies for next generation CPU thermal designs. Electronics Cooling, Vol. 11, No.1, February 2005.
[90] Duminy, S. Experimental investigation of pulsating heat pipes. Diploma Thesis, Institute of Nuclear Engineering and Energy Systems (IKE), University of Stuttgart, Germany, 1998.
[91] Khandekar, S., Charoensawan, P., Groll, M. and Terdtoon, P. Closed loop pulsating heat pipes Part B: visualization and semi-empirical modeling, Appl. Therm. Eng., Vol. 23, 2021–2033, 2003.

[92] Akachi, H., Polasek, F. and Stulc, P. Pulsating heat pipes. Proceedings of the 5th International Heat Pipe Symposium, Melbourne, Australia, 1996, ISBN 0-08-042842-8, pp 208–217.

[93] Cornwell, K. and Kew, P.A. Boiling in small parallel channels, Proceedings of CEC Conference on Energy Efficiency in Process Technology, Elsevier Applied Sciences, Paper 22, pp 624–638, 1992.

[94] Khandekar, S., Dollinger, N. and Groll, M. Understanding operational regimes of closed loop pulsating heat pipes: an experimental study, Appl. Therm. Eng., Vol. 23, pp 707–719, 2003.

[95] Zhang, X.M., Xu, J.L. and Zhou. Z.Q. Experimental study of a pulsating heat pipe using FC-72, ethanol, and water as working fluids, Exp. Heat Trans., Vol. 17, pp 47–67, 2004.

[96] Sakulchangsatjatai, P., Terdtoon, P., Wongratanaphisan, T., Kamonpet P. and Murakami, M. Operation modeling of closed-end and closed-loop oscillating heat pipes at normal operating condition, App. Therm. Eng. Vol. 24, pp 995–1008, 2004.

[97] Katpradit, T., Wongratanaphisan, T., Terdtoon, P., Kamonpet, P., Polchai, A. and Akbarzadeh, A. Correlation to predict heat transfer characteristics of a closed end and oscillating heat pipe at critical state, Appl. Therm. Eng., Vol. 25, pp 2138–2151, 2005.

[98] Karimi, G. and Culham, J.R. Review and assessment of pulsating heat pipe mechanism for high heat flux electronic cooling, Inter Society Conference on Thermal Phenomena, 2004.

[99] Stenger, F.J. Experimental feasibility of water-filled capillary-pumped heat-transfer loops, NASA TM X-1310, NASA, Washington DC 1966.

[100] Gerasimov, Y.F., Maydanik, Y.F., Shchogolev, G.T. et al. Low-temperature heat pipes with separate channels for vapour and liquid, Eng.-Phys. J., Vol. 28, No. 6, pp 957–960 (in Russian) 1975.

[101] Delil, A.A.M. Research issues on two-phase loops for space applications, National Aerospace Laboratory Report NLR-TP-2000-703, 2000.

[102] Maydanik, Y.F. Loop heat pipes. Appl. Therm. Eng., Vol. 25, pp 635–657, 2005.

[103] Nikitin M. and Cullimore, B. CPL and LHP technologies, what are the differences, what are the similarities, SAE Paper 981587, pp 400–408, 1998.

[104] Murakoa, I., Ramos, F.M. and Vlassov, V.V. Analysis of the operational characteristics and limits of a loop heat pipe with porous element in the condenser. Int. J. Heat Mass Transf., Vol. 44, pp 2287–2297, 2001.

[105] Bazzo, E. and Riehl, R.R. Operation characteristics of a small-scale capillary pumped loop, Appl. Therm. Eng., Vol. 23, pp 687–705, 2003.

[106] Hamdan, M. et al. Loop heat pipe (LHP) development by utilizing coherent porous silicon (CPS) wicks. Inter Society Conference on Thermal Phenomena, IEEE, 2002.

[107] Pouzet, E. et al, Dynamic response of a capillary pumped loop subjected to various heat load transients. Int. J. Heat Mass Transf., Vol. 47, pp 2293–2316, 2004.

[108] Khodabandeh, R. Heat transfer and pressure drop in a thermosyphon loop for cooling of electronic components Doctoral Thesis, KTH, Royal Institute of Technology, Stockholm, 2004.

[109] Khodabandeh R. Pressure drop in riser and evaporator in an advanced two-phase thermosyphon loop, Int. J. Refrigeration, Vol. 28, No. 5, pp 725–734, 2005.

[110] Lockhart, R.W. and Martinelli, R.G., Proposed correlations for isothermal two-phase two-component flow in pipes. Chem. Eng. Prog., Vol. 45, 39–48, 1949.

[111] Bowers, M.B. and Mudawar, I. Two-phase electronic cooling using mini-channel and macro-channel heat sinks – part II, flow rate and pressure drop constraints, ASME J. Electronic Packag., Vol. 116, pp. 298–305, 1994.

[112] Khodabandeh R Heat transfer in the evaporator of an advanced two-phase thermosyphon loop. Int. J. Refrigeration. Vol. 28, No. 2, pp 190–202, 2005.

[113] Dube, V., Akbarzadeh, A. and Andrews, J. The effects of non-condensable gases on the performance of loop thermosyphon heat exchangers, Appl. Therm. Eng., Vol. 24, pp 2439–2451, 2004.

[114] Dube, V. The development and application of a loop thermosyphon heat exchanger for industrial waste heat recovery and determination of the influence of non-condensable gases on its performance. Ph.D. Thesis, RMIT, University of Melbourne, Australia, 2003.

7

APPLICATIONS OF THE HEAT PIPE

The heat pipe has been, and currently is being, studied for a wide variety of applications, covering almost the complete spectrum of temperatures encountered in heat transfer processes. These applications range from the use of liquid helium heat pipes to aid target cooling in particle accelerators, to cooling systems for state-of-the-art nuclear reactors and potential developments aimed at new measuring techniques for the temperature range 2000–3000 °C.

The decade since the last edition of *Heat Pipes* has seen mass production of heat pipes on a scale not envisaged before – millions of units per month being fabricated for thermal management (generally cooling) of the processors in desktop and notebook computers. The position of heat pipes in spacecraft has been challenged by developments in lightweight mechanical pumps (for fluid pumping) and by some thermal storage technologies. Nevertheless, the concurrent development and use of loop heat pipes (see also Chapters 6 and 8) has ensured that the 'heat pipe' solution remains at or near the top of the list of preferred options [1].

7.1 BROAD AREAS OF APPLICATION

In general, the applications come within a number of broad groups, each of which describes a property of the heat pipe. These groups are:

(i) Separation of heat source and sink
(ii) Temperature flattening, or isothermalisation
(iii) Heat flux transformation
(iv) Temperature control
(v) Thermal diodes and switches

The high effective thermal conductivity of a heat pipe enables heat to be transferred at high efficiency over considerable distances. In many applications where component cooling is required, it may be inconvenient or undesirable thermally to dissipate the heat via a heat sink or radiator located immediately adjacent to the component. For example, heat dissipation from a high power device within a module containing other temperature-sensitive components would be affected by using the heat pipe to connect the component to a remote heat sink located outside the module. Thermal insulation could minimise heat losses from intermediate sections of the heat pipe. Most uses of tubular heat pipes in electronics thermal control are directed at separating the heat source from the sink.

The second property listed above, temperature flattening, is closely related to source–sink separation. As a heat pipe, by its nature, tends towards operation at a uniform temperature, it may be used to reduce thermal gradients between unevenly heated areas of a body. The body may be the outer skin of a satellite, part of which is facing the sun, the cooler section being in shadow; this is illustrated diagrammatically in Fig. 7.1. Alternatively, heat pipes 'immersed' in a batch chemical reactor could assist uniform reaction rates by taking heat from more exothermic regions to less active parts of the reactants.

Heat flux transformation has attractions for other reactor technologies. In thermionics, for example, the transformation of a comparatively low heat flux, as generated by radioactive isotopes, into sufficiently high heat fluxes capable of being utilised effectively in thermionic generators was an early application.

The fourth area of application, temperature control, is best carried out using the variable conductance heat pipe (VCHP). This type can be used to control accurately

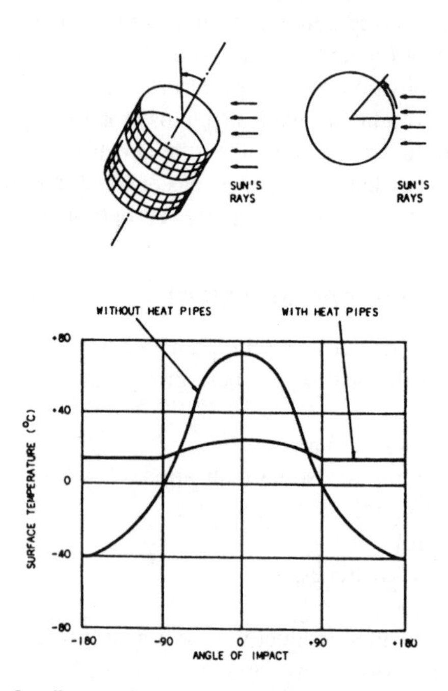

Fig. 7.1 Satellite isothermalisation (Courtesy Dornier Review).

the temperature of devices mounted on the heat pipe evaporator section. While the VCHP found its first major applications in spacecraft, it has now become widely accepted in many more mundane applications, ranging from temperature control in electronics equipment to ovens and furnaces. One such example is shown in Fig. 7.2, a VCHP.

The thermal diode, described in Chapter 6, has a number of specialised applications where heat transport in one direction only is a prerequisite.

As with any other device, the heat pipe must fulfil a number of criteria before it becomes fully acceptable in applications in industry. For example, in the diecasting and injection moulding industries, the heat pipe has to be

(i) Reliable and safe
(ii) Satisfy a required performance
(iii) Cost-effective
(iv) Easy to install and remove

Obviously, each application must be studied in its own right, and the criteria vary considerably. A feature of the moulding processes, for example, is the presence of high-frequency accelerations and decelerations. In these processes, therefore, heat pipes should be capable of operating when subjected to this motion, and this necessitates development work in close association with the potential users.

Fig. 7.2 VCHP used for temperature control (Courtesy Themacore International, Inc.).

7.2 HEAT PIPES IN ENERGY STORAGE SYSTEMS

By their nature, many energy storage systems should lose or gain as little heat as possible during 'inactive' periods, while also delivering or taking in heat (or 'coolth') as predetermined rates, some of which may be rather high, when required to function actively. The nature of the chemicals used in some phase change storage media, in particular, such as their low thermal conductivity, gives heat pipes an opportunity to enhance performance. One sensible heat 'store' that has benefited considerably from heat pipes is the ground. Use of the ground as either a heat source or a heat sink – well known to heat pump users – has been used to deice roads using heat pipes and, as discussed below, as a sensible heat sink for underground train thermal management. The philosophy of thermal management of nuclear waste and some reactors, where heat pipes have been considered, is similar.

Heat pipes have been used extensively in a variety of energy storage systems. Heat pipes are suited to thermal storage systems, in particular, in the role of heat delivery and removal, because of their high effective thermal conductivity and their passive operation. As aids to temperature stratification in hot water storage tanks, to their incorporation in stores for heat or 'coolth' using phase change materials (PCMs), the unique properties of heat pipes can permit systems to operate in a manner not generally possible using conventional heat exchangers. Heat pipes and thermosyphons have also been applied where heat transfer from or into the ground has been required – for example, in the preservation of permafrost and for deicing roads, as described later in this chapter (see Section 7.7). These were among the earliest mass-produced heat pipe applications. The safety aspect of heat pipes, due to the two walls intervening between the evaporator and the condenser, has also encouraged their use for heat removal from nuclear fuel stores and reactors themselves – see below and Appendix 3 and Bibliography (heat pipes in chemical reactors are discussed in Section 7.3).

7.2.1 Why use heat pipes in energy storage systems

The limitations of some thermal storage systems, be they for storing heat or 'coolth', tend to be strongly dependent on the properties of the heat storage medium used, such as specific or latent heats, density and thermal conductivity. For the reader interested in this specific topic, many excellent reviews have been written on the subject, for example, Dincer and Rosen [2] and Zalba et al. [3]. Cost bears strongly on the choice of storage medium, and unfortunately low-cost materials tend to require the largest storage volume per watt-hour of heat stored. The materials that undergo a phase change, thus releasing latent heat – as in a heat pipe, but in this case normally changing from solid to liquid – tend to have the smallest storage volumes, but are generally more expensive, and may require special encapsulation materials, due to corrosion or toxicity. This can be a limiting factor in applications in occupied buildings, for example.

While heat may be stored at any temperature from just above ambient to in excess of 1000 °C, for the storage of 'coolth' for air-conditioning applications – an important energy-saving opportunity, the temperature range is the modest. The storage medium may be expected to operate mainly within the 10–25 °C band. Although the use of heat pipes for the storage at cryogenic temperatures is less known, there is no reason why heat pipes using, for example, nitrogen as the working fluid should not be employed.

A major disadvantage of many potential heat storage candidates is the poor thermal conductivity, whatever phase they are in. It is the role of heat pipes (and other 'enhanced' heat transfer devices such as compact fin assemblies) which has allowed the practical use of heat storage systems to extend into areas where limitations on internal conduction have inhibited the performance in the past. Often the heat pipe is critical to the successful operation of the unit, both in charging and discharging modes. In a number of applications, it additionally allows a compact modular unit to be developed and helps ensure separation of reactive storage media from occupied spaces, an important health and safety factor – not just for the nuclear stores!

7.2.2 Heat pipes in sensible heat storage devices

One of the most common uses for heat pipes associated with storage is to absorb solar energy and transfer it to water, either static or flowing. Solar collectors employing heat pipes are made by several manufacturers. The concept is described in one form by Azad et al. [4]. The use of individual heat pipes linking to water stores is also cited by Polasek [5].

Work on heat pipes and their terrestrial applications in the Former Soviet Union (FSU) was, and in some CIS members continues to be, perhaps more prolific than anywhere else in the world. One of the laboratories most involved with such uses is the Luikov Heat and Mass Transfer Institute in Minsk, Belarus. Many years ago, Vasiliev [6] and his team (see e.g. Caruso et al. [7]) examined the performance of a heat store that used horizontal heat pipes to transfer heat into and out of the store. The store was charged with dry sand or pebbles – used in the FSU in houses and greenhouses, often located under the building to capture solar heat or heat in warm air or warm water. The heat pipes were found to be an effective way for heat transfer during both charge and discharge processes.

The mean energy transfer of the pipes was 200 Wh/m, and a $6 \times 5 \times 2$ m tank had ten 6 m long heat pipes 1 m apart. Each discharge interval gave about 100 W per pipe.

A second system developed in Belarus [8], illustrated in Fig. 7.3, employed electric heating elements, using 'off-peak' electricity, to raise the temperature of storage bricks within the unit to around 500 °C. The heat pipes, with evaporator sections in the lower half of the unit and condensers in the central finned section above the core, allowed heat discharge to take place over a 24–48-h period, with a boost being provided by fan-assisted heat transfer.

Heat pipes and heat stores have also employed the 'thermal diode' aspect of heat pipes – the ability to transfer heat in one direction only, thus minimising heat losses – see Section 6.2 in Chapter 6. Sodha et al. [9] referred to work at a European

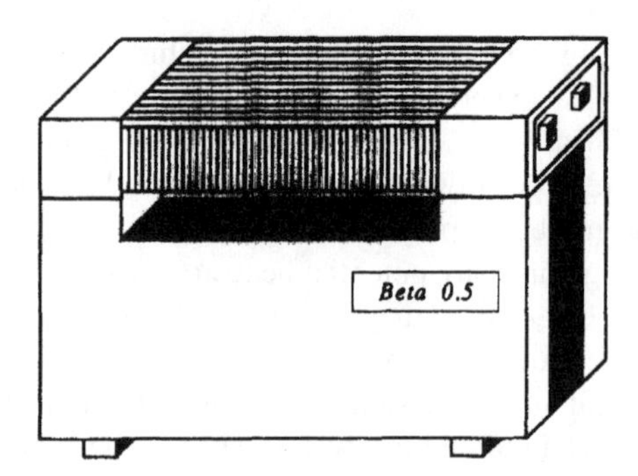

Fig. 7.3 The heat storage unit, employing heat pipes, developed at the Luikov Institute, Minsk.

Commission Laboratory (ISPRA) in the late 1970s on using heat pipes to take heat into a 'storage wall' of a building.

The heat pipe is able to transfer heat effectively from within thermal stores employing solid or liquid storage media (without phase change – see the section below on reactor decay heat removal). However, its practical use in conjunction with PCMs in the context of building environment conditioning is relatively new, and it may help the reader to innovate further in this area if we review some of the approaches taken to storing heat or 'coolth' using PCMs. Following this, details of a system developed during the past 5 years in the United Kingdom will be given.

7.2.3 Tunnel structures and earth as a heat 'sink'

The London underground railway system – the 'tube' – was designed and largely constructed in the Victorian era. Tunnels in many cases are deep and small in diameter, passenger loads are increasing and modern air-conditioning systems are rarely used – in fact air conditioning underground can of course lead to local heat gains, depending upon the location of the condenser. The New York City Transit Authority calculated that the operation of underground railway systems can generate sufficient heat to raise the tunnel and station temperatures by 8–11 K above the ambient. In London, where ambients can reach 30 °C or more, temperatures of over 37 °C have been recorded on some trains, making passenger comfort difficult to achieve.

London South Bank University (LSBU) [15, 16] has identified heat pipes as one option for removing heat from the tunnels. The tunnel structure and the surrounding earth tend to have a moderating influence on the underground railway air temperatures, taking in or rejecting heat, depending upon the air temperatures in the tunnels. This is called the 'tunnel heat sink effect' and ways being investigated by the LSBU team of enhancing this effect include the use of heat pipes.

Heat pipes can enhance the tunnel heat sink effect by modifying the thermal conductivity of the ground surrounding the tunnel – analogous to the effect on sensible heat stores of other types. Figure 7.4 shows the effect of soil thermal

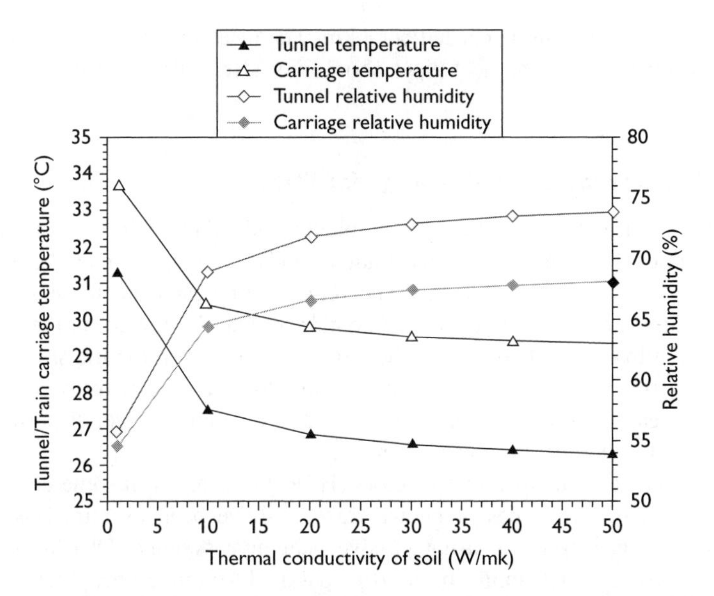

Fig. 7.4 The effect of soil thermal conductivity on tunnel and train carriage conditions [16].

conductivity on the conditions within the tunnel and the underground trains. It can be seen that if the thermal conductivity of the ground can be modified, by an order of magnitude, the temperatures in the tunnel and carriages can be lowered by 12 per cent. The number of heat pipes needed per kilometre of tunnel length can be obtained with reference to Fig. 7.5. This suggests that 2500 units of 130 W heat

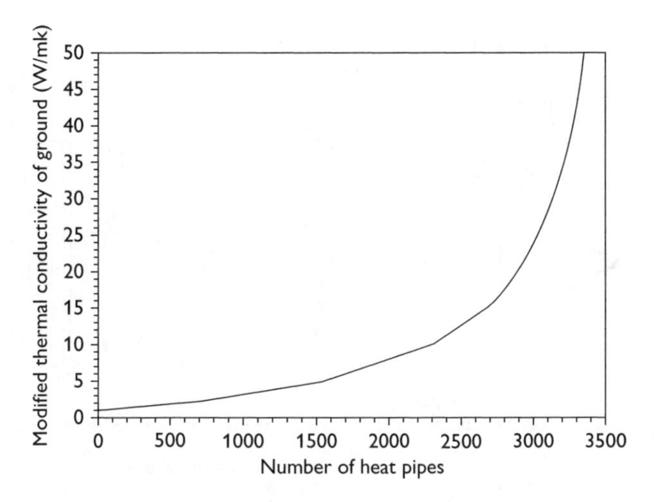

Fig. 7.5 The effect of the number of heat pipes on soil thermal conductivity. The heat pipes are rated at 130 W and the data relate to a 1 km length of 'tube' [16].

transfer capability would be needed per 1 km. The research team points out that the fitting and operation of the heat pipes should not affect the integrity of the tunnel structure.

7.2.4 Nuclear reactors and storage facilities

Radioactive waste from nuclear power plants and other facilities may generate modest amounts of heat over very long periods. The waste may be stored in containers or under water in tanks. The removal of heat from these stores may be done using 'active' methods, or the containers may be finned on their external surfaces, to allow natural convection heat transfer. However, the transfer of heat to the walls of the container suffers from the same thermal conductivity problems as in other stores, and the insertion of heat pipes can improve overall performance, minimising store temperature excursions.

The modular high temperature reactors (HTRs) of interest in some countries, as being safer than other types of nuclear reactor, have been examined as suitable candidates for heat removal using passive heat pipe systems. Ohashi et al. [17] examined how a VCHP might be used to passively remove decay heat from the core, should a depressurisation of the primary system occur or forced circulation coolant fail. The work involved modelling and experimental verification of the performance of the VCHP.

The concept is illustrated in Fig. 7.6 and shows a rather unusual 'loop' VCHP with a large condenser section and a separate condensate return pipe connected to

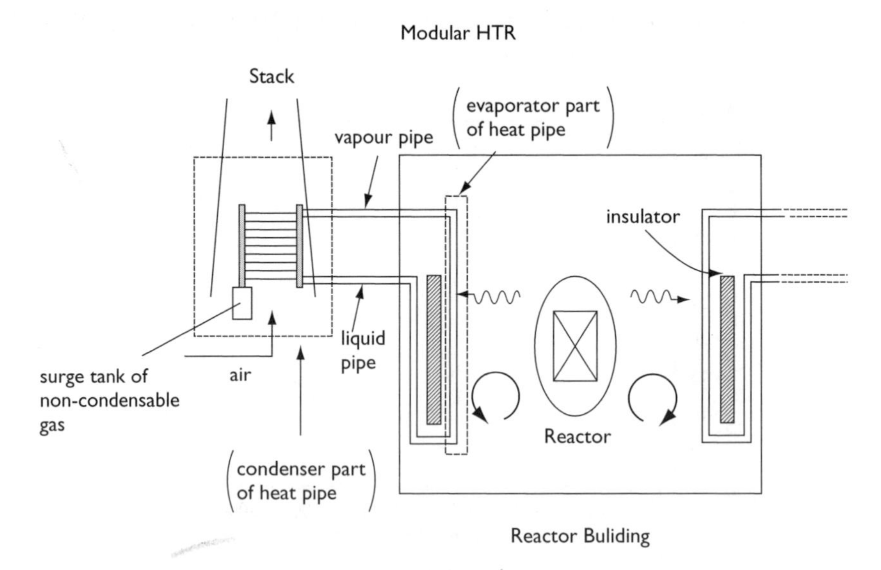

Fig. 7.6 Concept of VCHP decay heat removal system for the modular high temperature reactor [17].

the evaporator. It was calculated that a system using heat pipes could maintain the core at a temperature no higher than 1399 °C and the reactor vessel at a maximum of 415 °C, well below the safety limits. For the size of reactor considered, the heat removal rate was 560 kW.

7.2.5 Heat pipes in phase change stores (using PCMs)

The use of PCMs, like single-phase storage media, is beset by problems with poor thermal conductivity and unique freezing and melting profiles. Some laboratories have used metallic foils and compact heat exchanger structures to enhance the heat transfer in PCMs. A logical development is the introduction of heat pipes into PCMs. This section discusses some aspects of this and early proposals.

In the 1980s, Lee and Wu [10] reported on the heat transfer within PCMs when heat pipes were inserted in them. This research, at the University of Ottawa, was based upon paraffin wax as the PCM (the wax variant being Sun P-116). A water thermosyphon was used, its dimensions being 22.2 mm internal diameter and 610 mm length. It was concluded that an increase in the total heat transfer coefficient of the system was observed for the following changes in parameters:

- an increase in superheat of the PCM (allowing the PCM to exceed the saturation temperature)
- an increase in heat transfer coefficients at the thermosyphon evaporator and condenser sections (to be expected).

A review of the patent literature reveals further studies. US Patent 5386701, assigned to Cao Yiding (USA) and published in 1995, describes a system for human body cooling using a suit with heat pipes in it, these being connected to a PCM module. The heat pipes are flexible. With an aerospace theme, US Patent 4673030 published in 1987 and assigned to the Hughes Aircraft Company covers a thermal diode-type heat pipe allowing heat to be taken from electronic equipment into a PCM (*n*-heptadecane is proposed for 20 °C operation). A more recent reference, describing a study of solidification of a PCM in the presence of a heat pipe [11], considers the tools needed to predict the solidification process in the PCM when a finned heat pipe is used to extract the energy. The authors claim that the major advantage of heat pipes in PCM thermal stores occurs when simultaneously charging and discharging stores. Such a situation is unlikely to occur in practice with applications for buildings – most stores are charged during one period and discharged later. For building applications, the merits of heat pipes lie elsewhere.

7.2.5.1 The heat pipe in a passive cooling system for relieving air-conditioning loads

A system based upon the use of heat pipes to aid heat transfer into and out of PCMs (from and to, respectively, ambient air) has been developed over a number of years at Nottingham University in the United Kingdom by a team led by Dr David Etheridge.

The operation of this system, perfected in conjunction with PCM and heat pipe suppliers and an installer, is comparatively straightforward. During the night, cool air is used to 'freeze' the PCM, and during the day heat is extracted from the room air, which 'melts' the PCM. This cycle is repeated on a daily basis. The crucial process is the transfer of heat between the air and the PCM. Heat transfer coefficients need to be high, because the temperature differences between the air and the PCM are low, typically less than 6 °C.

As highlighted above, the main problem is achieving sufficient heat transfer into (and out of) a PCM, because the material essentially behaves as a solid and conduction is the only transfer mechanism – in common with most sensible heat stores. Moreover, direct contact between the air and the PCM is undesirable, because of odours and perceived health hazards, and the possibility of this happening should be minimised. The approach adopted was to use a heat pipe to provide an indirect but effective heat transfer path between the air and the PCM, with forced convection on the airside. This allows the PCM to be encased in a rigid sealed container.

Thus, a single module consists of a container of PCM into which one half of the heat pipe is embedded [12, 13]. The other half of the pipe is exposed to the air. Both halves of the pipe are equipped with finned heat exchangers. The direction of heat flow changes from day to night, so the heat pipe is designed for reversible operation and is mounted horizontally.

Hydrated Glauber's salt was the basic material, with borax as an additive to obtain the required transition temperature range (nominally 21–23 °C). The latent heat capacity is 198 kJ/kg and the density is 1480 kg/m^3. The PCM volume of the present module is 7.8 l, which corresponds to a latent heat storage capacity of 0.64 kWh per module.

The modules are installed in a floor-standing unit that is suitable for installation in both new and existing buildings. Seven modules were used in each unit, giving a latent cooling capacity of 4.4 kWh, e.g. 500 W of cooling for 8 h. The unit also houses the fan. Figure 7.7 shows the installation at one of the two test sites. The fan is mounted centrally and pulls air across the heat pipes. At night, the air is drawn through a duct that is connected to a motorised window vent. During the day, the vent is closed and air is drawn directly from the room through the open flap (also motorised). In a permanent installation, the vent would be installed in the wall rather than the window, to give a more compact installation. Both the vent and the flap were under automatic control [14].

7.2.5.2 Field trials of the cooling unit

Field tests were carried out in offices at two sites, the EcoHouse in the University of Nottingham – where heat-pipe-based solar collectors are also used, and the Pinxton premises of Building Product Design Ltd, who were also partners in the project. At

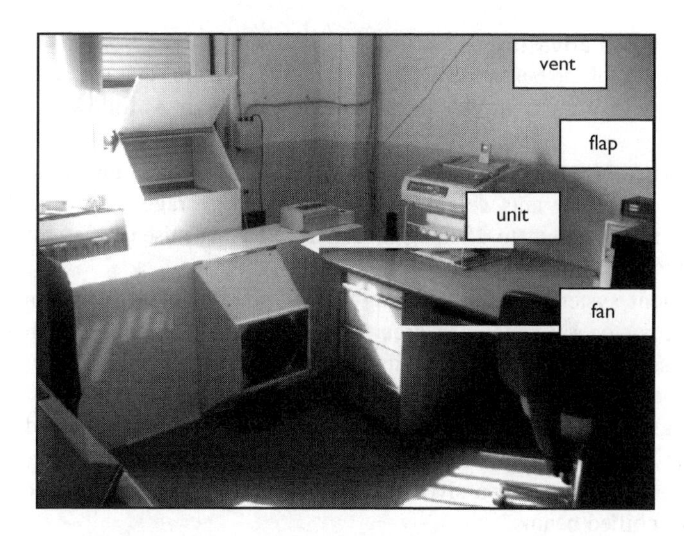

Fig. 7.7 System installation of the PCM–heat pipe cooler (Courtesy of the University of Nottingham).

the university, an unoccupied office (volume $16.4\,m^3$) hosted the unit. The effect of occupation was simulated using controllable heat sources. These provided a known and variable heat input and allowed the unit to be assessed under fairly extreme conditions (heat gains $>50\,W/m^2$). The test room at Pinxton was an occupied office (volume $26.3\,m^3$) and offered a true working environment. It has a south-facing window and is known to suffer from overheating in the summer months. The room had one occupant, one set of computer equipment, an overhead fluorescent light and three 60 W spotlights.

It was found that the system was capable of maintaining control over the room temperature and could respond quickly to changes in heat gains. In particular, the high thermal storage capacity of the PCM and the efficient heat transfer of the system allowed the room temperature to be controlled at a constant level in the same way as an air-conditioning system. This was recognised as being very impressive for what is basically a passive system. With forced nighttime cooling alone, the room temperature would have continued to rise resulting in a maximum temperature 2 °C or 3 °C higher. With only natural nighttime cooling of the room fabric, the rise would almost certainly have been even larger. These results therefore demonstrate not only that the system works in the manner intended but also that it offers significant benefits over other 'passive' systems.

The occupied office at Pinxton provided the opportunity to obtain the subjective opinion of the occupant. Although definitive conclusions should not of course be drawn from a sample size of one, the views expressed were favourable. The occupant felt that it improved the general freshness of his office and kept the temperature within a comfortable range.

7.2.5.3 System advantages

The advantages of the heat-pipe–PCM system over systems that use additional mass (e.g. concrete beams) to provide sensible heat storage are as follows:

- The PCM stores energy in the form of latent heat, which is much more efficient in terms of volume and weight than sensible heat storage. For example, 100 kg of PCM is equivalent to 2500 kg of concrete beams raised by a temperature of 2 K (a factor of 25).
- The present system is floor-standing and is more suitable for retrofitting. The potential for energy saving is greater for existing buildings than it is for new buildings, simply because of their greater numbers. The market is also much greater, leading to economies of scale.
- The system is modular and a specified cooling rate can be achieved simply by selecting the number of modular units. A small number of units gives a performance corresponding to passive concrete beams and a larger number corresponds to active chilled beams.
- The heat pipe allows the PCM to be used in a way that minimises the risk of contact with the room air arising from any leakage. In adopting this approach, it is important to ensure adequate heat transfer from the air to the PCM. The system enhances heat transfer in two ways. A heat pipe transfers heat into or out of the PCM and a fan enhances heat transfer between the air and the heat pipe.
- The success of the system is demonstrated by the fact that it offers control over the air temperature to a degree that is comparable to an air-conditioning unit. This fast response allows a control system to minimise the energy consumption of the fan, which is important in any system that aims to reduce carbon emissions (a fact that is sometimes overlooked).
- The fan also enhances the night cooling of the fabric. However, the results in the test rooms indicate that fabric cooling alone is relatively ineffective.

The industrial partners in the field test project were Building Product Design Ltd, Environmental Process Systems Ltd, Thermacore Europe and David Reay & Associates, and in 2005 a further award was made to allow the development of the system to continue.

7.3 HEAT PIPES IN CHEMICAL REACTORS

The selection of chemical reactors as a heat pipe application area allows us to highlight the benefits of the heat pipe based on isothermalisation/temperature flattening device and on a highly effective heat transfer unit. The VCHP can also be effective in reactor temperature control – an area yet to be fully exploited. (See Section 7.2.4 for reference to the use of the VCHP in nuclear reactors.)

Heat pipe technology offers a number of potential benefits to reactor performance and operation. While there remain some technical difficulties in a number of applications, in particular at very high temperatures, and examples of their use are few, there has in the last few years been increased interest in achieving uniformity of reactor temperature, and thus good product quality. The reason for this is that increased yield of high purity, high added value chemicals (such as pharmaceutical products) means less waste and higher profitability. Some conventional reactors, such as the 'stirred pot', can be characterised by highly nonuniform temperature distributions.

One of the first references relevant to heat pipe use in reactors was in the United States [18]. Although not directly employing heat pipes, the proposal was for a tube wall reactor in which a Raney catalyst is sprayed directly onto the heat transfer surface, and the energy of the reaction is transferred directly to boiling thermal fluid through the tube wall. This system was operated successfully.

One view of a heat pipe linked to a reactor [19] was to put a catalyst onto the surface of a heat pipe (or other heat exchanger) and carry out the reaction thereon. This was studied in the context of a catalytic incinerator, in which the catalyst was put on the tubes of a waste heat boiler, combining two unit operations (catalytic incinerator and waste heat boiler). Callinan and Burford [20] suggested that surface heat transfer coefficients on a finned tube coated with a catalyst could be of the order of 900 $W/m^2 K$, over an order of magnitude greater than a forced convection coefficient.

The catalyst need not, of course, be applied directly onto the heat pipe in order to give benefits. Reay and Ramshaw [19] listed the following positive effects the heat pipe might have on catalytic behaviour:

- Isothermal operation where the heat pipe would be inherently safe for the removal of reaction hotspots (and coldspots), and, as here, for uniform heat input and isothermalisation.
- Capability to deal with high heat fluxes, both radial and axial. Liquid metal heat pipes can be used in high-temperature chemical reactors.
- Heat pipes have much faster response time to a thermal input than, say, a solid conductor. They can therefore be used to speed up the heating of a catalyst support structure (and the catalyst) easing light-off in combustion, as an example.
- A range of heat pipe working fluids can be selected to cover all temperatures envisaged in catalysis. In Chapter 5, a sodium heat pipe destined for catalytic reforming plant is illustrated.
- Variable conductance heat pipes (see Chapter 6) can be designed to maintain the temperature of catalysts near or on their surface at a constant temperature, independent of heat generation rates. This opportunity for reactor control could be highly important in temperature critical reactions.
- Heat pipes have already been used to even out temperature excursions in reactors, thereby increasing yield and hence reactor efficiency.
- Heat pipe surfaces have been plasma-sprayed with materials such as alumina and are therefore readily treatable as catalyst support members.

Another area where heat pipes have been actively investigated is in adsorption reactions. These have been strongly promoted during the last 20 years for heat pumping and refrigeration duties and commonly use activated carbon as the adsorber bed and ammonia (or methanol) as the working fluid. As with all solid–gas reactions where a 'bed' is used, regeneration is generally a slow process due to the poor thermal conductivity of the bed. Metal inserts can be incorporated into the solid reactor to assist heating and/or cooling of the bulk material. A more effective way is believed to be to use variants of heat pipes – effectively thermal superconductors – to improve the heat transfer in solid–gas systems. Professor Robert Critoph, at Warwick University in the United Kingdom, is one of the principal exponents of enhanced adsorption cycle performance [21]. One concept is illustrated in Figs 7.8–7.10. The idea has also been applied in China to an adsorption cycle ice-making machine for fishing boats Ref [22], and [28] describes research, also in China, which has examined the use of loop heat pipes (see Chapter 6) in a silica gel–water adsorption chiller.

Foster Wheeler in the United States patented the use of heat pipes in a reforming reactor. US Patent 4315893 was filed on 17 December 1980 and is associated with the utilisation of the combustion gases from a gas turbine to provide a heat input to the reactor. Applications cited include the production of methanol, ammonia, as feed to a Fischer–Tropsch-type reactor and/or fuel cell gas.

The US heat pipe manufacturer Dynatherm has described an 'isothermal processing vessel' that can be used as a chemical reactor. It is claimed that hot or

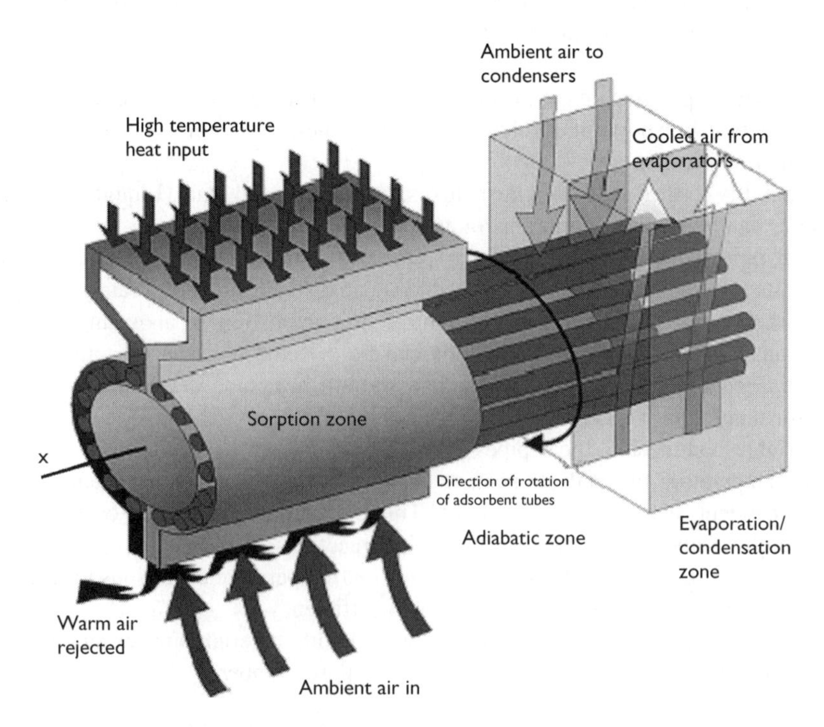

Fig. 7.8 The adsorption machine using heat-pipe-type tubular modules, developed at Warwick University [21].

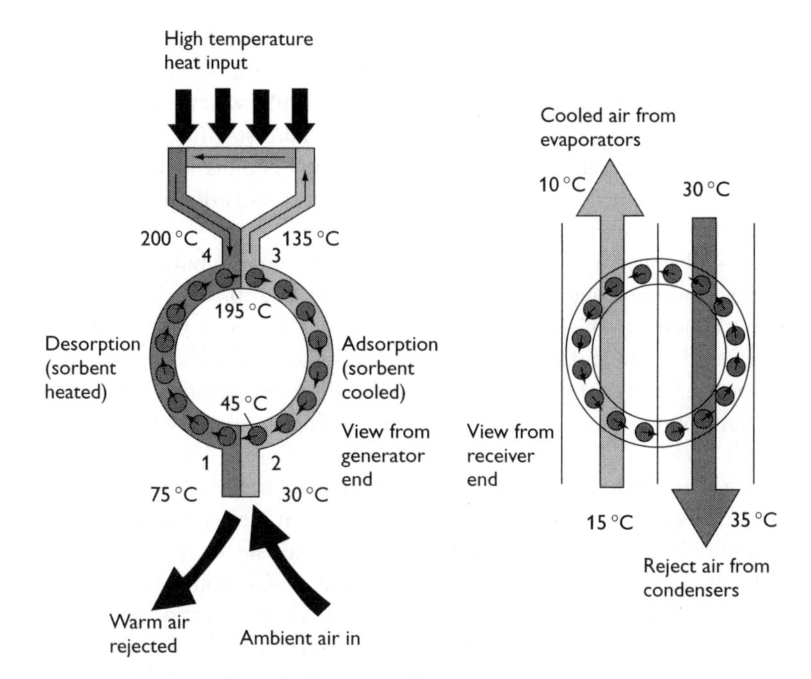

Fig. 7.9 Operation of the Warwick machine [21].

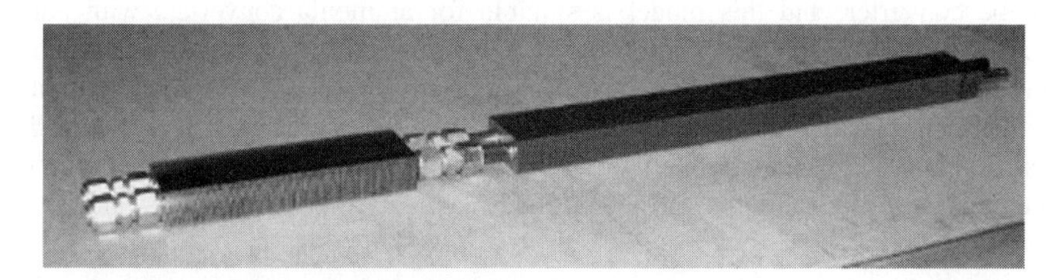

Fig. 7.10 A single module, the evaporation/condensation zone being on the RHS [21].

cold zones on reactor walls are eliminated. The reactor design is based upon the isothermal furnace muffles developed in the 1970s by several makers of liquid metal working fluid heat pipes. These were annular heat pipes, the inside of the chamber formed by the annulus being an excellent temperature-controlled furnace muffle, which can also be used for growing crystals under precise conditions.

The isothermal processing vessel uses the same configuration and it is claimed that large vessels with heat pipe walls can maintain temperatures within 0.5 °C. Size and configurations can range from small diameter long tube-flow reactors to large diameter batch or stirred tank reactors. Dimensions range from 2.5 cm to 1 m diameter and 30 cm to 3 m length.

Richardson et al. at the University of Houston [23] employed sodium heat pipes in their prototype steam reformer. Some years prior to this, another US research team studied the use of an annular heat pipe to cool a tube wall catalytic reactor [24]. It was found that the heat pipe gave better temperature stability, greater flexibility and increased productivity compared to conventional cooling methods. The impact of the more uniform heat load, caused by the inherent redistribution of heat created by the annular unit, was particularly beneficial.

Work is still underway on heat pipe reactors in China (China originating some of the early work on reactor isothermalisation using heat pipes). The research team at Nanjing University of Technology [25] believes that heat pipes can contribute towards raising the output and yield of reactors. The university has been aiming at the development of a heat pipe chemical reactor 'infinitely approaching' the optimal reaction temperature under high-temperature and high-pressure conditions (480 °C, 32 MPa).

Research was conducted on an ammonia converter used in fertiliser production. The aim is to optimise the temperature distribution in the reactor bed and increase the net ammonia yield without increasing the flow resistance.

The evaporator of a loop heat pipe is placed in the converter to extract the heat used to produce steam. This allows control of the reaction temperature, so it approaches the most suitable value, thereby increasing the net ammonia yield. Specifically, the structure is divided into a finite section model and an infinite approaching model, with the operation curves as shown in Fig. 7.11. With the so-called finite section model, there are a finite and fixed number of catalyst layers in the converter, and this model is suitable for ammonia converters with adiabatic reaction and indirect heat exchange between sections as well as axial–radial ammonia converters. The infinite approaching model is suitable for axial ammonia converters, in which the loop heat pipes are placed in the catalyst layers, and the reaction heat is extracted while the reaction is going on, so that the temperature of the catalyst layers can approach the optimal reaction temperature curve.

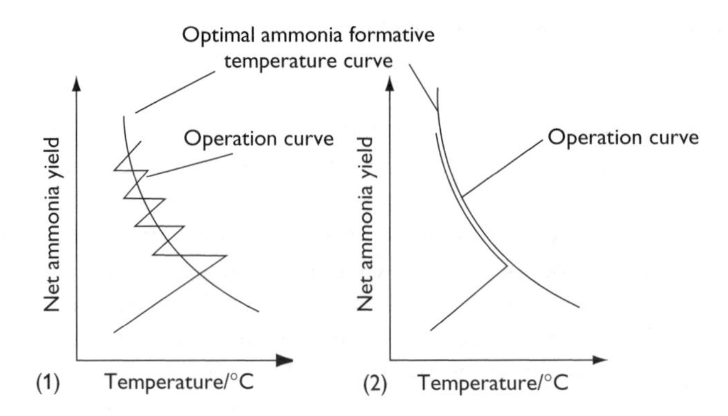

Fig. 7.11 Heat pipe reactor operation curves (China).

Simulation software calculations revealed that the net ammonia yield is over 19 per cent in all cases, 6 per cent higher than the present values in ammonia converters (about 13 per cent), leading the researchers to recommend hardware development of the heat pipe ammonia converter.

The trend towards increasing the use of renewable energy has not completely bypassed the chemical reactor area. A study in Morocco [26] examined linking a parabolic solar concentrator to a dehydrogenation reactor using a heat pipe.

It was determined at an early stage that for an effective reaction, good heat transfer was necessary, hence the interest in heat pipes. A bundle of heat pipes, in a heat exchanger, would be used in practice, as illustrated in Fig. 7.12c. The heat is taken from the parabolic collector as in Fig. 7.12d. With an operating temperature range that suggested mercury as the heat pipe working fluid (up to 730 K), and a duty of 20 kW, it was concluded that an efficiency of 86 per cent was feasible, the solar collector providing sufficient energy to allow the reaction to proceed effectively.

The chemical heat pipe, described in the earlier editions of *Heat Pipes*, is worthy of mention here. The term was introduced by the US General Electric in the mid-1970s to describe a chemical pipeline system for heat distribution [27]. The chemical heat pipe links endothermic and exothermic reactions, but a pump is used to transfer the chemicals along the two separate pipelines.

7.4 SPACECRAFT

Heat pipes, certainly at vapour temperatures up to 200 °C, have probably gained more from the developments associated with spacecraft applications than from any other area. The VCHP is a prime example of this 'technological fallout'.

The investment in spacecraft technology has moved onto the worldwide stage in recent years, with China, Japan, India and Brazil [39] investing in rocket and satellite technologies, in addition to the established countries and consortia such as the European Space Agency. This has led to duplication of effort, rather than great innovation, thus much of the data in earlier editions of *Heat Pipes* remain valid, in part as useful background information and also as a tutorial for those with a new interest in spacecraft thermal control. However, the loop heat pipe, some cryogenic developments and the demands of micro- and nanosatellites will ensure that there is enough to interest inquisitive researchers in satellite thermal control. The demands of

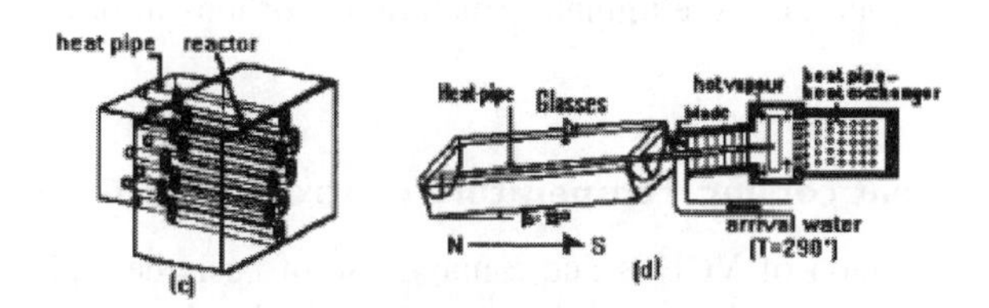

Fig. 7.12 Illustrations of components of the solar-powered dehydrogenation reactor [26].

electronics cooling can be greater in some avionics uses, compared to spacecraft use, but in programmes such as those of the US Air Force and its space-related activities, [40] high heat fluxes of over $100\,W/cm^2$ are projected (see also Chapter 8).

However, as discussed in Chapter 8 and highlighted by Swanson and Birur of NASA and JPL, respectively [1], heat pipes cannot satisfy all thermal requirements in spacecraft. The use of mechanical pumps in systems for moving liquid around is one option. While conventional heat pipes are limited in their performance, loop types have given heat pipes in spacecraft a new lease of life. The next development may well include active as well as passive components.

7.4.1 Spacecraft temperature equalisation

Temperature equalisation in spacecraft, whereby thermal gradients in the structure can be minimised to reduce the effects of external heating, such as solar radiation, and internal heat generation by electronics components or nuclear power supplies, has been discussed by Savage [47] in a paper reviewing several potential applications of the basic heat pipe. The use of a heat pipe connecting two vapour chambers on opposite faces of a satellite is analysed with proposals for reducing the temperature differences between a solar cell array facing the sun and the cold satellite face. If the solar array was two-sided, Savage proposed to mount the cells on a vapour chamber, using one side for radiation cooling, in addition to connecting the cell vapour chamber to extra radiators via heat pipes. Katzoff [48] proposed as an alternative the conversion of the tubular structural members of a satellite body into heat pipes.

The use of heat pipes to improve the temperature uniformity of nonuniformly irradiated skin structures has been proposed by Thurman and Mei [49], who also discussed production of near-isothermal radiator structures utilising heat pipes to improve the efficiency of waste rejection, the heat originating in a reactor/thermionic converter system. Conway and Kelley [50] studied the feasibility of a continuous circular heat pipe having many combinations of evaporator and condenser surfaces. This was constructed in the form of a toroid with eight heat sources and eight heat sinks. They concluded that a continuous heat pipe properly integrated with a spacecraft can be a highly effective means for reducing temperature differentials.

Kirkpatrick and Marcus [51] proposed the use of heat pipes to implement structural isothermalisation in the National Space Observatory and the Space Shuttle. It is particularly important to eliminate structural distortions in orbiting astronomy experiments.

7.4.2 Component cooling, temperature control and radiator design

The widest application of VCHPs and a major use of basic heat pipe units are in the removal of heat from electronic components and other heat-generating devices on satellites.

The variable conductance pipe offers temperature control within narrow limits, in addition to the simple heat transport function performed by basic heat pipes. One special requirement cited by Savage [47] would arise when it was required to maintain a particular subsystem at a temperature lower than that of its immediate environment.

Katzoff [48] proposes covering a complete instrument with a wicking material and then sealing the instrument within the evaporator section of a heat pipe that transfers the heat to a radiator. He also suggests that the heat pipe would be the most useful in cases where high-intensity-localised cooling is required for only a very short period, where instruments may reach peak power loads for short parts of a much longer time cycle. Travelling wave tubes (TWTs) are subject to varying internal heat distribution profiles, most of the dissipation being required at the collector. One possible arrangement is shown in Fig. 7.13, studied by the authors some years ago.

An interesting form of heat pipe was developed by Basiulis [54]. The 'unidirectional' heat pipe permits heat flow in one direction but acts as a thermal insulator in the opposite direction. Multiple wicks are used to produce a dry evaporator by limiting fluid return in one direction. By having a greater number of wicks in the preferred active evaporator, some of which only extend over the evaporator length, any heat input at the other end of the pipe results in rapid wick dryout because a considerable portion of the condensate enters those wick sections that cannot feed liquid back to the unwanted heat input section. A dielectric heat pipe was used to remove heat from the TWT in this application, taking heat unidirectionally to external radiators.

One of the most comprehensive studies on using the VCHP concept for electronics temperature control has been carried out by Kirkpatrick and Marcus [51]. They designed and manufactured a variable conductance pipe, referred to as the Ames Heat Pipe Experiment (AHPE), in which the heat pipe provided temperature

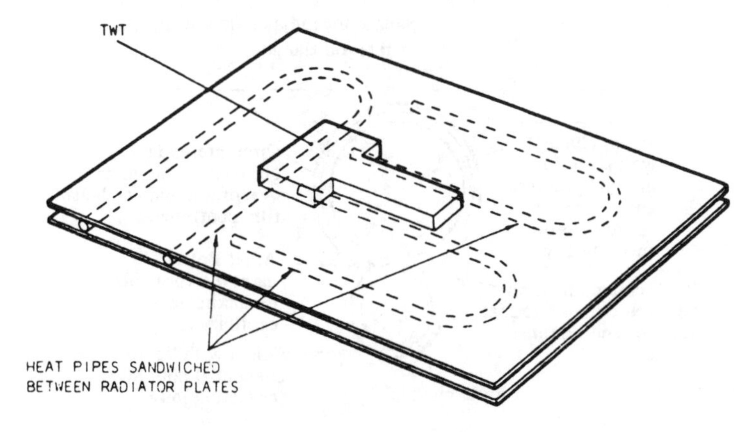

Fig. 7.13 Heat pipes for cooling a TWT mounted on a radiator plate.

stability for an onboard processor (OBP), by maintaining the OBP platform/AHPE interface at $17 \pm 3\,°C$. Power dissipation from the electronics processor varied between 10 and 30 W.

Grumman Aerospace [52] investigated other potential spacecraft applications of heat pipes. This included their use to control the liquid temperature in a closed circuit environmental control system, using water as the coolant. The maximum load to be dissipated was 3.82 kW. A second application to thermal control of equipment was in heat dissipation from a 1-kW power converter module. Heat pipes were brazed to the rectifier mounting plates. This resulted in a 15 per cent weight saving and a near-isothermal interface mounting plate. A required dissipation rate of 77 W was achieved.

Kirkpatrick and Marcus quote some very interesting figures on the power densities encountered in current and proposed systems. For example, Apollo power densities were typically 3 W/linear inch. Present state-of-the-art electronics for the International Space Station (ISS) and Shuttle are of the order of 30 W/linear inch, an order of higher magnitude. Typically, a fluid-(liquid)-cooled cold rail should handle the power densities appropriate to Apollo, but to meet the current requirements, an augmented heat pipe cold rail could cope with 60 W/linear inch. In the Space Shuttle, it was proposed that heat pipes be used to aid heat transfer from the lubricant fluid used for the auxiliary power unit.

A number of high-performance heat pipe configurations dedicated to spacecraft applications have been developed. These include the monogroove heat pipe, illustrated in Fig. 7.14, which has essentially separate vapour and liquid flow passages, and the advanced trapezoidal axially grooved heat pipe. The monogroove unit has given horizontal heat transport capabilities (ammonia at 21 °C) of 3300 W-M, while the related tapered artery unit can have a performance of 5200 W-M. The trapezoidal groove unit is designed for 30 000 W-M at 20 °C in a horizontal mode, again using ammonia. The diameter is approximately 50 mm [53].

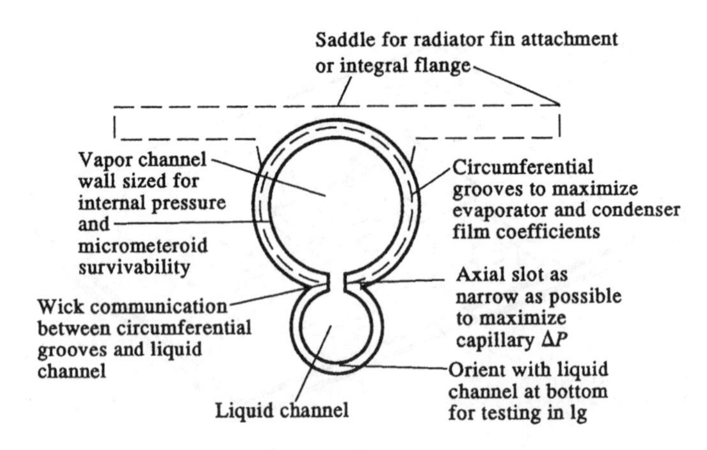

Fig. 7.14 Monogroove heat pipe configuration.

Turner [55], discussing applications for the VCHP, stressed the important advantage that these (and other heat pipes) offer in permitting direct thermal coupling of internal spacecraft components to radiators. While with RCA, he developed a VCHP capable of managing power inputs varying between 1 and 65 W. This advantage offered by heat pipes is also mentioned by Edelstein and Hembach [56]. Many current electronic packages, when integrated, rely solely on thermal radiation between the heat source (package) and the heat sink (space) using an intermediate radiating skin that need not have the components to be cooled in contact with it. Because of the fixed nature of the thermal coupling, wide ranges of equipment heat generation and/or environment heat loads will result in large temperature variations. Turner proposed the VCHP for directly linking the sources of heat to the radiator plate, which would, in turn, be made isothermal.

Scollon [57], working at General Electric, built a full-scale thermal model of a spacecraft and applied heat pipe technology to many of the thermal problems. He also selected the Earth Viewing Module of a communications and navigation satellite as one potential area for the application of these devices. The high solar heat loading on the east and west panels of the satellite necessitated the use of superinsulation to maintain these surfaces below the specified upper temperature limits. The north and south faces were the primary areas for heat rejection. By a system of internal and circumferential heat pipes, thermal control within the specification was achieved, dealing with 380 W of internal dissipation and 170 W of absorbed solar energy.

The ISS and modern communications satellites as well as prestigious projects, such as the Hubble Space Telescope, all have challenging thermal control problems. Companies such as Swales Aerospace in the United States and Alcatel Space in France, not to mention Russian organisations such as the Lavochkin Association, are all active in spacecraft thermal control using heat pipes.

Swales Aerospace has manufactured heat pipe radiators for the ISS, the unit shown in Fig. 7.15, having a plan size of 6.8×1.6 m.

The work of Swales Aerospace in integrating loop heat pipes with spacecraft and other 'objects in space' is worthy of mention. Illustrated in Fig. 7.16, the loop heat pipe can incorporate a flexible section, making deployment easier once in orbit. The Swales loop heat pipes were first demonstrated some 8 years ago in Shuttle flights STS-83 and STS-94.

The Hubble Space Telescope, one of the most successful and most internationally recognised and useful scientific devices in space, uses Swales heat pipes of the form shown in Fig. 7.17.

As pointed out by Swanson and Birur [1], thermal control requirements will be determined, if they are not already, by the needs of dimensional stability of large structures and thermal control of detectors at low cryogenic temperatures. Loop heat pipes will have a role to play here, as might nonheat pipe technologies.

Spacecraft is an established area of heat pipe use. As with terrestrial electronics thermal control, the increasing heat fluxes and absolute heat transport duties may increasingly challenge heat pipe capabilities. The heat pipe community is addressing

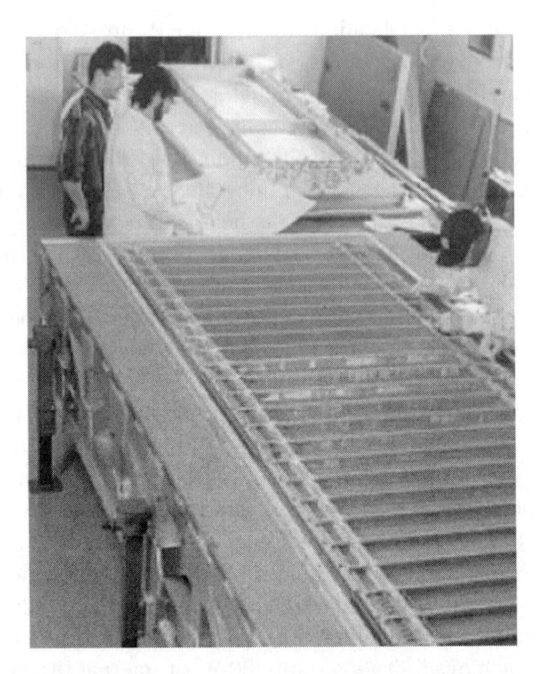

Fig. 7.15 ISS heat pipe radiator (Courtesy Swales Aerospace).

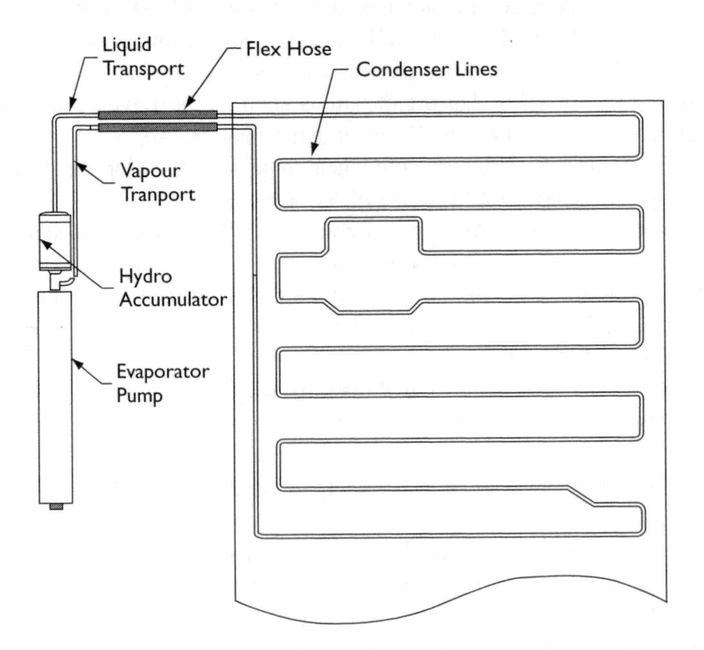

Fig. 7.16 Swales Aerospace loop heat pipe with flexible section (Courtesy Swales Aerospace).

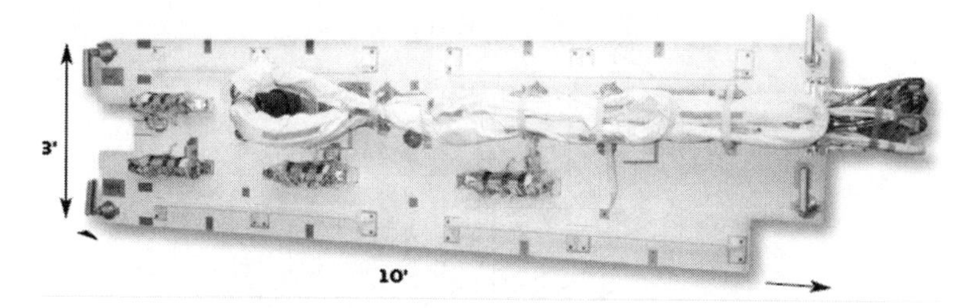

Fig. 7.17 Hubble Space Telescope Radiator with four capillary pumped loops and long flexible transport lines (Courtesy Swales Aerospace).

this, but it will result in opportunities for other thermal control systems, perhaps of a more 'active' nature.

7.5 ENERGY CONSERVATION AND RENEWABLE ENERGY

It is interesting to note that the preceding statement remains as valid today as it was over 10 years ago. Energy prices are even higher now, and the case for energy efficiency is even greater. The use of heat pipes in energy-efficient systems has not been fully exploited in the intervening period. This is due in part to the perceived high cost of some heat pipes (sometimes a realistic concern when looking at large heat recovery systems), a lack of appreciation of their capabilities and limitations and an aversion to risk-taking by industry and potential users.

A gas–gas heat pipe exchanger consists of a bundle of externally finned tubes that are made up as individual heat pipes of the type described above. The heat pipe evaporation/condensation cycle affects the transfer of heat from the 'evaporators', located in the duct carrying the countercurrent gas stream from which heat is required to be recovered, to the 'condensers' in the adjacent duct carrying the air which is to be preheated, as illustrated in Fig. 7.18.

Figure 7.19 shows units of this type constructed in the United Kingdom.

In the heat pipe heat exchanger, the tube bundle may be horizontal or vertical with the evaporator sections below the condensers. The angle of the heat pipes may be adjusted 'in situ' as a means of controlling the heat transport. This is a useful feature in air-conditioning applications.

Features of heat pipe heat exchangers which are attractive in industrial heat recovery applications are

(i) No moving parts and no external power requirements, implying high reliability.
(ii) Cross-contamination is totally eliminated because of a solid wall between the hot and cold gas streams.
(iii) A wide variety of sizes are available, and the unit is in general compact and suitable for all except the highest temperature applications.

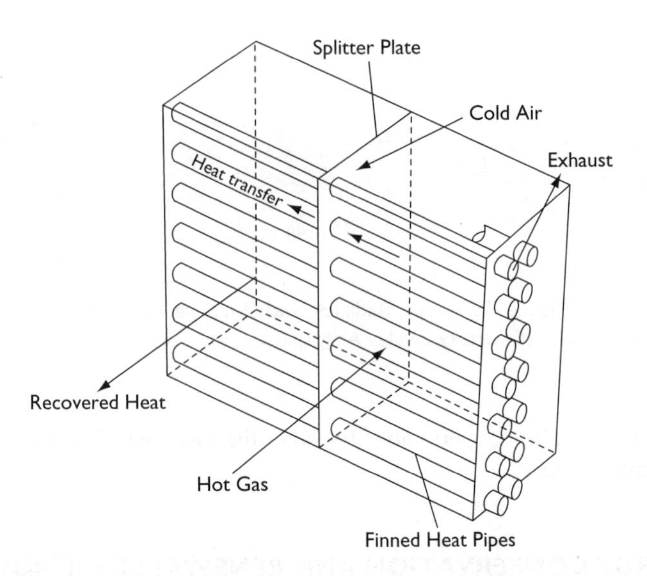

Fig. 7.18 Layout of a heat pipe heat exchanger showing means of heat transfer.

(iv) The heat pipe heat exchanger is fully reversible, i.e. heat can be transferred in either direction.
(v) Collection of condensate in the exhaust gases can be arranged, and the flexibility accruing to the use of a number of different fin spacings can permit easy cleaning if required.

The application of these devices fall into three main categories:

(i) Heat recovery in air-conditioning systems, normally involving comparatively low temperatures and duties, but including evaporative cooling, see Fig. 7.20.
(ii) Recovery of heat from a process exhaust stream to preheat air for space heating.
(iii) Recovery of waste heat from processes for reuse in the process, e.g. preheating of combustion air. This area of application is the most diverse and can involve a wide range of temperatures and duties.

The materials and working fluids used in the heat pipe heat recovery unit depend to a large extent on the operating temperature range, and as far as external tube surface and fins are concerned, on the contamination in the environment in which the unit is to operate. The working fluids for air conditioning and other applications where operating temperatures are unlikely to exceed 40 °C include HFCs. Moving up the temperature range, water is the best fluid to use. For hot exhausts in furnaces and direct gas-fired air circuits, higher temperature organics can be used.

In most instances, the tube material is copper or aluminium, with the same material being used for the extended surfaces. Where contamination in the gas is

Fig. 7.19 A heat pipe exchanger manufactured in the United Kingdom (Courtesy Isoterix Ltd – now Thermacore Europe).

Fig. 7.20 A heat pipe heat exchanger with spray bars (visible on the RHS). These are used on the exhaust side for indirect evaporative cooling, in addition to sensible heat transfer (Courtesy Deschamps Technologies).

likely to be acidic, or at higher temperature where a more durable material may be required, stainless steel is generally selected.

The tube bundle may be made up using commercially available helically wound finned tubes, or may be constructed like a refrigeration coil, the tubes being expanded into plates forming a complete rectangular 'fin' running the depth of the heat exchanger. The latter technique is preferable from a cost point of view.

Unit size varies with the air flow, a velocity of about 2–4 m/s being generally acceptable to keep the pressure drop through the bundle to a reasonable level.

Small units having a face size of 0.3 m (height) × 0.6 m (length) are available. The largest single units have face areas of over 100 m^2, for flue gas desulphurisation applications.

7.5.1 Heat pipes in renewable energy systems

Conventional energy efficiency still holds many opportunities for the use of heat pipe technology. However, it is in the area of renewables that the concept of a passive heat transfer system, with features such as flux transformation, has caught the imagination of researchers around the world. Starting with solar collectors, moving to concentrators and passive heating and cooling systems for buildings, the incorporation of turbines is now introducing an 'active' element into heat pipes and thermosyphons in renewable energy generation. A few examples are briefly discussed in the next sections.

7.5.1.1 The heat pipe turbine – power from low grade heat

The heat pipe turbine, sometimes called the thermosyphon Rankine engine, has been studied for power generation using solar, geothermal or other available low-grade heat sources. It is a heat pipe or thermosyphon, modified to incorporate a turbine in the adiabatic region. The basic configuration is a closed vertical cylinder functioning as an evaporator, an insulated section and a condenser. In the configuration described and developed by Johnson and Akbarzadeh [29], the turbine is placed in the upper end between the insulated section and the condenser section, and a plate is installed to separate the high-pressure region from the low-pressure region in the condenser. Conversion of enthalpy to kinetic energy is achieved through the nozzles. The mechanical energy developed by the turbine can be converted to electrical energy by direct coupling to an electrical generator. A schematic of one variant [30] is shown in Fig. 7.21.

The heat pipe turbine (Fig. 7.22) is able to convert high-velocity vapour energy circulated in the heat pipe to mechanical or electrical energy. The heat pipe turbine is not equipped with a pump and compression work is generated by gravity.

The fifth prototype of the unit developed by Akbarzadeh and colleagues in Australia and Japan was designed for 100 kW thermal input at 55 °C at the evaporator and 25 °C at the condenser. The specification was determined by the characteristics of a geothermal borehole at Portland, Victoria, Australia. The electrical output anticipated was about 3 kW, and it is intended that this will be the basis of a full-scale installation at Portland. Another unit is proposed for Kyushu in Japan (see Fig. 7.23), where the temperature of the geothermal water is substantially higher than that at the Australian site.

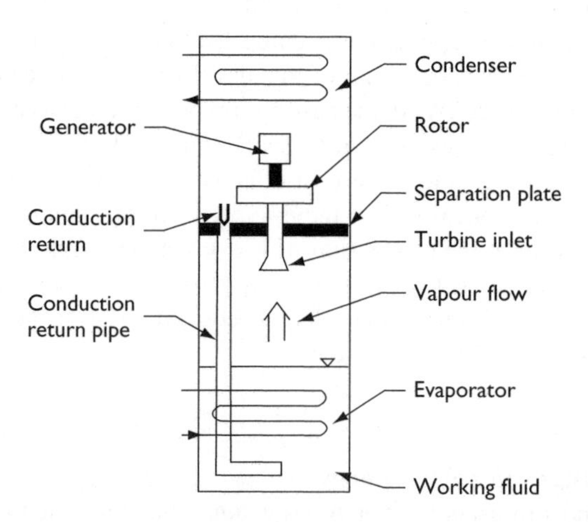

Fig. 7.21 The thermosyphon Rankine engine developed by Akbarzadeh et al.[30].

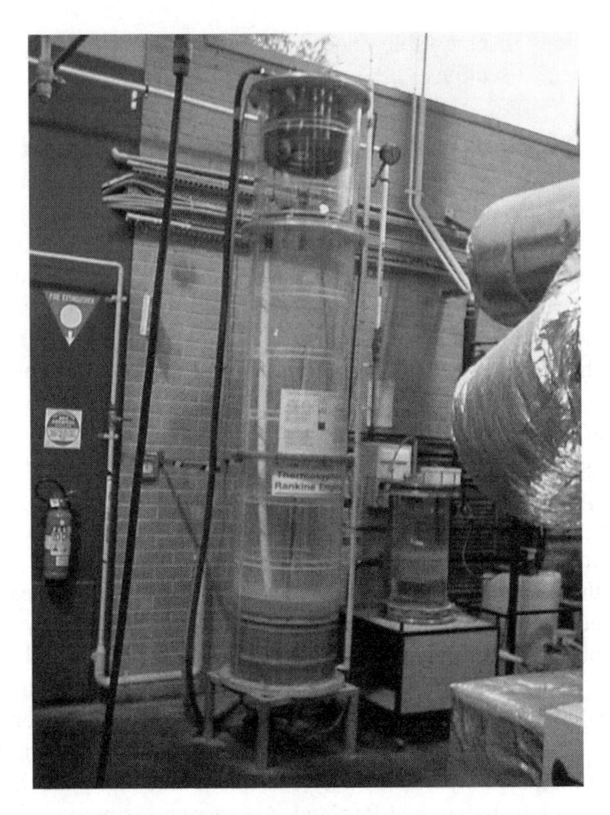

Fig. 7.22 Heat pipe turbine in the laboratory (Courtesy Prof. A. Akbarzadeh).

There is scope for further improvement and development of the heat pipe turbine: for example, integrating the generator and nozzle arrangement, both to decrease cost and to reduce friction. A chosen aim of this development is to keep the cost of the engine below A $2000 per kW. In the new prototype, this aim has not been achieved, and future research work must look into cheaper materials and manufacture to make the heat pipe Rankine engine more cost-effective. The simplicity of the machine and also the fact that it has very few moving parts offers potential for the heat pipe turbine to play a role in the conversion of low-grade heat from renewable sources into power. The heat pipe turbine will, if it proves to be cost-effective, find its application fields in hybrid thermal electric converters using geothermal energy, solar heat and waste heat as a heat source.

7.5.1.2 Heat pipes in solar energy

There are a number of users for heat pipes in solar-energy-related fields. A range of commercial manufacturers now market heat-pipe-based solar water heaters that use

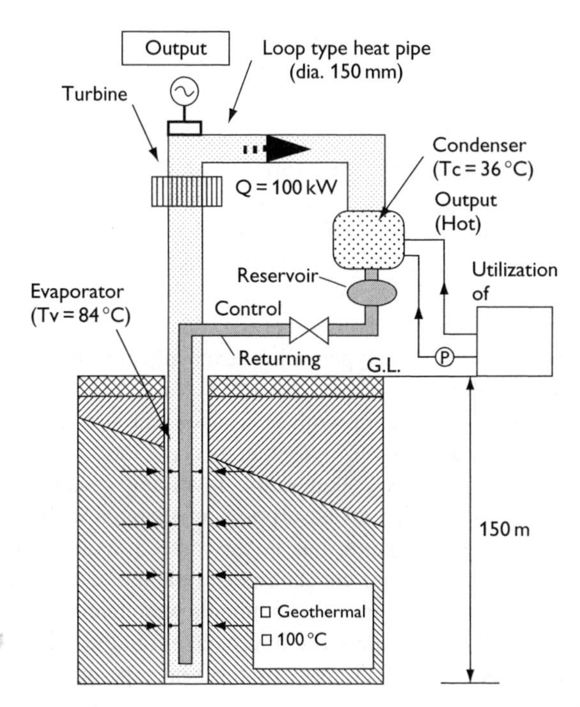

Fig. 7.23 Schematic of a loop heat pipe turbine assembly for power generation at Kyushu.

the heat pipes to transfer heat from the collector to the water store or circulation system. These are well-documented elsewhere. However, capillary pumped two-phase loops (loop heat pipes) have been investigated more recently by a number of workers for use in solar collectors. Work in Brazil [33] in conjunction with flat plate solar collectors showed that the capillary pumped loop could be used effectively to overcome the limitations of conventional solar collector systems. These are the need to have an electric pump to circulate the fluid, unless a gravity-fed system for natural circulation can be suitably positioned. Thus, the concept could be applied in rural areas where grid connections for electricity did not exist.

At the same laboratory as the development of the heat pipe turbine, research has been undertaken using thermosyphons to remove heat from solar photovoltaic cells. (Cooling of these cells can benefit conversion efficiency.) As well as cooling the cells, the recovered heat may be used for domestic water heating, for example, [31].

In Taiwan (as has been done in France some decades ago), heat pipes have been used in conjunction with heat pump water heaters in order to boost the performance of heat pump system. It is widely recognised that heat pumps can be inefficient energy deliverers when using ambient energy sources, in particular air. This is because if heating is required and the air temperature drops, for example, in the evening, the compressor of the heat pump has to do more work increasing the electricity (or gas) demand. This lowers the 'coefficient of performance' (COP) – the ratio of useful energy output to the energy needed to power the system.

The work in Taiwan [32] showed that by using the solar-heated water using the heat pipe water heater, the COP of the hybrid heat pipe/heat pump system could be improved by almost 30 per cent to 3.32, compared to the system when operating without the warmed water source. As the heat source temperature increases, it becomes economical to operate in the heat-pump-only mode.

Similar or greater improvements to heat pump performance in air–air systems can be achieved where heat recovery is used between the exhaust air and the incoming air, thus preheating the latter. The types of heat exchanger discussed in Section 7.5.1 would be used for this duty.

Renewable energy in conjunction with heat pipes can also be used to boost the performance of other thermodynamic cycles, in this case the adsorption refrigeration cycle (see also Section 7.3). A collaboration involving researchers in Belgium, Spain and Morocco [34] studied an adsorption refrigerator with activated carbon and ammonia as the working pairs, but in this case a stainless steel/water heat pipe was used to connect the solar concentrator (a parabolic solar collector) to the adsorption cycle generator (reactor). (No comments were made on the possibility of inert gas generation in the heat pipe.) The improvement in the heat transfer between the collector and the generator facilitated by the heat pipe has encouraged further research on the system.

Energy efficiency and renewable energy are areas of great importance. Heat pipes have a role to play in both conventional and RE technologies, in order to improve efficiencies. Cost-effectiveness is important, however, and in some applications, such as heat recovery and solar collectors, there have to be good reasons for using heat pipe technology.

7.6 PRESERVATION OF PERMAFROST

One of the largest contracts for heat pipes[1] was placed with McDonnell Douglas Corporation by Alyeska Pipeline Service Company for nearly 100 000 heat pipes for the Trans-Alaska pipeline. The value of the contract was approximately $13 000 000 and is of course now complete.

The function of these units is to prevent thawing of the permafrost around the pipe supports for elevated sections of the pipeline. Diameters of the heat pipes used are 5 and 7.5 cm and lengths vary between 9 and 18 m.

Several constructional problems are created when foundations have to be on or in a permafrost region. Frost heaving can cause differential upward motion of piles resulting in severe structural distortions. On the other hand, if the foundations are in soil overlying permafrost, downward movements can occur.

[1] Now dwarfed in size (in terms of number of units) by computer chip cooling uses – see Chapter 8.

If the thermal equilibrium is distorted by operation of the pipeline, thawing of the permafrost could occur, with the active layer (depth of annual thaw) increasing each summer until a new thermal equilibrium is achieved. This will of course affect the strength of the ground and the integrity of any local foundation structure. Insulation, ventilation and refrigeration systems have already been used to maintain permafrost around foundations. The use of heat pipes and thermosyphons has also been studied by several laboratories [58, 59].

The system developed by McDonnell Douglas [58] uses ammonia as the working fluid, heat from the ground being transmitted upwards to a radiator located above the ground level. Several forms of 'Cryo-anchor' have been tested and two types are illustrated in Fig. 7.24. As the heat pipe is very long and is operating in the vertical position, the only means for transferring heat from the atmosphere into the ground will be conduction along the solid wall.

Results have shown that rapid soil cooling occurs in the autumn after installation and as air temperatures increased in the spring, Cryo-anchor cooling ceased. The temperature rose fairly rapidly as heat flowed from the surrounding less cold permafrost, with large radial temperature gradients. As the gradients became smaller, the temperature rise slowed through the summer months.

At the end of the thaw season, the permafrost remained almost 0.5 °C below its normal temperature. This cooling would permit a 10–12 m reduction in pile length required to support the pipe structure. The system is now installed on the pipeline.

In the context of the recent developments during the construction of a railway line between China and Tibet, where parts cross permafrost areas, Cheng [42] reviews previous work using thermosyphons for permafrost preservation. The first use, he states, of thermosyphons to maintain permafrost below railway embankments was in 1984, in Alaska. More recently, the Alaska Department of Transportation experimented with hairpin thermosyphons, which the author suggests are more suitable for roadway engineering than the simple vertical thermosyphons piloted

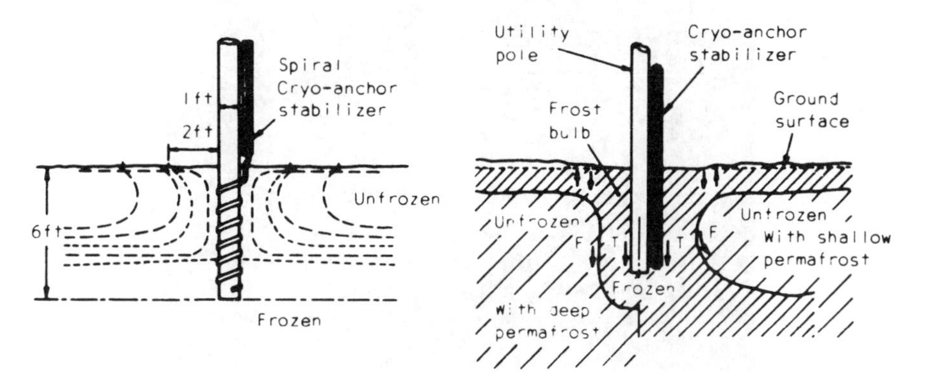

Fig. 7.24 Types of 'Cryo-anchor' for permafrost preservation (Courtesy McDonnell Douglas Corp.).

for the Tibet–Qinghai railway [43]. For those interested in the effect of global warming on permafrost, the Chinese study of vertical thermosyphons assumed an air temperature increase of 2 °C over the next 50 years.

The preservation of permafrost will become increasingly difficult as global warming progresses. Many transport structures and buildings rely upon permafrost foundations, and it is possible that there will be a growth of interest in thermosyphon and related technologies as means for preserving the status quo.

7.7 SNOW MELTING AND DEICING

An interesting area of application, and one in which work in Japan, Russia and the USA has been particularly intense, has been the use of heat pipes to melt snow and prevent icing. Work carried out by Mitsubishi Electric Corporation in collaboration with the Japanese Ministry of Construction on heat pipe deicing and snow melting has involved a large number of test sites [36].

The operating principle of the heat pipe snow melting (or deicing) system is based upon the use of heat stored in the ground as the heat input to the evaporators of the heat pipes. A typical test panel, illustrated during assembly in Fig. 7.25 would measure 20 m × 3.5 m, and contain 28 carbon steel/ammonia heat pipes, set 5 cm below the surface. The concrete panel, as shown in Fig. 7.26 was 25 cm thick. The sizes and location of the heat pipes were varied so that the effect of this variation could be studied and an optimum location selected.

The system was found to be capable of delivering 20–130 W/m^2, and heat extracted during winter operation using the ground source heat pipes was found to be regenerated the following October. Deicing of ships has also been carried out [37].

Fig. 7.25 Heat pipe snow melting panel during installation (Courtesy Mitsubishi Electric Company).

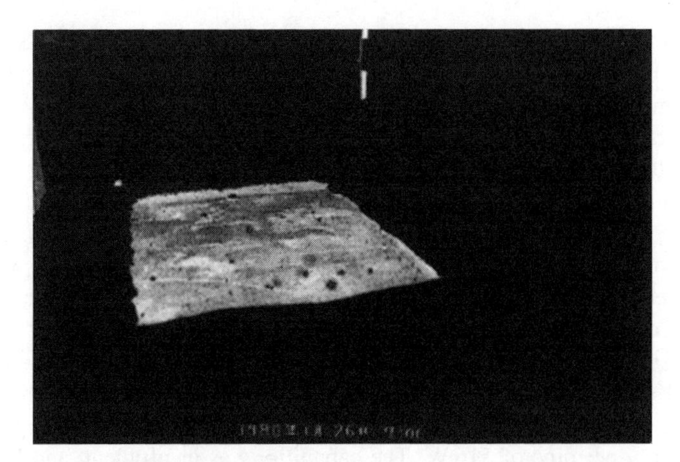

Fig. 7.26 The test panel during operational trials (Courtesy Mitsubishi Electric Company).

Interest in snow melting and deicing of pavements and roads has grown in the United States. An excellent review of systems, including the use of heat pipes, was carried out in 2000 at the Oregon Institute of Technology [35]. This covered uses of the various technologies in Japan, Argentina and Switzerland as well as in several US states. The author of the review points out that as well as the safety benefits of the passive melting systems, the elimination of snow and slush clearing reduced labour costs.

Not covered in depth in other references, the author cites the ASHRAE Handbook data (1995) for allowing calculation of the energy needs for snow melting. This is a function of several variables, notably snow fall rate, air temperature and relative humidity and the wind velocity. Lund, the report author, quoted the heat inputs required to melt snow in several states of the United States and for several classifications in terms of importance (residential being low, hospital entrances being high). The installation heat inputs varied from about $270\,W/m^2$ for low classification areas in less cold states (Portland, Oregon) to more than $1\,kW/m^2$ (Chicago) for area requiring maximum attention. The relatively high heat input needs for some areas suggest that passive systems may have difficulty in meeting the demand, unless very large 'heat flux transformation' is undertaken, with heat pipe evaporators being spread over massive underground areas.

This is borne out by the examples of early US installations, where the evaporator sections were typically 10 times longer than the condensers, the latter being below the pavement where melting is required. Satisfactory performance was achieved with one unit of 177 heat pipes with 30 m long evaporators warming almost $1000\,m^2$ of pavement. (In the case of the bridge heating system in Virginia, USA described below, a conventional propane boiler was used to provide an energy input to the heat pipe evaporators.)

Lund, in concluding that the two main systems for snow and ice melting of pavements are those based on heat pipes and the direct use of geothermal hot water, gives useful cost data. The geothermal systems costs typically $215/m^2 of pavement, plus the well and pumping system, while heat pipes work out at $375/m^2 installed on a highway bridge deck (see below).

It was in 1995 that the Virginia Department of Transportation in the United States started installing a heated bridge across the Buffalo River in Amherst County, as shown in Fig. 7.27. The cost of the bridge was $665 000, of which $181.5 was for the heating system that was based upon heat pipes [38]. The heating system, based upon a hybrid system involving a boiler unit heating a circulating ethylene glycol system, feeding heat to the evaporators of a heat pipe assembly that then warmed the road crossing the bridge, as illustrated in Fig. 7.28.

A total of 241 gravity-assisted heat pipes were used, of length 12 m and internal diameter 13 mm. Interestingly, a number of different working fluids were used. The pipes were initially charged with R123, an HCFC refrigerant, and this gave a heat transport duty per pipe of 516 W. The capability was insufficient, and higher heat transport (1050 W) per pipe was achieved with the HFC, R134a. Ethanol was also tried as a working fluid before ammonia was ultimately selected. Ammonia, unsurprisingly, boosted performance to 2.5 kW per pipe, at a retrofit cost of $18.73/m^2. Running costs for the gas were $18/h and maintenance $500/year.

The subsequent assessment of the performance of the ammonia system, reported during an evaluation in 2001 [46], concluded that while heat pipe technology could be effective in this type of application, a reliable bridge deck heating system was still 'work in progress'. Correct working fluid selection was important, and maintenance of the active mechanical parts of the system was time-consuming.

Fig. 7.27 The heat pipe heated bridge in Virginia, USA.

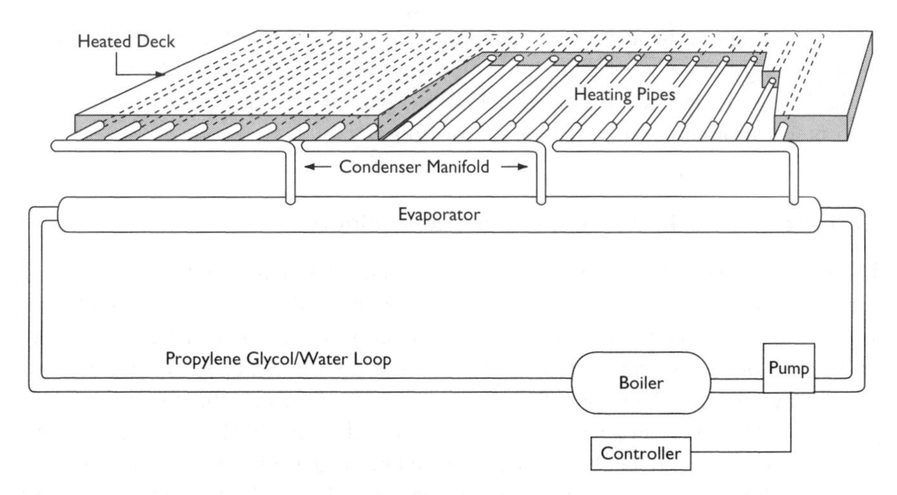

Fig. 7.28 The heating system, showing the boiler (heated by propane gas) supplying energy to the heat pipe system.

However, compared to an alternative hydronic heating system, the operating costs were almost halved.

The use of heat pipes or thermosyphons for deicing and snow melting has some economic benefits. Obviously, in order to use the full passive benefits, substantial evaporator sections need to be buried. However, the civil engineering can be reduced by introducing an active heating system, as was done in Virginia.

7.8 HEAT PIPES IN THE FOOD INDUSTRY

The food sector has been a major growth area in recent years. There are demands for improved energy efficiency in food cooling and refrigeration, more convenient and rapid ways for cooking and chilling food and increased interest in renewable energy for, for example, foodstuff drying and preservation.

The 'heat pipe cooking pin' pioneered by a company in the United States some 35 years ago, and also marketed in the United Kingdom, was perhaps the first large-scale domestic application of heat pipe technology (although Gaugler's patent – see Chapter 1 – described a refrigerator employing a heat pipe). In spite of being subject to long-term gas generation because of the container material, the concept was highly effective in allowing joints of meat to be cooked in a shorter time. The heat pipe was, of course, a highly effective meat skewer, and could lead to up to 50 per cent reductions in cooking time. One of the most effective claims was that one could, by sticking the 'cooking pin' only halfway into the joint, have well-done meat at one end and rare meat at the other!

For an unknown reason, its availability on the market seems to have waned. Perhaps the idea of putting a small pressure vessel into a hot oven caught the attention of safety authorities, or perhaps a chef of less intelligence than normal included it in a microwave oven when preparing a meal! Nevertheless, in these days of high-energy prices, a cooking pin comeback would be welcome! Some research in the United Kingdom, discussed below, suggests that at least at the commercial food preparation level, this may be a possibility.

7.8.1 Heat pipes in chilled food display cabinets

The modern retail cabinets used for chilling and displaying food relies solely on convective heat transfer and has changed little, since the first designs introduced around 1965. In a conventional multishelf supermarket cabinet, the food is stored on shelves or decks and is cooled by cold air supplied from within the cabinet. Axial flow fans circulate the cold air, and this is directed into the display area as a jet, which also forms an air curtain at the front to the cabinet (Fig. 7.29). The air is cooled by passing it over the refrigerant evaporator, the refrigerant being typically at −7 °C. Professor Maidment and colleagues, working at the London South Bank University [41, 60] found that gains from outside the cabinet, by radiation for example, were only a fraction of the energy delivered by the cabinet cooling system. This had only to maintain the food chilled, as it was normally delivered in a cooled state to the cabinets. The low heat transfer rate leads to excess energy

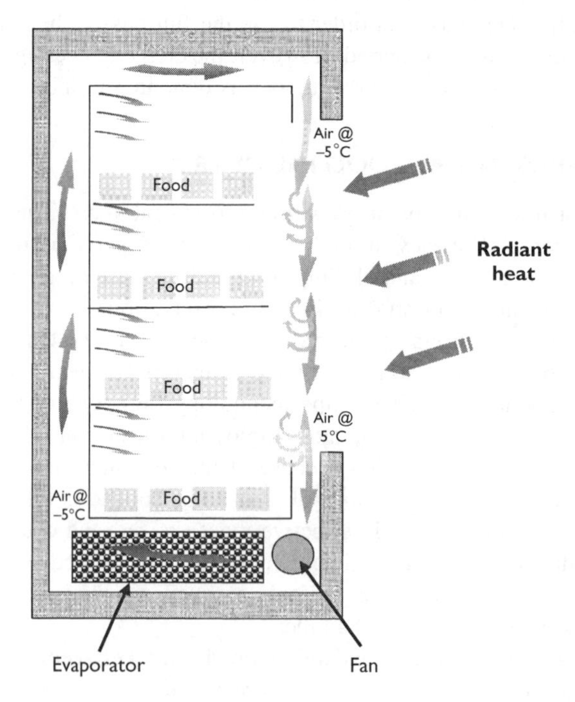

Fig. 7.29 Principle of operation of a conventional chilled food cabinet [60].

use, excessive equipment sizing and reduced food shelf life. Recorded temperature distributions showed wide variations within the cabinet.

Before moving to the heat pipe solution, it was shown that putting food in contact with a cooled shelf (part of the evaporator) in a delicatessen cabinet provided substantial additional cooling. This conductive heat transfer component makes such a significant contribution to cooling and food quality that overall energy use was much reduced. However, the solution would not be practical in most chilled cabinets, as adjustable shelving is used. The heat pipe solution was then proposed.

Illustrated in Fig. 7.30, the unit employs shelves that as well as displaying the food, take heat from the food to the cabinet cold air supply duct. The shelf is constructed as a heat pipe.

Modelling of the system allowed comparison of the conventional and heat-pipe-assisted solutions. The predicted data using heat pipe shelves suggested that food core temperatures 2.5–3.5 K less than those achievable conventionally would be possible. This could lead to energy and capital cost reductions as well as improvements in the quality of the food. Electric defrost might even be avoided.

7.8.2 Cooking, cooling and defrosting meat

Cooking, cooling and defrosting of meat is the major users of energy in the food sector, domestic, commercial and industrial, and improvements in the rate of these

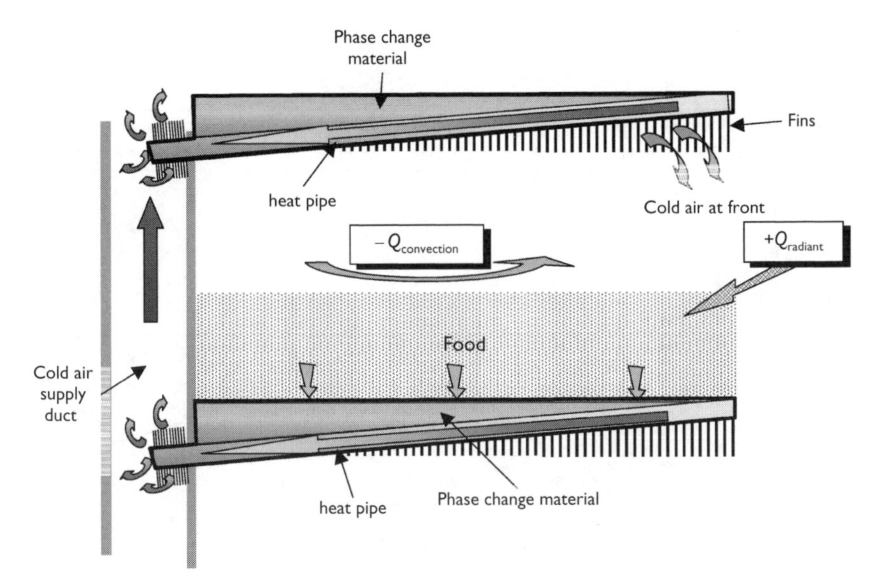

Fig. 7.30 The operation of heat pipe cooling shelves fitted to a supermarket food cabinet [60].

processes need a way of overcoming the poor thermal conductivity of the 'product'. While microwave ovens can be used in some domestic cooking/defrosting processes, on a larger scale the equipment is less popular and can have health hazards. The heat pipe, as a 'super thermal conductor', has supplanted the conventional meat skewer in some kitchens, as mentioned above, but a scientific analysis of the benefits across a range of food processing stages has, until recently, been lacking. This has been rectified by research at the Food Refrigeration and Process Engineering Centre (FRPERC) of Bristol University in the United Kingdom.

One of the first projects at FRPERC involved investigating freezing and defrosting of meat joints (beef topside). Two sets of experimental data are shown in Figs 7.31 and 7.32 for freezing and cooling, respectively. (Joints were typically 0.9–3 kg in weight.)

The graphs show data for two experiments. These include plots of the temperature profile using the heat pipe 'pin', a control case, where the temperature inside the joint was recorded without a heat pipe, and measurement of the ambient air temperature (in the freezer compartment in Fig. 7.31, and in ambient air for data in Fig. 7.32).

Much more data than that shown were obtained, and the researchers concluded that during the initial trials reductions of up to 42 per cent in freezing time and up to 54.5 per cent in thawing time were achieved. The heat pipe used in these cases was 600 mm long, 9.5 mm in diameter and used ethanol as the working fluid. A later series of tests using a smaller heat pipe (300 mm × 6.5 mm) gave average freezing and thawing time reductions of 18 per cent and 20 per cent, respectively.

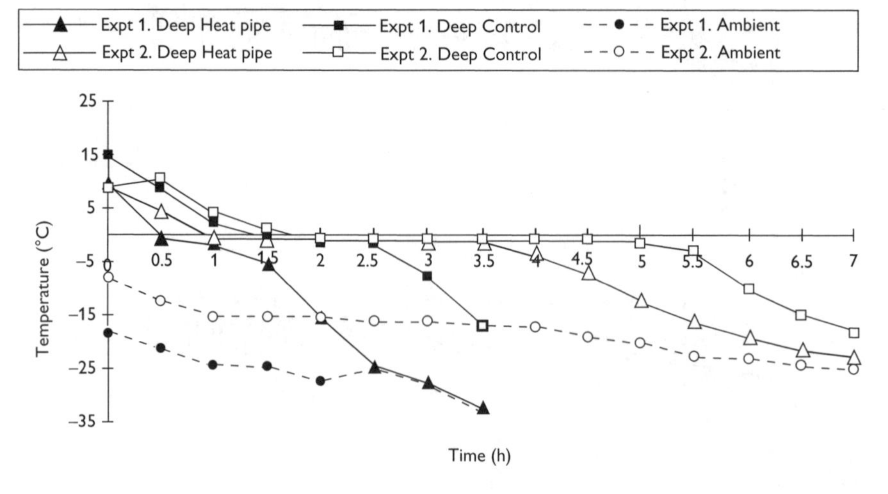

Fig. 7.31 Freezing curve for joint in an ambient in the freezing room reducing from −18 °C to −30 °C (Courtesy Food Refrigeration and Process Engineering Research Centre [45]).

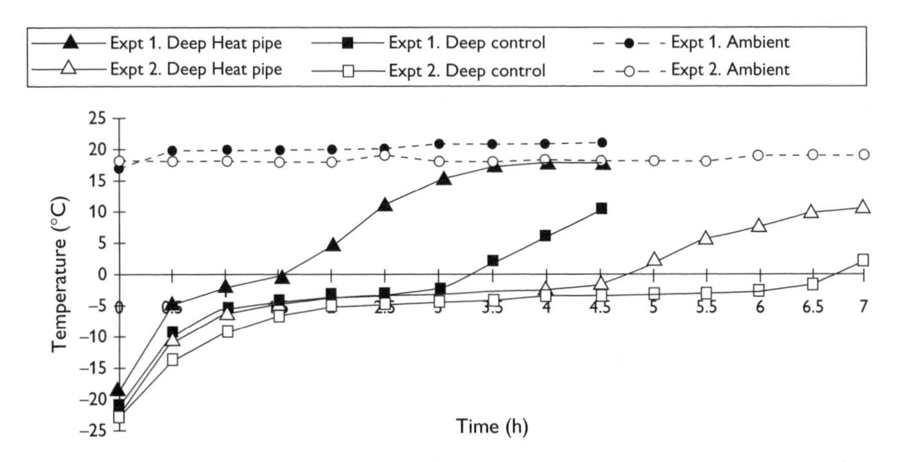

Fig. 7.32 Thawing curve for joint in an ambient rising from 17 °C to 21 °C (Courtesy Food Refrigeration and Process Engineering Research Centre [45]).

The authors of the report suggest that there are cost disadvantages in that the heat pipes have to be left in the joint after freezing, until it thaws, and some damage is done to the joint on insertion/removal. However, they point out that in specific cases, such as to meet European Union requirement on carcass chilling, they are likely to be cost-effective.

More recently the same team has reported on using heat pipes to enhance the cooking and cooling of joints of bacon [44]. In this series of experiments, heat pipes with greater diameters were used than in the earlier freezing work. Three sizes were tested, diameters varying between 12 and 19 mm, with the length constant at 330 mm. The heat pipes used methanol as the working fluid and were finned on the airside.

The results showed that the use of heat pipes in cooking and cooling could halve the cooking time and help ensure that the joints were cooked to a safe internal temperature. For cooling, the time was reduced by 25 per cent. The energy-saving implications of the reduced cooking time are important – a 37–43 per cent reduction in use being attained.

Food preservation by freezing or chilling and food preparation involving heating or cooling are major and growing uses of energy. The few examples mentioned above show how significant contributions to energy saving can be made by heat pipes. The field is wide open for further exploitation.

7.9 MISCELLANEOUS HEAT PIPE APPLICATIONS

To assist the reader in lateral thinking, a number of other applications of heat pipes are listed below.

Heat pipe roll-bond panels for warming bathroom floors (Japan).
Heat-pipe-cooled dipstick for cooling motor bike engine oil (Japan).

Passive cooling of remote weather station equipment (Canada).
Cooling of drills (Russia)
Thermal control of thermoelectric generators (USA).
Cooling of gas turbine blades (Czechoslovakia).
Cooling of semi-automatic welding equipment (Russia).
Cooling of soldering iron bit (UK).
Cooling of bearings for emergency feedwater pumps (UK).
Cooling of targets in particle accelerators (UK).
Thermal control in electric batteries.
Car passenger compartment heating.
Thermal control of injection moulding and diecasting processes.
Preservation of ice 'island' drilling platforms (Russia).

7.10 HEAT PIPE APPLICATIONS – BIBLIOGRAPHY

The range of heat pipe applications, both proposed and practised, is considerable. The reader is advised to consult the Bibliography – Appendix 3 – for additional references related to heat pipe (and thermosyphon) applications.

REFERENCES

[1] Swanson, T.D. and Birur, G.C. NASA thermal control technologies for robotic spacecraft. Appl. Therm. Eng., Vol. 23, pp 1055–1065, 2003.
[2] Dincer, I. and Rosen, M.A. Thermal Energy Storage — Systems and Applications. John Wiley, Chichester, 2001.
[3] Zalba, B., Marin, J.M., Cabeza, L.F. and Mehling, H. Review on thermal energy storage with phase change: materials, heat transfer analysis and applications. Appl. Therm. Eng., Vol. 23, pp 251–283, 2003.
[4] Azad, E. et al. Solar water heater using gravity-assisted heat pipe. Heat Recov. Syst. CHP, Vol. 7, No. 4, pp 343–350, 1987.
[5] Polasek, F. Heat pipe research and development in East European countries. Heat Recovery Syst CHP, Vol. 9, No. 1, pp 3–18, 1989.
[6] Vasiliev, L.L. Heat pipe research and development in the USSR. Heat Recovery Syst. CHP, Vol. 9, No. 4, pp 313–333, 1989.
[7] Caruso, A. et al. Heat pipe heat storage performance. Heat Recovery Syst. CHP, Vol. 9, No. 5, pp 407–410, 1989.
[8] Vasiliev, L.L., Boldak, I.M., Domorod, L.S., Rabetsky, M.I. and Schirokov, E.I. Experimental device for the residential heating with heat pipe and electric heat storage blocks. Heat Recovery Syst. CHP, Vol. 12, No. 1, pp 81–85, 1992.
[9] Sodha et al. Solar Passive Building: Science and Design. Pergamon, Oxford, 1986.
[10] Lee, Y. and Wu, C.-Z. Solidification heat transfer characteristics in presence of two-phase closed thermosyphons in latent heat energy storage systems. Proceedings of 6th International Heat Pipe Conference. 25–29 May, Grenoble, 1987.
[11] Horbaniuc, B., Dumitrascu, G. and Popescu, A. Mathematical models for the study of solidification within a longitudinally finned heat pipe latent heat thermal storage system. Energy Conver. Manage., Vol. 40, pp 1765–1774, 1999.

[12] Turnpenny, J.R., Etheridge, D.W. and Reay, D.A. Novel ventilation cooling system for reducing air conditioning in buildings: 1 Testing and theoretical modelling. Appl. Therm. Eng., Vol. 20, pp 1019–1038, 2000.

[13] Turnpenny, J.R., Etheridge, D.W. and Reay, D.A. Novel ventilation cooling system for reducing air conditioning in buildings: Part II: Testing of prototype. Appl. Therm. Eng., Vol. 22, pp 1203–1217, 2001.

[14] Etheridge, D., Murphy, K. and Reay, D.A. A novel ventilation cooling system for reducing air conditioning in buildings: Review of options and report on field tests. Building Serv. Eng. Res. Technol., Vol. 27, No. 1, pp 27-39, 2006.

[15] Ampofo, F., Maidment, G. and Missended, J. Underground railway environment in the UK Part 1: Review of thermal comfort. Appl. Therm. Eng., Vol. 24, pp 611–631, 2004.

[16] Ampofo, F., Maidment, G. and Missenden, J. Underground railway environment in the UK Part 3: Methods of delivering cooling. Appl. Therm. Eng., Vol. 24, pp 647–659, 2004.

[17] Ohashi, K., Hayakawa, H., Yamada, M., Hayashi, T. and Ishii, T. Preliminary study on the application of the heat pipe to the passive decay heat removal system of the modular HTR. Prog. Nucl. Energy, Vol. 32, No. 3–4, pp 587–594, 1998.

[18] Ralston, T.D. et al. Tube wall methanation reactors with combined diffusion and kinetic resistance. Bureau of Mines, US Department of Interior Report of Investigations 7941, 1974.

[19] Reay, D.A. and Ramshaw, C. An investigation into catalytic combustion. Report for ETSU, Rolls-Royce, NEL, ICI Chemicals and Polymers, EA Technology British Gas, IMI Marston and Heatric, August 1992.

[20] Callinan, J.P. and Burford, D.L. The analysis of finned catalytic heat exchangers. ASME Report 77-HT-67, ASME, New York, 1977.

[21] Critoph, R.E. Multiple bed regenerative adsorption cycle using the monolithic carbon-ammonia pair. Appl. Therm. Eng., Vol. 22, pp 667–677, 2002.

[22] Wang, L.W., Wang, R.Z., Wu, J.Y., Xia, Z.Z. and Wang, K. A new type adsorber for adsorption ice maker on fishing boats. Energy Conver. Manage., Vol. 46, pp 2301–2316, 2005.

[23] Richardson, J.T., Paripatyadar, S.A. and Shen, J.C. Dynamics of a sodium heat pipe reforming reactor. AIChEJ, Vol. 34, No. 5, pp 743–752, May 1988.

[24] Parent, Y.O., Caram, H.S. and Coughlin, R.W. Tube-wall catalytic reactor cooled by an annular heat pipe. AIChE J., Vol. 29, No. 3, pp 443–451, 1983.

[25] Zhang, H. and Zhuang, J. Research, development and industrial application of heat pipe technology. Proceedings of 12 International Heat Pipe Conference, Moscow, 2002. Appl. Therm. Eng., Vol. 23, pp 1067–1083, 2003.

[26] Aghbalou, F. et al. A parabolic solar collector heat pipe heat exchanger reactor assembly for cyclohexane's dehydrogenation: a simulation study. Renewable Energy, Vol. 14, No. 1–4, pp 61–67, 1998.

[27] Anon. Energy storage and transmission by chemical heat pipe. Heat Air Cond. J., pp 12, 14, 16, June 1979.

[28] Wang, D.C. et al. Study of a novel silica gel-water adsorption chiller. Part 1. Design and performance prediction. Int. J. Refrigeration, Vol. 28, pp 1073–1083, 2005.

[29] Johnson, P. and Akbarzadeh, A. Thermosyphon Rankine engine for solar energy and waste heat applications. Proceedings of 7th International Heat Pipe Conference, Minsk, USSR, 1990.

[30] Akbarzadeh, A. et al. Formulation and analysis of the heat pipe turbine for production of power from renewable sources. Appl. Therm. Eng., Vol. 21, pp 1551–1563, 2001.
[31] Akbarzadeh, A. and Wadowski, T. Heat pipe-based cooling systems for photovoltaic cells under concentrated solar radiation. Appl. Therm Eng., Vol. 16, No. 1, pp 81–87, 1996.
[32] Huang, B.J., Lee, J.P. and Chyng, J.P. Heat-pipe enhanced solar-assisted heat pump water heater. Solar Energy, Vol. 78, pp 375–381, 2005.
[33] Bazzo, E. and Nogoseke, M. Capillary pumping systems for solar heating applications. Appl. Therm Eng., Vol. 23, pp 1153–1165, 2003.
[34] Aghbalou, F. et al. Heat and mass transfer during adsorption of ammonia in a cylindrical adsorbent bed: Thermal performance study of a combined parabolic solar collector, water heat pipe and adsorber generator assembly. Appl. Therm. Eng., Vol. 24, pp 2537–2555, 2004.
[35] Lund, J.W. Pavement snow melting. Report of the Geo-Heat Center, Oregon Institute of Technology, Klamath Falls, Oregon, USA, 2001.
[36] Tanaka, O. et al. Snow melting using heat pipes. Advances in Heat Pipe Technology, Proceedings of IV International Heat Pipe Conference, Pergamon Press, Oxford, 1981.
[37] Matsuda, S. et al. Test of a horizontal heat pipe deicing panel for use on marine vessels.
[38] Hoppe, E.J. Evaluation of Virginia's first heated bridge. Report VTRC 01-R8, Virginia Transportation Research Council, December 2000.
[39] Bazzo, E. and Riehl, R.R. Operation characteristics of a small-scale capillary pumped loop. Appl. Therm. Eng., Vol. 23, pp 687–705, 2003.
[40] Lin, Lanchao, et al. High performance miniature heat pipe. Int. J. Heat Mass Transf., Vol. 45, pp 3131–3142, 2002.
[41] Maidment, G.G. et al. An investigation of a novel cooling system for chilled food display cabinets. Proceedings of IMechE Vol. 219, Part E: J. Process Mechanical Engineering, pp 157–165, 2005.
[42] Cheng, G. et al. A roadbed cooling approach for the construction of Qinghai-Tibet railway. Cold Reg. Sci. Technol., Vol. 42, pp 169–176, 2005.
[43] Zhi, W. et al. Analysis on effect of permafrost protection by two-phase closed thermosyphon and insulation jointly in permafrost regions. Cold Reg. Sci. Technol., Vol. 43, No. 3, pp 105–163, December 2005.
[44] James, C. et al. The heat pipe and its potential for enhancing the cooking and cooling of meat joints. Int. J. Food Sci. Technol., Vol. 40, pp 419–423, 2005.
[45] James, C. and James, S.J. The heat pipe and its potential for enhancing the freezing and thawing of meat in the catering industry. Int. J. Refrigeration, Vol. 22, pp 414–424, 1999.
[46] http://www.virginiadot.org/vtrc/briefs/01-r8rb/01-r8rb.htm (Evaluation of Virginia's first heated bridge).
[47] Savage, C.J. Heat pipes and vapour chambers for satellite thermal balance. RAE Tech. Report 69125, June 1969.
[48] Katzoff, S. Heat pipes and vapour chambers for thermal control of spacecraft. AIAA Paper 67–310, April 1967.
[49] Thurman, J.L. and Mei, S. Application of heat pipes to spacecraft thermal control problems. Tech. Note AST-275, Brown Engineering (Teledyne), July 1968.
[50] Conway, E.C. and Kelley, M.J. A continuous heat pipe for spacecraft thermal control. General Electric Space Systems, Pennsylvania (undated).
[51] Kirkpatrick, J.P. and Marcus, B.D. A variable conductance heat pipe experiment. AIAA Paper 71–411, 1971.

[52] Roukis, J. et al. Heat pipe applications for space vehicles. Grumman Aerospace, AIAA paper 71–412, 1971.

[53] Dobran, F. Heat pipe research and development in the Americas. Heat Recovery Systems & CHP, Vol. 9, No.1, pp 67–100, 1989.

[54] Basiulis, A. Uni-directional heat pipes to control TWT temperature in synchronous orbit. NASA Contract NAS-3-9710, Hughes Aircraft Co., California, 1969.

[55] Turner, R.C. The constant temperature heat pipe — a unique device for the thermal control of spacecraft components. AIAA 4th Thermo-physics Conference, Paper 69–632, (RCA), June 1969.

[56] Edelstein, F. and Hembach, R.J. Design, fabrication and testing of a variable conductance heat pipe for equipment thermal control. AIAA Paper 71–422, 1971.

[57] Scollon, T.R. Heat pipe energy distribution system for spacecraft thermal control. AIAA Paper 71–412, 1971.

[58] Waters, E.D. Arctic tundra kept frozen by heat pipes. Oil Gas J. (US), pp 122–125, August 26, 1974.

[59] Larkin, B.S. and Johnston, G.H. An experimental field study of the use of two-phase thermosyphons for the preservation of permafrost. Paper at 1973 Annual Congress of Engineering, Inst of Canada, Montreal, October 1973.

[60] Wang, F., Maidment, G.G., Missenden, J.F., Karayianns, T.G. and Bailey, C. A novel superconductive food display cabinet. Submitted to Int. J. Refrigeration, 2005.

8

COOLING OF ELECTRONIC COMPONENTS

In the previous editions of *Heat Pipes*, electronics cooling applications (and closely related discussions of heat pipes in electrical plant) have been considered as part of the wider application field. Now, however, with heat pipes in this application being orders of magnitude more numerous than elsewhere, and the challenges for the future of heat pipes and their derivatives here being considerable, a dedicated chapter is warranted.

As pointed out by Thermacore Inc. [1], the major manufacturer of heat pipes for electronics thermal control, the microelectronics, telecommunications, power electronics and to some extent the electrical power industries are constantly striving towards miniaturisation of devices that inevitably results in greater power densities. Therefore, there is a challenge to develop efficient management of heat removal from these high flux devices.

Current systems consist of rack-mounted units with total electronic chip powers of up to 120 W per printed circuit board (PCB) and total rack powers of approximately 5 kW. From the point of view of convenience and compatibility, it is desirable to maintain the standard sizes of casings and electronic connections in the back of each case. Therefore, more components are inserted onto the same size of PCB creating higher packing densities, facilitated by the miniaturisation of components.

However, this miniaturisation, together with increasing processing speeds decreases the heat transfer surface area and increases power. This generates very high heat fluxes resulting in large temperature rises. Therefore, to maintain the chips within operating conditions, more heat must be removed and traditional methods such as simple forced air convection become inadequate. Thermal management is now believed by many to be the limiting factor in the development of higher power electronic devices and new, preferably low-cost, methods of cooling are required. The nature of the heat concentration in very small volumes/areas of surface challenges innovative thermal engineers, with solutions such as jet impingement and microrefrigerators being considered.

There are some steps that can alleviate the thermal control problem. For example, the move from bipolar to complementary metal oxide semiconductor and the change in voltage from 5 V to a voltage lower than 1.5 V have extended the lifetimes of standard cooling systems, at least for low-power electronics (PC, phone chip, etc.). However, the International Technology Road map indicates no slowdown in the rate of increase in cooling demands for the next generations of chips. Predictions, made about 2 years ago, for 2006 are of peak heat fluxes around 160 W/cm^2.

Previously mentioned in Chapter 7 in the section on Spacecraft applications, the challenges of miniaturisation have similar implications. The current status in that sector is neatly summarised by Swanson and Birur at NASA [2] as: 'While advanced two-phase technology such as capillary pumped loops and loop heat pipes (LHPs) offer major advantages over traditional thermal control technology, it is clear that this technology alone will not meet the needs of all future scientific spacecraft ...'

The need to handle much higher heat fluxes has prompted much research in two-phase heat transfer, much of it directly or indirectly related to heat pipes. For example, Thermacore and its parent, Modine, are industrial supporters of a major UK-university-based project on boiling and condensation in microchannels that started at the end of 2005, involving the authors. This project, one of the largest in the area funded in Europe (>£1 million) is supported partially based on the demands of chip cooling.

Much of the research will be relevant to micro-heat pipes, because, as highlighted earlier, as power ratings are increased and sizes are decreased, multiphase heat transfer in channels of very small cross section becomes increasingly attractive, compared to single-phase cooling. Multiphase pressure drop imposes a constraint on the lower limit of cross-sectional dimension of $\sim$50 μm and microchannels in the range of practical interest of 50–1000 μm may either be components of heat pipe capillary structures, or at the extreme, the whole heat pipe vapour and liquid cross section. As an example, cooling by evaporation in microchannels embedded in a silicon chip is a strong contender for cooling new electronic devices at heat fluxes exceeding 100 W/cm^2 – and these microchannels may of course be an array of micro-heat pipes (see also Chapter 6).

In the interim, work by researchers at UES and the US Air Force Research Laboratory has reported heat flux capabilities greater than 140 W/cm^2 using novel capillary fin type structures [6]. (This approaches the 2006 figure mentioned above.) However, at high heat fluxes (some exceeding 280 W/cm^2), the temperature difference between the evaporator wall and the vapour temperature was rather high – around 70 °C.

For the reader who wishes to consult an overview of the many methods for managing heat in microelectronics, including systems that compete with heat pipes, the review by Garimella is interesting [7]. He covers microchannels, micropumps, jet impingement and phase change energy storage (like the systems in Chapter 7 but smaller) as well as micro-heat pipes. Ellsworthy and Simons of IBM Corporation, in *Electronics Cooling* in 2005 [13], suggest that alternative chip cooling methods, such as water-cooled cold plates and water jet impingement, can achieve cooling fluxes of 300–450 W/cm^2. While these are not needed yet, heat pipe researchers will be challenged to meet such requirements, should they arise.

8.1 FEATURES OF THE HEAT PIPE

As outlined at the beginning of Chapter 7, it is convenient to describe in broad terms how the heat pipe might be used in this area, based upon its main properties. Three are particularly important here

- Separation of heat source and sink
- Temperature flattening
- Temperature control

The system geometry may now be considered, and this may conveniently be divided into five major categories, each representing a different type of heat pipe system

- tubular (including individual micro-heat pipes)
- flat plate (including vapour chambers)
- micro-heat pipes and arrays
- LHPs
- direct contact systems

8.1.1 Tubular heat pipes

In its tubular form, with round, oval, rectangular or other cross sections, two prime functions of the heat pipe may be identified

- transfer of heat to a remote location
- production of a compact heat sink

It is possible to add heat flux transformation to this, although in electronics thermal management it would be difficult to find applications that do not involve taking a high heat flux and dissipating it in the form of a low heat flux, over a larger area.

By using the heat pipe as a heat transfer medium between two isolated locations (and as shown in Fig. 8.1, these may be isolated electrically as well as physically), recognisable applications become evident. It becomes possible to connect the heat pipe condenser to any of the following:

- a solid heat sink (including finned heat sinks, of course)
- immersion in another cooling medium

- a separate part of the component or component array
- another heat pipe
- the wall of the module containing the component(s) being cooled

8.1.1.1 Electrically isolated heat pipes

Heat pipes can be used for applications requiring electrical isolation. Figure 8.1 shows a heat pipe assembly that incorporates ceramic isolated inserts in the adiabatic section of the heat pipe. These inserts can hold off kilovolts of electricity. This

Fig. 8.1 An electrically isolated heat pipe assembly (Courtesy of Thermacore International Inc.).

particular assembly is used on high-speed trains to cool power electronics. The internal working fluid is a dielectric too.[1]

In applications where size and weight need to be kept a minimum (this applying to most applications in the twenty-first Century), the near-isothermal operation of the heat pipe may be used to raise the temperature of fins or other forms of extended surface. This leads to higher heat transfer to the ultimate heat sink (for example, air in the case of laptop computers), and the advantage may be used to uprate the device or reduce the size and weight of the metallic heat sink.

There are two possible ways of using the heat pipe here

- mount the component to be cooled directly onto the heat pipe;
- mount the component onto a plate into which the heat pipes are inserted.

8.1.2 Flat plate heat pipes

The second of the five categories of heat pipes likely to be the most useful in electronics cooling is the flat plate unit. It is often used for 'heat spreading' or 'temperature flattening', and is highly effective in this role. The applications of the flat plate unit include, but are not limited to

- multicomponent array temperature flattening;
- multicomponent array cooling;
- doubling as a wall of a module or mounting plate;
- single-component temperature flattening.

The ability of flat plate heat pipes (and vapour chambers) to double as structural components opens up many possibilities and of course not just related to electronics thermal management. The example shown in Fig. 8.2, although not strictly a flat plate heat pipe, rather heat pipes in a flat plate, is the first step towards producing the near-isothermal surface.

8.1.2.1 Embedded heat pipes

It is very common to take the standard heat pipes and embed them into the base of a heat sink as shown in Fig. 8.3. This allows the use of the high volume, lower cost heat pipes in a heat sink design. The heat pipes are flattened, soldered or epoxied into the heat sink and then machined to create a flat mounting surface to the electronics being cooled.

[1] Electrically insulated heat pipes have been examined in a number of electrical engineering areas too. One of the authors was involved many years ago with the A. Reyrolle Switchgear Company, testing heat pipes for use at up to 33 kV potential, examining the suitable dielectric fluids and wall materials. As in many application areas, including cooling buried cable junctions, judicious use of heat pipes to remove hotspots can allow upgrading of equipment at minimum extra cost.

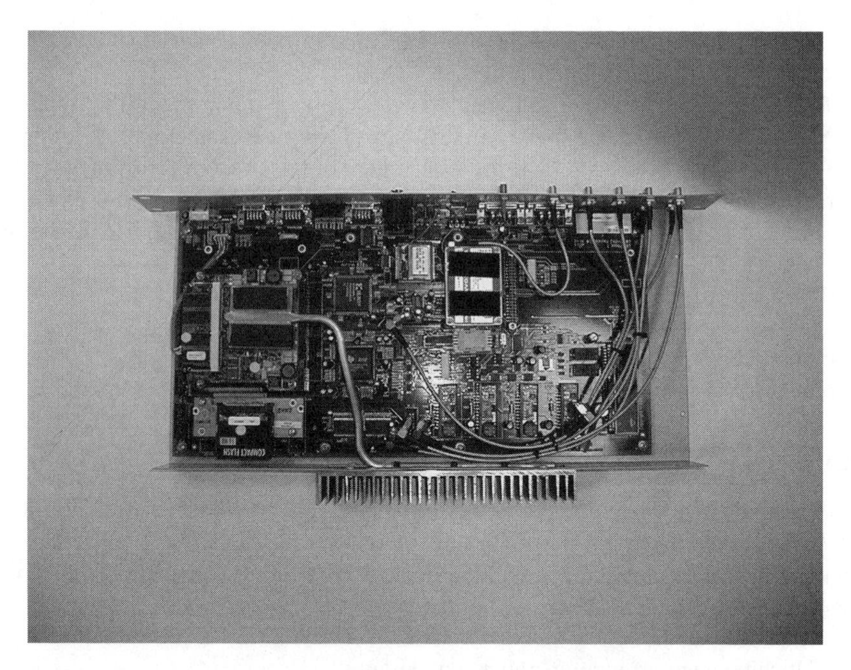

Fig. 8.2 Component (10 W processor chip) on evaporator block (centre left) with heat dissipated from natural convection heat sink (centre bottom). Courtesy of Thermacore International Inc. (See also the case study in Section 8.2.2 and Fig. 8.25.)

Fig. 8.3 Heat pipes embedded in a heat sink – 'approaching' a flat plate heat pipe (Courtesy of Thermacore International Inc.).

8.1.2.2 Vapour chamber heat pipe

A flat plate heat pipe is shown in Fig. 8.4. This flat plate heat pipe design is hollow on the inside and the vapour is free to travel throughout. It is either soldered or epoxied to the heat sink.

Fig. 8.4 A vapour chamber flat plate heat pipe (Courtesy of Thermacore International Inc.).

Fig. 8.5 The Therma Base flat plate heat pipe/vapour chamber. In this variant, the dimples show where penetrations (to locate components) can be safely made (Courtesy of Thermacore International Inc.).

8.1.3 Micro-heat pipes and arrays

Micro-heat pipes are becoming increasingly important as the components requiring thermal control become smaller and heat fluxes increase. Moon et al. in South Korea [8, 10] has reported on miniature (2–4 mm diameter) and micro-heat pipes (down to less than 1 mm in terms of characteristic dimensions, as shown in Fig. 8.6). The units of triangular cross section can transport up to 7 W, the performance as a function of inclination angle being shown in Fig. 8.7. The location within a small notebook computer is illustrated in Fig. 8.8.

Note that in Fig. 8.7 the fill ratio is quoted. In such small volumes, it is important to ensure that the correct amount of working fluid is inserted, so as to allow 'heat pipe functioning' to take place.

The response time of heat pipes can be critical in some electronics cooling applications. As shown in Fig. 8.9 [4], this can be rapid – a few seconds to achieve full mass flow of vapour.

Micro-heat pipes are also discussed in Chapter 6.

8.1.4 Loop heat pipes

Loop heat pipes (LHPs) – see also Chapter 6 – are becoming increasingly popular for electronics thermal management. With all the main features of conventional heat

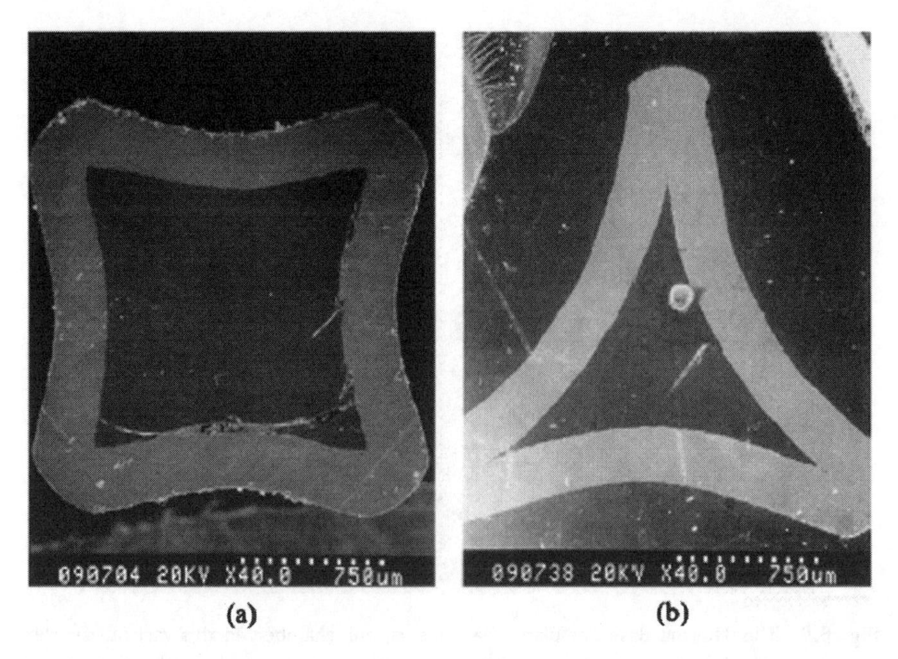

Fig. 8.6 Micro-heat pipe cross sections. The wall material is copper and working fluid is water [10].

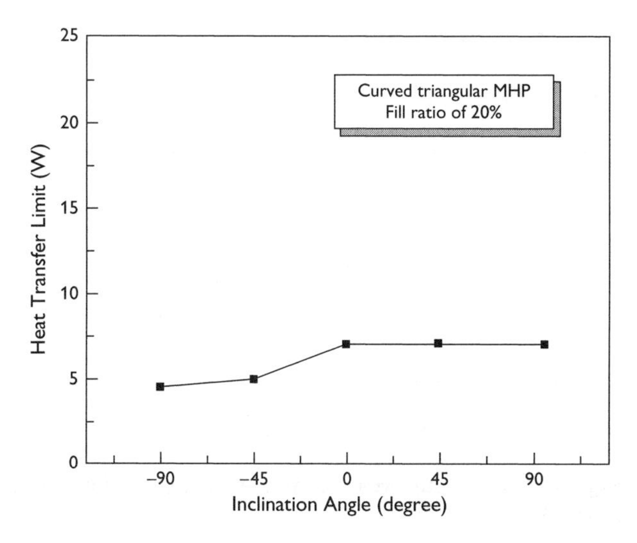

Fig. 8.7 The performance of the triangular cross-section micro-heat pipe, as a function of inclination angle [10].

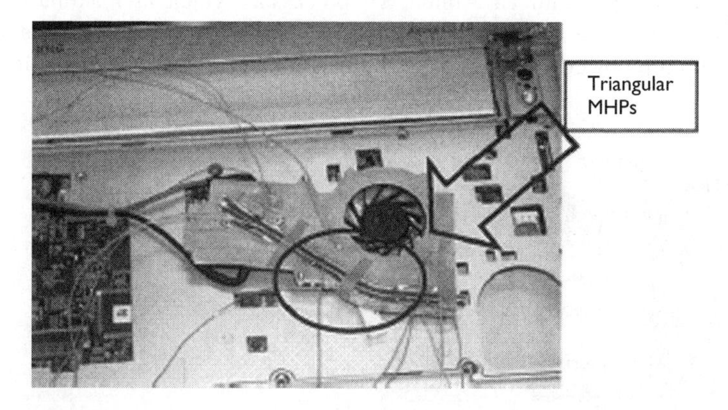

Fig. 8.8 Location of two triangular cross-section micro-heat pipes in a sub-notebook personal computer. Water is the working fluid [10].

pipes, they are additionally not constrained by the distance limitations of their close relations. Highly effective heat transfer over distances of metres against gravity, or over tens of metres horizontally, open up new fields of application. Multiple evaporators and condensers are also possible [3].

Computers, in particular laptop PCs, are new spheres of LHP applications. First experience was reported, according to Maydanik in 2001, when a number of compact coolers for central processing units (CPUs) with a mass of about 50 g was created based on LHPs. They dissipated about 25–30 W. The scheme and the external view

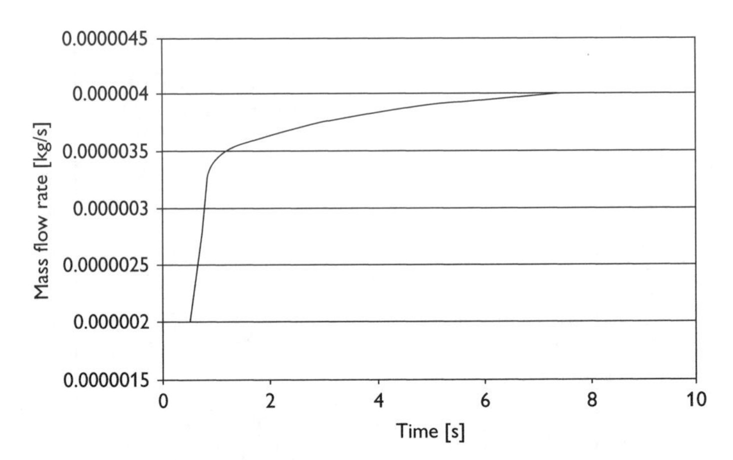

Fig. 8.9 Vapour mass flow rate evolution at the evaporator for a step variation of evaporator power [4].

of such coolers are shown in Fig. 8.10. He reported that a cooler based on a copper water LHP with an evaporator diameter of 6 mm was being tested. It was intended for a notebook PC with an Athlon XP processor, which at maximum loading dissipates about 70 W. It is suggested that LHPs may be the only heat-pipe-based solution if chip powers increase much further in laptops and similar small units.

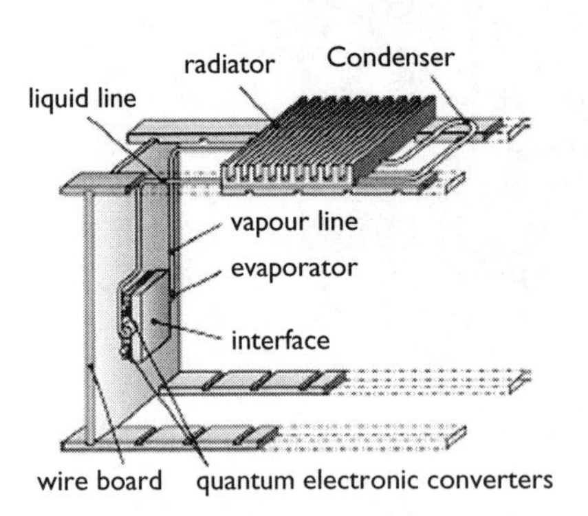

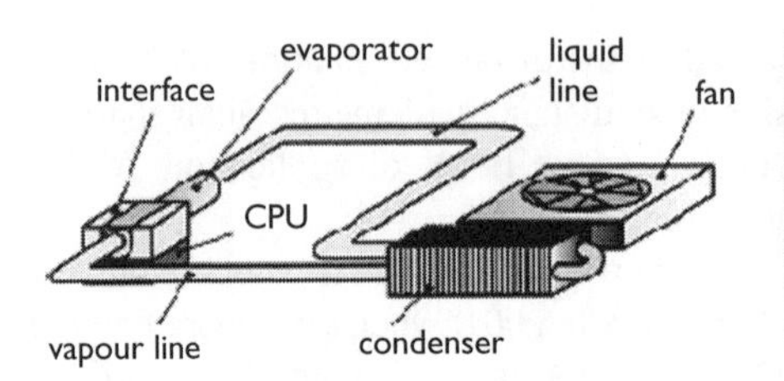

Fig. 8.10 Early loop heat pipe coolers for computer CPUs [3].

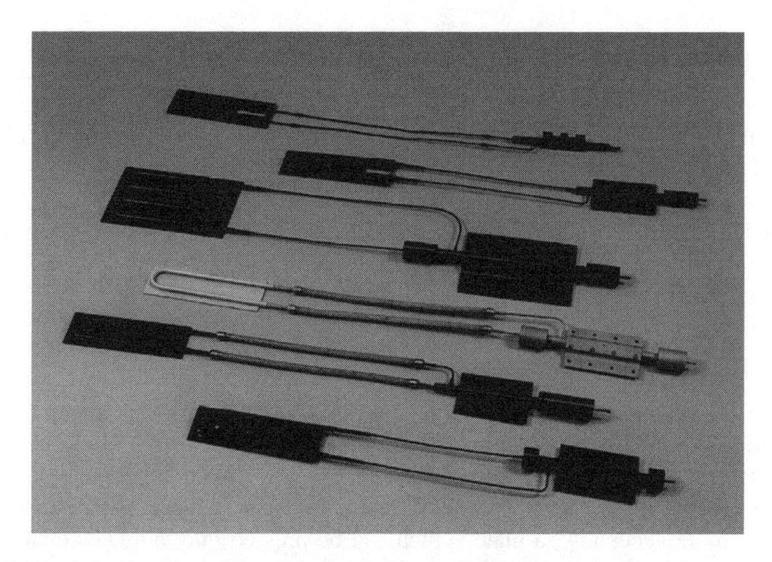

Fig. 8.11 A family of loop heat pipes, some having flexible sections, others being bendable (Courtesy of Thermacore International Inc.).

Figure 8.11 shows a family of LHPs. Two of the LHPs pictured have flexible tubing sections for increased flexibility in application – see applications later. The others have straight tubing sections. This tubing can be coiled or bent to route through complex mounting applications.

A version of the thermacore loop thermosyphon (LTS) with horizontal transport lines is shown in Fig. 8.12. To reduce the fabrication cost, the LTS was designed using the companies' standard vapour chamber parts as building blocks. The flat copper evaporator with a footprint of 65 mm × 90 mm had a sintered copper capillary structure on the inner surfaces that was 1.5 mm thick with an effective pore radius of about 50 μm. The condenser consisted of two flat copper boxes, 65 mm × 90 mm each, without any internal structure. The two boxes were connected using two short tubes one above the other. The smooth-wall transport lines were 60 cm long.

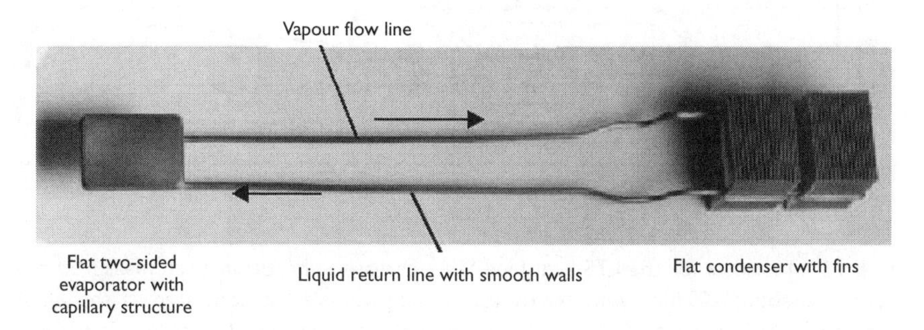

Fig. 8.12 A thermacore loop thermosyphon (LTS) with horizontal transport lines [9].

The LTS was filled with 75 cc of pure ethanol and tested in the horizontal orientation as shown in Fig. 8.12. Two flat aluminium blocks with two cartridge heaters each and a central hole for the thermocouple measuring the heated wall temperature were used to heat the evaporator from both sides. The condenser with folded aluminium fins on the surface was air-cooled.

A selection of test data is shown in Fig. 8.13. Thermocouple number 6 was located on the top portion of the far end of the condenser. The LTS transported up to 200 W with an air velocity across the condenser of 1 m/s and the heated wall temperature below 70 °C.

8.1.5 Direct contact systems

One of the problems with integrating heat pipes and the electronic components is mounting the devices and minimising interface resistances. Two ways for easing this problem have been proposed and patented in the United Kingdom.

The first of these, developed at Marconi, involves using a conformable heat pipe, or more accurately, a plate, which can be pressed into intimate contact with heat-generating components, with a minimal thermal interface between these and the wick. The second method illustrated in Fig. 8.14 involves removal of the heat pipe wall altogether and the use of a liquid reservoir to feed wicks that cover the components to be cooled. In this case, the module would be a sealed unit with provision for heat extraction located externally.

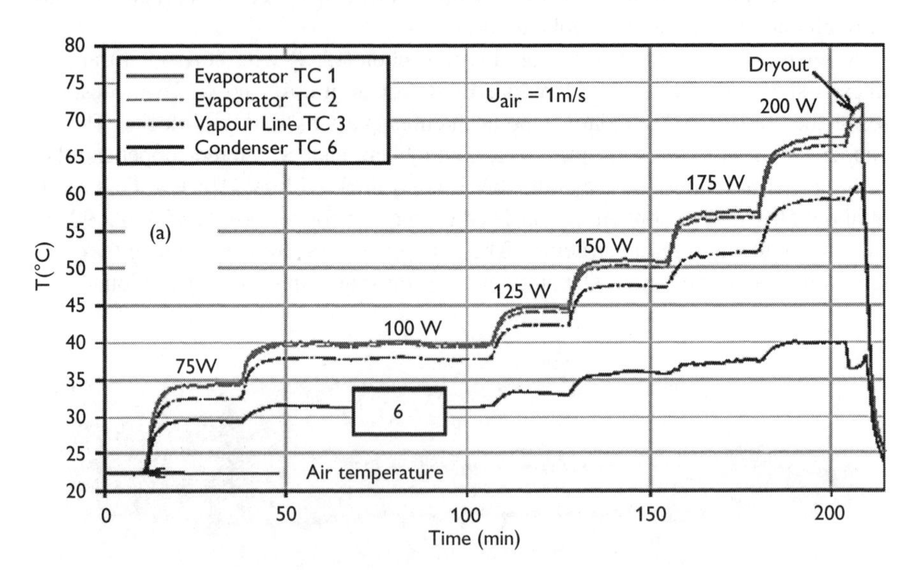

Fig. 8.13 Test data for the LTS heat pipe. The increase in heat transport, measured over a period of about 200 min, with steady state being achieved at each step, is shown. As the capability is exceeded, dryout occurs and heat transport breaks down (Courtesy of Thermacore International Inc.).

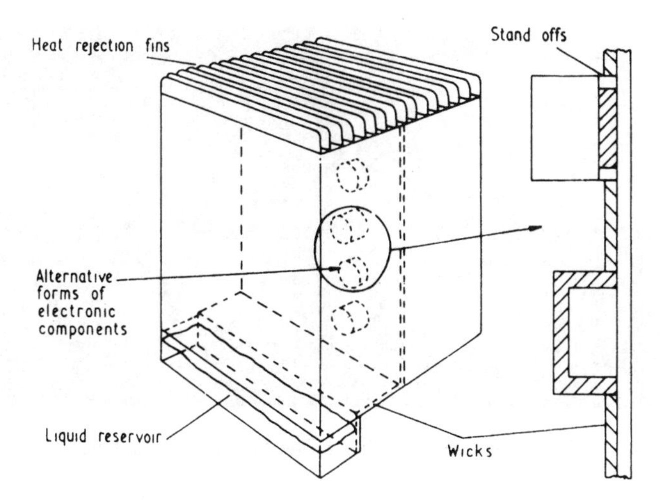

Fig. 8.14 Cooling electronic components by direct contact.

There are few subsequent examples of direct contact systems in commercial use. Compatibility and electrical properties limit the number of potential fluids, unless full encapsulation of the electronics packages can be guaranteed. Access for component replacement is limited, of course, but with the trend towards replacing complete electronics packages, rather than components, a sealed system may not be a problem.

Direct contact cooling is used in other industrial sectors, and jets of air can be used to remove heat from chips. Use of two-phase heat transfer in such systems can create safety problems due to chemical dispersal, if total loss units are used.

8.1.6 Sheet heat pipes

Furukawa Electric in Japan has developed an innovative sheet-type heat sink, a vapour chamber claimed to be the thinnest in the world. This is directed at the cooling of semiconductor chips in mobile electronic equipment such as notebook PCs, cellular phones and personal digital assistants as well as charge-coupled devices, these being imaging units used in digital cameras. The heat sink can also be used for equalizing the heat distribution within the casings.

The Furukawa argument is that as equipment (such as laptops and phones) gets lighter, thinner and smaller, conventional rigid heat pipes, even at the miniature scale, become difficult to install due to the limited space allowed for mounting. The perceived need was to develop an efficient heat conductor that is as thin and as flexible as paper, while providing heat-dissipating and isothermalising functions.

Furukawa points out that although heat-dissipating sheets called 'thermal sheets' using carbon graphite and the like having high heat conductivity have recently been used in practical applications, their effectiveness is not adequate for current

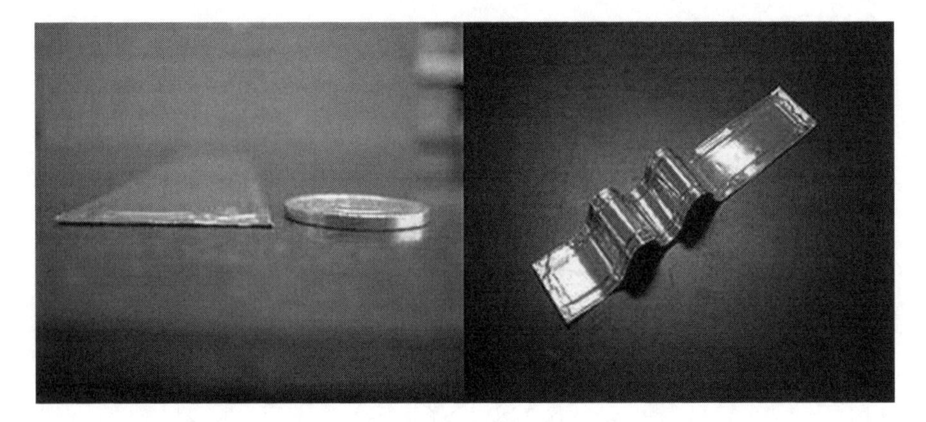

Fig. 8.15 The Furukawa Electric vapour chamber.

demands. The company addressed this challenge by developing a sheet-type heat sink with a thickness of less than 1 mm (Fig. 8.15). The width, length and thickness can be selected to match the product to the design requirements of individual electronic equipment with regard to heat-dissipating performance and shape.

With regard to performance, a unit 0.6 mm in thickness, 20 mm in width and 150 mm in length can dissipate up to 10 W and several tens of W if the thickness is increased to 1 mm. Furukawa estimated 2 years ago that production would rapidly reach 5 million units per annum.

8.1.7 Miscellaneous systems

There are a number of other fluid flow mechanisms used in heat pipes, which can benefit their application in electronics thermal management. Two mechanisms that have been studied by Thermacore include the pulsating heat pipe, more specifically the combined pulsating and capillary transport (CPCT) system, and the use of graded wicks [11]. The CPCT-type has the potential of achieving heat flux capabilities of over 300 W/cm^2.

The CPCT unit, a variant of the pulsating heat pipes described in Chapter 6, uses both pulsating fluid motion and capillary forces to take the working fluid to high heat flux regions. The grooves serve as channels for the pulsating fluid and supplementary sintered wicks are provided for extra nucleation sites – a 'belt-and-braces approach'. (Another pulsating heat pipe design is described in Ref. [12], this Canadian research leading to optimum fill ratio criteria.)

The graded wick concept, a derivative of composite wicks described in Chapters 3 and 5, use relatively open pores for liquid transport between the evaporator and the condenser but fine pores (sintered) close to the heat input and output walls of the heat pipe. Thus, the pressure drop along most of the pipe is modest. Performance was slightly better than conventional units but pore size optimisation could lead to larger improvements. Ref. [11] includes theoretical analyses of the systems, for those wishing to follow up the concepts in depth.

8.2 APPLICATIONS

In this section, a small number of applications are used to illustrate the features of several heat pipe types, as supplied by Thermacore Inc. A case study is also given, showing in simple format the procedures from concept to application of a heat pipe for cooling a processor.

8.2.1 Flexible heat pipes

Figure 8.16 is a photograph of a family of flexible heat pipes. The adiabatic section is made from flexible hose. This allows freedom when mounting the heat pipe to the device being cooled and the heat sink. In addition, the flexible section can accommodate relative motion between the heat source and the sink. These heat pipes were developed for aircraft use. The evaporator section mounts to electronics on an actuator and the condenser attaches to the aircraft structure. The actuator moves while the condenser remains stationary.

In collaboration with Intel Corporation, in 1999, Thermacore looked at some solutions for current and future laptop computers [5], the first heat pipe being used in such a computer in 1994. Conventional units use a heat pipe to take heat from a CPU to the edge of the laptop, normally the side or rear panel, where a finned heat

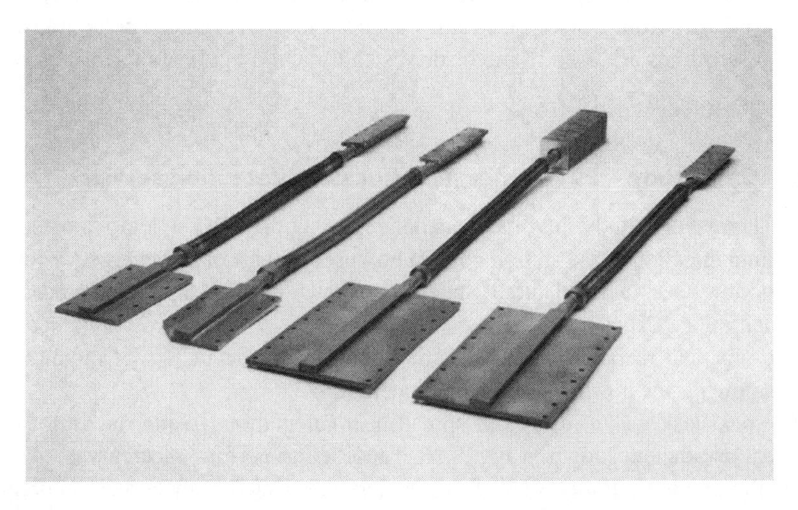

Fig. 8.16 A family of flexible heat pipes, used in aircraft (Courtesy of Thermacore International Inc.).

sink on the condenser section can be forced air-cooled, as illustrated in Fig. 8.2 earlier in this chapter. Thermacore proposed that a 'thermal hinge' be used to allow the heat pipe to be extended into the lid of the laptop (which of course houses the screen). The lid could thus become a good natural convection-cooled heat sink, allowing the fan to be discarded.

The thermal hinge differs from the flexible heat pipe 'joints' shown above by being of rigid construction, as a hinge, and thus having to conduct heat effectively while being subjected to a large number of cycles (opening and closing of the lid). Thermacore tests showed that effective performance was maintained over 30 000 cycles. The design needs to maintain the torque at the hinge, in order to allow consistent heat transfer over thousands of cycles.

(When discussing flexible heat pipes, the reader may also come across reference to 'bendable' heat pipes. As illustrated in one of the case studies in Chapter 4, the ability to bend a heat pipe, perhaps only once, in order to fit it within a piece of equipment, can also be a specific requirement.)

8.2.1.1 Heat pipes to cool a concentrated heat source

Figure 8.17 shows a heat pipe assembly design to remove heat from a concentrated high heat flux source and spread it to a large finned area where the heat can be more effectively removed.

8.2.1.2 Multi-kilowatt heat pipe assembly

Figure 8.18 is a photograph of a heat pipe assembly designed to remove up to 10 kW of heat from power electronics such as insulated gate bipolar transistors. These assemblies are used in motor drives and traction applications.

8.2.2 Case study – E911 emergency location detection service

The Thermacore study involved challenges to cool a 2.3 W chip located in a telecommunications base station control box used as part of an emergency location detection service in the United States. As shown in Fig. 8.19, the processor is mounted on a PCB.

The unit had to be natural convection-cooled and the PCB was mounted in a rack, allowing for horizontal heat pipe operation.

The next task was to carry our spreadsheet calculations, using the various geometrical parameters shown in Fig. 8.20. These led to performance curves, allowing the optimum configuration of surfaces, etc. to be selected to meet the specification.

The conceptual solution, shown in Fig. 8.21, involves finned condenser heat sink that is a 200 mm aluminium extrusion. A 4 mm diameter copper heat pipe with a

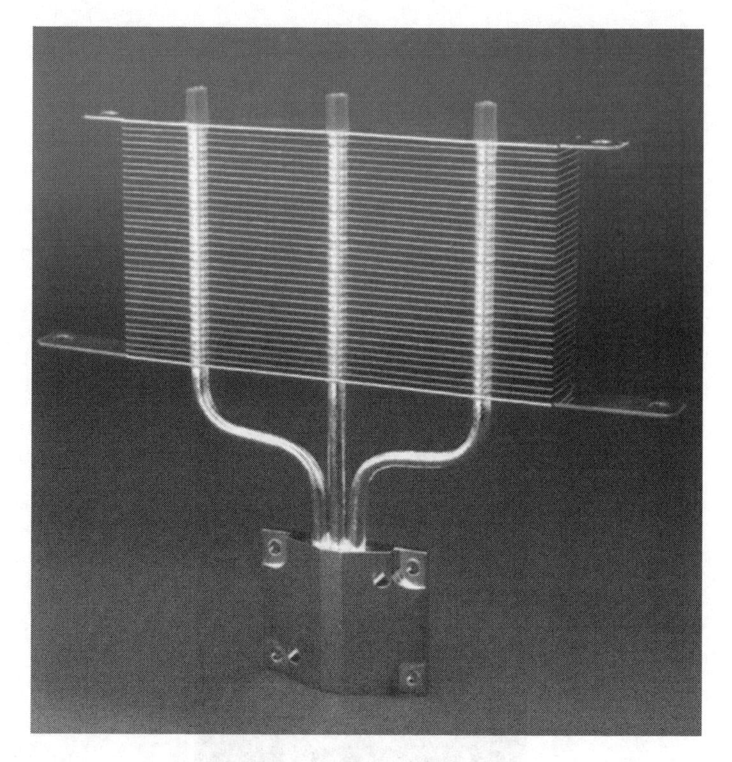

Fig. 8.17 Heat pipe assembly for removing heat from a concentrated heat source (Courtesy of Thermacore International Inc.).

sintered wick, and a 49 mm × 49 mm copper block into which the evaporator section of the heat pipe is fitted.

Once this has been fixed, the details of locating the components onto the heat pipe need to be examined. In this case, as illustrated in Fig. 8.22, thermally conductive double-sided tape is used to minimise the thermal gradient across the evaporator block junction, and at the condenser floating mounting bushes will be used.

Figure 8.23 shows the prototype unit, where the heat pipe is mechanically fixed to the extrusion used for heat rejection from the evaporator, and the evaporator block is soldered to the heat pipe.

Figure 8.24 shows the laboratory test set-up, where parameters can be adjusted to ensure that the unit meets the customer requirements. Then a unit will be constructed for field trials, as shown in Fig. 8.25. Their finned heat sink can be seen on the outer wall of the package, bottom centre, while the evaporator, over the processor, is seen centre left.

Fig. 8.18 Multi-kilowatt heat pipe assembly (Courtesy of Thermacore International Inc.).

Fig. 8.19 The location of the processor on the PCB (Courtesy of Thermacore International Inc.).

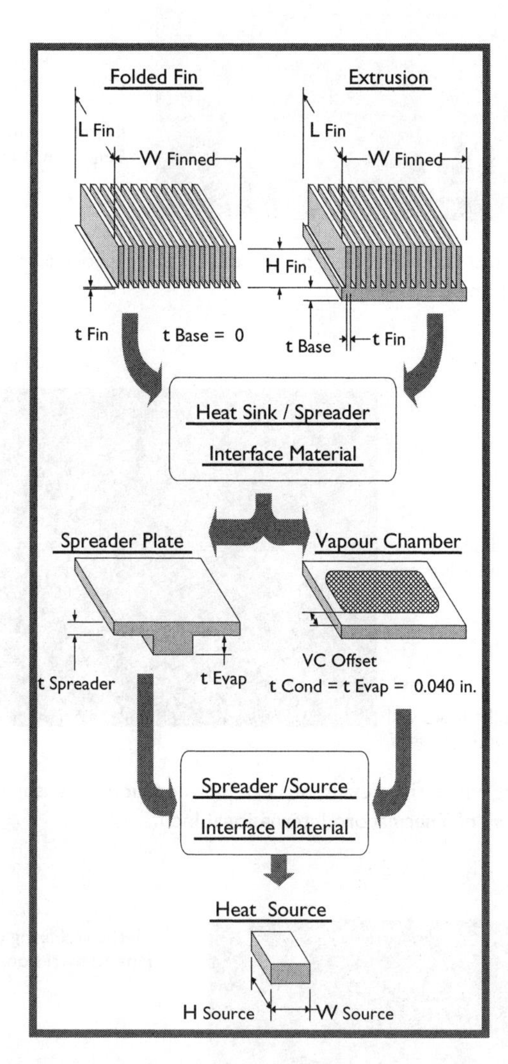

Fig. 8.20 Illustration of some of the parameters used in the heat pipe design spreadsheets (Courtesy of Thermacore International Inc.).

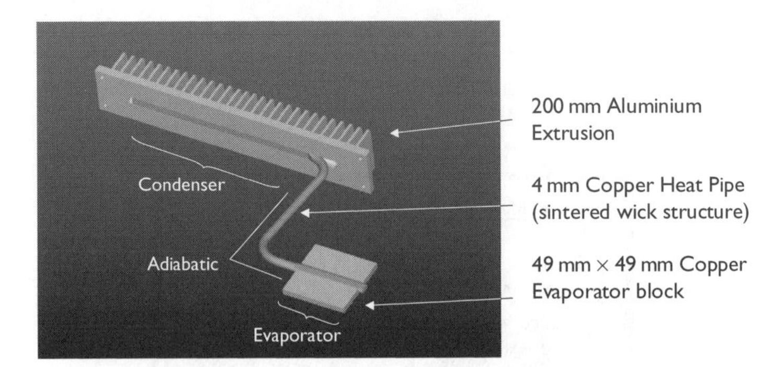

Fig. 8.21 Conceptual solution based on the optimisation procedure (Courtesy of Thermacore International Inc.).

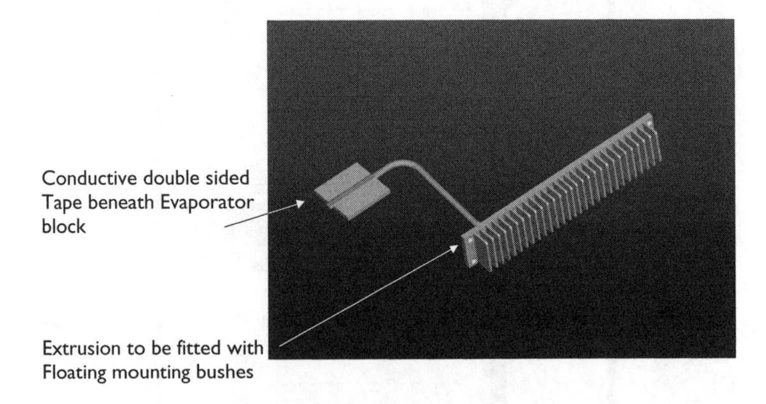

Fig. 8.22 Deciding upon the details of heat pipe location to ensure minimum thermal resistances (Courtesy of Thermacore International Inc.).

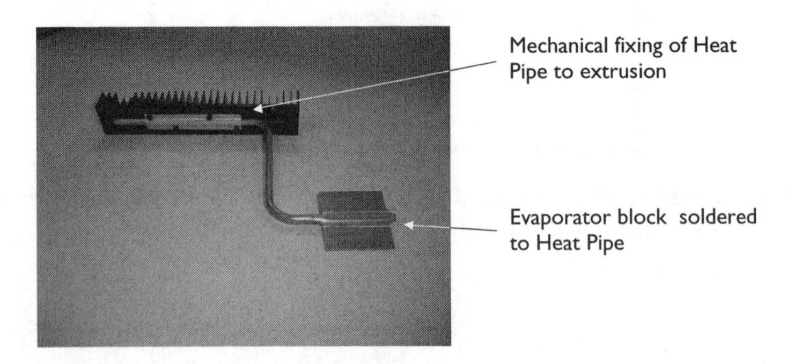

Fig. 8.23 The prototype unit (Courtesy of Thermacore International Inc.).

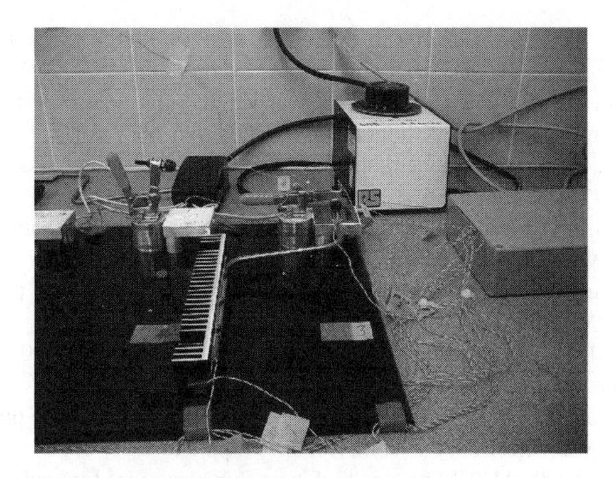

Fig. 8.24 The test facility used to check performance and to make sure it merits customer specifications (with adjustments to construction/assembly where necessary).

Fig. 8.25 The finished product, ready for field trials (Courtesy of Thermacore International Inc.).

REFERENCES

[1] McGlen, R., Jachuck, R. and Lin, S. Integrated thermal management techniques for high power electronic devices. Proceedings of 8th UK National Heat Transfer Conference, Oxford University, 2003.

[2] Swanson, T.D. and Birur, G.C. NASA thermal control technologies for robotic spacecraft. Appl. Therm. Eng., Vol. 23, pp 1055–1065, 2003.

[3] Maydanik, Yu.F. Loop heat pipes. Appl. Therm. Eng., Vol. 25, pp 635–657, 2005.

[4] Murer, S. et al. Experimental and numerical analysis of the transient response of a miniature heat pipe. Appl. Therm. Eng., Vol. 25, pp 2566–2577, 2005.
[5] Ali, A., DeHoff, R. and Grubb, K. Advanced thermal solutions for higher power notebook computers, 1999. Available at: www.thermacore.com.
[6] Lin, L. et al. High performance miniature heat pipe. Int. J. Heat Mass Transf., Vol. 45, pp 3131–3142, 2002.
[7] Garimella, S.V. Advances in mesoscale thermal management technologies for microelectronics. Microelectron. J., (*in press*).
[8] Moon, S.H. et al. Improving thermal performance of miniature heat pipe for notebook PC cooling. Microelectron. Reliability, Vol. 42, pp 135–140, 2002.
[9] Khrustalev, D. Loop thermosyphons for cooling of electronics, 2002. Available at: www.thermacore.com.
[10] Moon, S.H. et al. Improving thermal performance of miniature heat pipe for notebook PC cooling. Microelectron. Reliability, Vol. 44, pp 315–321, 2004.
[11] Zuo, Z.J. and North, M.T. Miniature high heat flux heat pipes for cooling of electronics. Available at: www.thermacore.com.
[12] Nikkanen, K. et al. Pulsating heat pipes for microelectronics cooling. Proceedings of 2005 AIChE Spring National Meeting, 2005, pp 2421–2428.
[13] Ellsworthy, M.J. and Simons, R.E. High powered chip cooling – air and beyond. Electron. Cooling Online, 2005.

For the reader interested in following up other methods of thermal management, and challenges, the following short bibliography may be of use:

Amon, C.H. et al. MEMS-enabled thermal management of high-heat-flux devices EDIFICE: embedded droplet impingement for integrated cooling of electronics. Exp. Therm. Fluid Sci., Vol. 25, No. 5, pp 231–242, 2002.

Azar, K. The history of power dissipation. Electron. Cooling, Vol. 6, No. 1, 2000.

Chen, Y. and Cheng, P. Heat transfer and pressure drop in fractal tree-like micro channel nets. Int. J. Heat Mass Transf., Vol. 45, pp 2643–2648, 2002.

Dumont, G. et al. Water-cooled electronics. Nucl. Instrum. Methods Phys. Res. A., Vol. 440, pp 213–223, 2000.

Groll, S. et al. Thermal control of electronic equipment by heat pipes. Rev. Gen. Therm., Vol. 37, pp 323–352, 1998.

Gromoll, B. Micro cooling systems for high density packaging. Rev. Gen. Therm., Vol. 37, pp 781–787, 1998.

Intel Corp. Silicon Moore's Law. Intel Corporation, 2003.

Kang, S.-W. et al. Fabrication and test of radial grooved micro heat pipes. Appl. Therm. Eng., Vol. 22, pp 1559–1568, 2002.

Lin, K.S. et al. Prospects of confined flow boiling in thermal management of microsystems. Appl. Therm. Eng., Vol. 22, pp. 825–837, 2002.

Moore, G.E. Cramming more components onto integrated circuits. Electronics, Vol. 38, No. 8, 1965.

Park, H. et al. Fabrication of a microchannel integrated with inner sensors and the analysis of its laminar flow characteristics. Sens. Actuators A Phys., Vol. 103, pp 317–329, 2003.

Qu, W. et al. Experimental and numerical study of pressure drop and heat transfer in a single-phase micro-channel heat sink. Int. J. Heat Mass Transf., Vol. 45, pp 2549–2565, 2002.

Scott Downing, R. et al. Single and two-phase pressure drop characteristics in miniature helical channels. Exp. Therm. Fluid Sci., Vol. 26, pp 535–546, 2002.

Cheng, Y.T. et al. Localised silicon fusion and eutectic bonding for MEMS fabrication and packaging. J. Microelectromech. Syst., Vol. 9, No. 1, pp 3–8, 2000.

[illegible]

[illegible]

[illegible]

[illegible]

APPENDIX 1

WORKING FLUID PROPERTIES

Note that the CFCs have been omitted from these tables. The replacement refrigerants such as R134a are as yet largely untried in heat pipes.

Fluids listed: (in order of appearance)		
	Helium	Heptane
	Nitrogen	Water
	Ammonia	Flutec PP9
	High-temperature organics	Pentane
	Mercury	Caesium
	Acetone	Potassium
	Methanol	Sodium
	Flutec PP2	Lithium
	Ethanol	

Properties listed:		
	Latent heat of evaporation	Vapour dynamic viscosity
	Liquid density	Vapour pressure
	Vapour density	Vapour specific heat
	Liquid thermal conductivity	Liquid surface tension
	Liquid dynamic viscosity	

Helium

Temp °C	Latent heat kJ/kg	Liquid density kg/m³	Vapour density kg/m³	Liquid thermal conductivity W/m °C × 10^{-2}	Liquid viscos. cP × 10^2	Vapour viscos. cP × 10^3	Vapour press. Bar	Vapour specific heat kJ/kg °C	Liquid surface tension N/m × 10^3
−271	22.8	148.3	26.0	1.81	3.90	0.20	0.06	2.045	0.26
−270	23.6	140.7	17.0	2.24	3.70	0.30	0.32	2.699	0.19
−269	20.9	128.0	10.0	2.77	2.90	0.60	1.00	4.619	0.09
−268	4.0	113.8	8.5	3.50	1.34	0.90	2.29	6.642	0.01

Nitrogen

Temp °C	Latent heat kJ/kg	Liquid density kg/m³	Vapour density kg/m³	Liquid thermal conductivity W/m °C	Liquid viscos. cP × 10^1	Vapour viscos. cP × 10^2	Vapour press. Bar	Vapour specific heat kJ/kg °C	Liquid surface tension N/m × 10^2
−203	210.0	830.0	1.84	0.150	2.48	0.48	0.48	1.083	1.054
−200	205.5	818.0	3.81	0.146	1.94	0.51	0.74	1.082	0.985
−195	198.0	798.0	7.10	0.139	1.51	0.56	1.62	1.079	0.870
−190	190.5	778.0	10.39	0.132	1.26	0.60	3.31	1.077	0.766
−185	183.0	758.0	13.68	0.125	1.08	0.65	4.99	1.074	0.662
−180	173.7	732.0	22.05	0.117	0.95	0.71	6.69	1.072	0.561
−175	163.2	702.0	33.80	0.110	0.86	0.77	8.37	1.070	0.464
−170	152.7	672.0	45.55	0.103	0.80	0.83	1.07	1.068	0.367
−160	124.2	603.0	80.90	0.089	0.72	1.00	19.37	1.063	0.185
−150	66.8	474.0	194.00	0.075	0.65	1.50	28.80.	1.059	0.110

Ammonia

Temp °C	Latent heat kJ/kg	Liquid density kg/m^3	Vapour density kg/m^3	Liquid thermal conductivity W/m °C	Liquid viscos. cP	Vapour viscos. $cP \times 10^2$	Vapour press. Bar	Vapour specific heat kJ/kg °C	Liquid surface tension $N/m \times 10^2$
−60	1343	714.4	0.03	0.294	0.36	0.72	0.27	2.050	4.062
−40	1384	690.4	0.05	0.303	0.29	0.79	0.76	2.075	3.574
−20	1338	665.5	1.62	0.304	0.26	0.85	1.93	2.100	3.090
0	1263	638.6	3.48	0.298	0.25	0.92	4.24	2.125	2.480
20	1187	610.3	6.69	0.286	0.22	1.01	8.46	2.150	2.133
40	1101	579.5	12.00	0.272	0.20	1.16	15.34	2.160	1.833
60	1026	545.2	20.49	0.255	0.17	1.27	29.80	2.180	1.367
80	891	505.7	34.13	0.235	0.15	1.40	40.90	2.210	0.767
100	699	455.1	54.92	0.212	0.11	1.60	63.12	2.260	0.500
120	428	374.4	113.16	0.184	0.07	1.89	90.44	2.292	0.150

Pentane

Temp °C	Latent heat kJ/kg	Liquid density kg/m^3	Vapour density kg/m^3	Liquid thermal conductivity W/m °C	Liquid viscos. cP	Vapour viscos. $cP \times 10^2$	Vapour press. Bar	Vapour specific heat kJ/kg °C	Liquid surface tension $N/m \times 10^2$
−20	390.0	663.0	0.01	0.149	0.344	0.51	0.10	0.825	2.01
0	378.3	644.0	0.75	0.143	0.283	0.53	0.24	0.874	1.79
20	366.9	625.5	2.20	0.138	0.242	0.58	0.76	0.922	1.58
40	355.5	607.0	4.35	0.133	0.200	0.63	1.52	0.971	1.37
60	342.3	585.0	6.51	0.128	0.174	0.69	2.28	1.021	1.17
80	329.1	563.0	10.61	0.127	0.147	0.74	3.89	1.050	0.97
100	295.7	537.6	16.54	0.124	0.128	0.81	7.19	1.088	0.83
120	269.7	509.4	25.20	0.122	0.120	0.90	13.81	1.164	0.68

Acetone

Temp °C	Latent heat kJ/kg	Liquid density kg/m^3	Vapour density kg/m^3	Liquid thermal conductivity W/m °C	Liquid viscos. cP	Vapour viscos. cP × 10^2	Vapour press. Bar	Vapour specific heat kJ/kg °C	Liquid surface tension N/m × 10^2
−40	660.0	860.0	0.03	0.200	0.800	0.68	0.01	2.00	3.10
−20	615.6	845.0	0.10	0.189	0.500	0.73	0.03	2.06	2.76
0	564.0	812.0	0.26	0.183	0.395	0.78	0.10	2.11	2.62
20	552.0	790.0	0.64	0.181	0.323	0.82	0.27	2.16	2.37
40	536.0	768.0	1.05	0.175	0.269	0.86	0.60	2.22	2.12
60	517.0	744.0	2.37	0.168	0.226	0.90	1.15	2.28	1.86
80	495.0	719.0	4.30	0.160	0.192	0.95	2.15	2.34	1.62
100	472.0	689.6	6.94	0.148	0.170	0.98	4.43	2.39	1.34
120	426.1	660.3	11.02	0.135	0.148	0.99	6.70	2.45	1.07
140	394.4	631.8	18.61	0.126	0.132	1.03	10.49	2.50	0.81

Methanol

Temp °C	Latent heat kJ/kg	Liquid density kg/m^3	Vapour density kg/m^3	Liquid thermal conductivity W/m °C	Liquid viscos. cP	Vapour viscos. cP × 10^2	Vapour press. Bar	Vapour specific heat kJ/kg °C	Liquid surface tension N/m × 10^2
−50	1194	843.5	0.01	0.210	1.700	0.72	0.01	1.20	3.26
−30	1187	833.5	0.01	0.208	1.300	0.78	0.02	1.27	2.95
−10	1182	818.7	0.04	0.206	0.945	0.85	0.04	1.34	2.63
10	1175	800.5	0.12	0.204	0.701	0.91	0.10	1.40	2.36
30	1155	782.0	0.31	0.203	0.521	0.98	0.25	1.47	2.18
50	1125	764.1	0.77	0.202	0.399	1.04	0.55	1.54	2.01
70	1085	746.2	1.47	0.201	0.314	1.11	1.31	1.61	1.85
90	1035	724.4	3.01	0.199	0.259	1.19	2.69	1.79	1.66
110	980	703.6	5.64	0.197	0.211	1.26	4.98	1.92	1.46
130	920	685.2	9.81	0.195	0.166	1.31	7.86	1.92	1.25
150	850	653.2	15.90	0.193	0.138	1.38	8.94	1.92	1.04

Flutec PP2

Temp °C	Latent heat kJ/kg	Liquid density kg/m^3	Vapour density kg/m^3	Liquid thermal conductivity W/m °C	Liquid viscos. cP	Vapour viscos. cP × 10^2	Vapour press. Bar	Vapour specific heat kJ/kg °C	Liquid surface tension N/m × 10^2
−30	106.2	1942	0.13	0.637	5.200	0.98	0.01	0.72	1.90
−10	103.1	1886	0.44	0.626	3.500	1.03	0.02	0.81	1.71
10	99.8	1829	1.39	0.613	2.140	1.07	0.09	0.92	1.52
30	96.3	1773	2.96	0.601	1.435	1.12	0.22	1.01	1.32
50	91.8	1716	6.43	0.588	1.005	1.17	0.39	1.07	1.13
70	87.0	1660	11.79	0.575	0.720	1.22	0.62	1.11	0.93
90	82.1	1599	21.99	0.563	0.543	1.26	1.43	1.17	0.73
110	76.5	1558	34.92	0.550	0.429	1.31	2.82	1.25	0.52
130	70.3	1515	57.21	0.537	0.314	1.36	4.83	1.33	0.32
160	59.1	1440	103.63	0.518	0.167	1.43	8.76	1.45	0.01

Ethanol

Temp °C	Latent heat kJ/kg	Liquid density kg/m^3	Vapour density kg/m^3	Liquid thermal conductivity W/m °C	Liquid viscos. cP	Vapour viscos. cP × 10^2	Vapour press. Bar	Vapour specific heat kJ/kg °C	Liquid surface tension N/m × 10^2
−30	939.4	825.0	0.02	0.177	3.40	0.75	0.01	1.25	2.76
−10	928.7	813.0	0.03	0.173	2.20	0.80	0.02	1.31	2.66
10	904.8	798.0	0.05	0.170	1.50	0.85	0.03	1.37	2.57
30	888.6	781.0	0.38	0.168	1.02	0.91	0.10	1.44	2.44
50	872.3	762.2	0.72	0.166	0.72	0.97	0.29	1.51	2.31
70	858.3	743.1	1.32	0.165	0.51	1.02	0.76	1.58	2.17
90	832.1	725.3	2.59	0.163	0.37	1.07	1.43	1.65	2.04
110	786.6	704.1	5.17	0.160	0.28	1.13	2.66	1.72	1.89
130	734.4	678.7	9.25	0.159	0.21	1.18	4.30	1.78	1.75

Heptane

Temp °C	Latent heat kJ/kg	Liquid density kg/m³	Vapour density kg/m³	Liquid thermal conductivity W/m °C	Liquid viscos. cP	Vapour viscos. cP × 10²	Vapour press. Bar	Vapour specific heat kJ/kg °C	Liquid surface tension N/m × 10²
−20	384.0	715.5	0.01	0.143	0.69	0.57	0.01	0.83	2.42
0	372.6	699.0	0.17	0.141	0.53	0.60	0.02	0.87	2.21
20	362.2	683.0	0.49	0.140	0.43	0.63	0.08	0.92	2.01
40	351.8	667.0	0.97	0.139	0.34	0.66	0.20	0.97	1.81
60	341.5	649.0	1.45	0.137	0.29	0.70	0.32	1.02	1.62
80	331.2	631.0	2.31	0.135	0.24	0.74	0.62	1.05	1.43
100	319.6	612.0	3.71	0.133	0.21	0.77	1.10	1.09	1.28
120	305.0	592.0	6.08	0.132	0.18	0.82	1.85	1.16	1.10

Water

Temp °C	Latent heat kJ/kg	Liquid density kg/m³	Vapour density kg/m³	Liquid thermal conductivity W/m °C	Liquid viscos. cP	Vapour viscos. cP × 10²	Vapour press. Bar	Vapour specific heat kJ/kg °C	Liquid surface tension N/m × 10²
20	2448	998.2	0.02	0.603	1.00	0.96	0.02	1.81	7.28
40	2402	992.3	0.05	0.630	0.65	1.04	0.07	1.89	6.96
60	2359	983.0	0.13	0.649	0.47	1.12	0.20	1.91	6.62
80	2309	972.0	0.29	0.668	0.36	1.19	0.47	1.95	6.26
100	2258	958.0	0.60	0.680	0.28	1.27	1.01	2.01	5.89
120	2200	945.0	1.12	0.682	0.23	1.34	2.02	2.09	5.50
140	2139	928.0	1.99	0.683	0.20	1.41	3.90	2.21	5.06
160	2074	909.0	3.27	0.679	0.17	1.49	6.44	2.38	4.66
180	2003	888.0	5.16	0.669	0.15	1.57	10.04	2.62	4.29
200	1967	865.0	7.87	0.659	0.14	1.65	16.19	2.91	3.89

Flutec PP9

Temp °C	Latent heat kJ/kg	Liquid density kg/m^3	Vapour density kg/m^3	Liquid thermal conductivity W/m °C	Liquid viscos. cP	Vapour viscos. cP × 10^2	Vapour press. Bar	Vapour specific heat kJ/kg °C	Liquid surface tension N/m × 10^2
−30	103.0	2098	0.01	0.060	5.77	0.82	0.00	0.80	2.36
0	98.4	2029	0.01	0.059	3.31	0.90	0.00	0.87	2.08
30	94.5	1960	0.12	0.057	1.48	1.06	0.01	0.94	1.80
60	90.2	1891	0.61	0.056	0.94	1.18	0.03	1.02	1.52
90	86.1	1822	1.93	0.054	0.65	1.21	0.12	1.09	1.24
120	83.0	1753	4.52	0.053	0.49	1.23	0.28	1.15	0.95
150	77.4	1685	11.81	0.052	0.38	1.26	0.61	1.23	0.67
180	70.8	1604	25.13	0.051	0.30	1.33	1.58	1.30	0.40
225	59.4	1455	63.27	0.049	0.21	1.44	4.21	1.41	0.01

High Temperature Organic (Diphenyl–Diphenyl Oxide Eutectic)

Temp °C	Latent heat kJ/kg	Liquid density kg/m^3	Vapour density kg/m^3	Liquid thermal conductivity W/m °C	Liquid viscos. cP	Vapour viscos. cP × 10	Vapour press. Bar	Vapour specific heat kJ/kg °C	Liquid surface tension N/m × 10^2
100	354.0	992.0	0.03	0.131	0.97	0.67	0.01	1.34	3.50
150	338.0	951.0	0.22	0.125	0.57	0.78	0.05	1.51	3.00
200	321.0	905.0	0.94	0.119	0.39	0.89	0.25	1.67	2.50
250	301.0	858.0	3.60	0.113	0.27	1.00	0.88	1.81	2.00
300	278.0	809.0	8.74	0.106	0.20	1.12	2.43	1.95	1.50
350	251.0	755.0	19.37	0.099	0.15	1.23	5.55	2.03	1.00
400	219.0	691.0	41.89	0.093	0.12	1.34	10.90	2.11	0.50
450	185.0	625.0	81.00	0.086	0.10	1.45	19.00	2.19	0.03

Mercury

Temp °C	Latent heat kJ/kg	Liquid density kg/m^3	Vapour density kg/m^3	Liquid thermal conductivity W/m °C	Liquid viscos. cP	Vapour viscos. cP $\times 10^2$	Vapour press. Bar	Vapour specific heat kJ/kg °C	Liquid surface tension N/m $\times 10^2$
150	308.8	13 230	0.01	9.99	1.09	0.39	0.01	1.04	4.45
250	303.8	12 995	0.60	11.23	0.96	0.48	0.18	1.04	4.15
300	301.8	12 880	1.73	11.73	0.93	0.53	0.44	1.04	4.00
350	298.9	12 763	4.45	12.18	0.89	0.61	1.16	1.04	3.82
400	296.3	12 656	8.75	12.58	0.86	0.66	2.42	1.04	3.74
450	293.8	12 508	16.80	12.96	0.83	0.70	4.92	1.04	3.61
500	291.3	12 308	28.60	13.31	0.80	0.75	8.86	1.04	3.41
550	288.8	12 154	44.92	13.62	0.79	0.81	15.03	1.04	3.25
600	286.3	12 054	65.75	13.87	0.78	0.87	23.77	1.04	3.15
650	283.5	11 962	94.39	14.15	0.78	0.95	34.95	1.04	3.03
750	277.0	11 800	170.00	14.80	0.77	1.10	63.00	1.04	2.75

Caesium

Temp °C	Latent heat kJ/kg	Liquid density kg/m^3	Vapour density $kg/m^3 \times 10^2$	Liquid thermal conductivity W/m °C	Liquid viscos. cP	Vapour viscos. cP $\times 10^2$	Vapour press. Bar	Vapour specific heat kJ/kg °C $\times$ 10	Liquid surface tension N/m $\times 10^2$
375	530.4	1740	0.01	20.76	0.25	2.20	0.02	1.56	5.81
425	520.4	1730	0.01	20.51	0.23	2.30	0.04	1.56	5.61
475	515.2	1720	0.02	20.02	0.22	2.40	0.09	1.56	5.36
525	510.2	1710	0.03	19.52	0.20	2.50	0.16	1.56	5.11
575	502.8	1700	0.07	18.83	0.19	2.55	0.36	1.56	4.81
625	495.3	1690	0.10	18.13	0.18	2.60	0.57	1.56	4.51
675	490.2	1680	0.18	17.48	0.17	2.67	1.04	1.56	4.21
725	485.2	1670	0.26	16.83	0.17	2.75	1.52	1.56	3.91
775	477.8	1655	0.40	16.18	0.16	2.28	2.46	1.56	3.66
825	470.3	1640	0.55	15.53	0.16	2.90	3.41	1.56	3.41

Potassium

Temp °C	Latent heat kJ/kg	Liquid density kg/m^3	Vapour density kg/m^3	Liquid thermal conductivity W/m °C	Liquid viscos. cP	Vapour viscos. cP × 10^2	Vapour press. Bar	Vapour specific heat kJ/kg °C	Liquid surface tension N/m × 10^2
350	2093	763.1	0.002	51.08	0.21	0.15	0.01	5.32	9.50
400	2078	748.1	0.006	49.08	0.19	0.16	0.01	5.32	9.04
450	2060	735.4	0.015	47.08	0.18	0.16	0.02	5.32	8.69
500	2040	725.4	0.031	45.08	0.17	0.17	0.05	5.32	8.44
550	2020	715.4	0.062	43.31	0.15	0.17	0.10	5.32	8.16
600	2000	705.4	0.111	41.81	0.14	0.18	0.19	5.32	7.86
650	1980	695.4	0.193	40.08	0.13	0.19	0.35	5.32	7.51
700	1969	685.4	0.314	38.08	0.12	0.19	0.61	5.32	7.12
750	1938	675.4	0.486	36.31	0.12	0.20	0.99	5.32	6.72
800	1913	665.4	0.716	34.81	0.11	0.20	1.55	5.32	6.32
850	1883	653.1	1.054	33.31	0.10	0.21	2.34	5.32	5.92

Sodium

Temp °C	Latent heat kJ/kg	Liquid density kg/m^3	Vapour density kg/m^3	Liquid thermal conductivity W/m °C	Liquid viscos. cP	Vapour viscos. cP × 10	Vapour press. Bar	Vapour specific heat kJ/kg °C × 10	Liquid surface tension N/m × 10^2
500	4370	828.1	0.003	70.08	0.24	0.18	0.01	9.04	1.51
600	4243	805.4	0.013	64.62	0.21	0.19	0.04	9.04	1.42
700	4090	763.5	0.050	60.81	0.19	0.20	0.15	9.04	1.33
800	3977	757.3	0.134	57.81	0.18	0.22	0.47	9.04	1.23
900	3913	745.4	0.306	53.35	0.17	0.23	1.25	9.04	1.13
1000	3827	725.4	0.667	49.08	0.16	0.24	2.81	9.04	1.04
1100	3690	690.8	1.306	45.08	0.16	0.25	5.49	9.04	0.95
1200	3577	669.0	2.303	41.08	0.15	0.26	9.59	9.04	0.86
1300	3477	654.0	3.622	37.08	0.15	0.27	15.91	9.04	0.77

Lithium

Temp °C	Latent heat kJ/kg	Liquid density kg/m^3	Vapour density kg/m^3	Liquid thermal conductivity W/m °C	Liquid viscos. cP	Vapour viscos. $cP \times 10^2$	Vapour press. Bar	Vapour specific heat kJ/kg °C	Liquid surface tension $N/m \times 10^2$
1030	20 500	450	0.005	67	0.24	1.67	0.07	0.532	2.90
1130	20 100	440	0.013	69	0.24	1.74	0.17	0.532	2.85
1230	20 000	430	0.028	70	0.23	1.83	0.45	0.532	2.75
1330	19 700	420	0.057	69	0.23	1.91	0.96	0.532	2.60
1430	19 200	410	0.108	68	0.23	2.00	1.85	0.532	2.40
1530	18 900	405	0.193	65	0.23	2.10	3.30	0.532	2.25
1630	18 500	400	0.340	62	0.23	2.17	5.30	0.532	2.10
1730	18 200	398	0.490	59	0.23	2.26	8.90	0.532	2.05

APPENDIX 2

THERMAL CONDUCTIVITY OF HEAT PIPE CONTAINER AND WICK MATERIALS

Material	Thermal conductivity (W/m °C)
Aluminium	205
Brass	113
Copper (0–100 °C)	394
Glass	0.75
Nickel (0–100 °C)	88
Mild steel	45
Stainless steel (type 304)	17.3
Teflon	0.17

APPENDIX 3

BIBLIOGRAPHY

This bibliography lists papers on heat pipes and thermosyphons, which are not generally cited as references in the main text. In the main, they are related to interesting and recent uses of heat pipes, but a number of other papers, some of an older vintage, are included, where the authors believe that they may be of particular interest or relevance.

There are of course many online databases that have good search capabilities for heat pipe papers. Some of these are by subscription, such as www.sciencedirect.com and the Scopus Search Alert, but many academic institutions will subscribe to these, as will major companies and research laboratories.

Another major source of information on heat pipes and their uses is in the lists of patents. There are a number of patent databases now accessible online. Some allow free access to at least the patent abstract, based on a keyword search. Searches in all databases can be done on the basis of keywords for the technology, patent numbers, (where known), companies or inventors. Care should be taken in ascertaining where 'free' access stops and charges begin, e.g. for ordering the full patent specification.

The UK Patent Office has its own web site, offering a variety of facilities. Free access to patent abstracts and other services is available: http://www.patent.gov.uk/

Once onto the UK Patent Office Home Page, directions to the patent search are given. On clicking on this, one is given access to the interface to the published patent application databases of the UK Patent Office, the European Patent Office and other European national patent offices. There is also access to the database of published patent applications: Esp@cenet

Full copies of the specification, drawings and claims can be viewed online, if they are available. Using a keyword search for compact heat exchangers, for example, the user will find patent abstracts from Eastern and Western Europe, as well as the USA and World patents.

Good data sources for patents, in terms of web accessibility and ease of searching, are those associated with United States Patents:

http://www.uspto.gov/ – This is the United States Patent and Trademark Office Home Page and is the official site for searching the US patent database.

A few patents are cited in the main text of the book, but the ease of access to patent databases makes a dedicated listing here less important than in previous editions.

BIBLIOGRAPHY

Abdel-Samad, S. et al. Deuterium heat pipes – cryogenic targets for COSY experiments. Nuclear Instruments and Methods in Physics Research, Section A. Vol. 550, No. 1, pp 61–69, 2005.

Amili, P. and Yortsos, Y.C. Stability of heat pipes in vapour-dominated systems. Int. J. Heat Mass Transf., Vol. 47, pp 1233–1246, 2004.

Borzenko, V.I. and Malyshenko, S.P. Investigations of rewetting phenomena at steam generation on surfaces with porous coatings. J. Phys., IV: JP, Vol. 9, No. 3, 1999.

Choi, M.K. SWIFT Burst Alert Telescope loop heat pipe thermal system characteristics and flight operation procedure. Proceedings of Intersociety Energy Conversion Engineering Conference, pp 357–360, 2002.

Cimbala, J.M. et al. Study of a loop heat pipe using neutron radiography. Appl. Radiat. Isot., Vol. 61, pp 701–705, 2004.

Cui, H. et al. Thermal performance analysis on unit tube for heat pipe receiver. Solar Energy. In press

Dobson, R.T. An open oscillatory heat pipe steam-powered boat. Int. J. Mech. Eng. Educ., Vol. 31, No. 4, pp 339–358, 2003.

Du, Shejiao et al. Thermal hydraulic tests for developing two-phase thermo-siphon loop of CARR-CNS. Physica B., Vol. 358, pp. 285–295, 2005.

El-Genk, M.S. and Tournier, J-M.P. SAIRS – Scalable AMTEC integrated reactor space power system. Prog. Nucl. Energy, Vol. 45, No. 1, pp 25–69, 2004.

Gilchrist, S. et al. Thermal modeling of a rotating heat pipe are-engine nose cone anti-icing system. SAE General Aviation Technology Conference and Exhibition, SAE 2004-01-1817, Wichita, USA, 2004.

Groll, M. et al. Thermal control of electronic equipment by heat pipes. Rev. Gen. Therm., Vol. 37, pp 323–352, 1998.

Gu, J. et al. Microgravity performance of micro pulsating heat pipes. Microgravity Sci. Technol., Vol. 15, No. 1, pp 181–185, 2005.

Hoang, Triem et al. Advanced loop heat pipes for spacecraft central thermal bus concept. Proceedings of 12th International Heat Transfer Conference, Vol. 4, pp 495–500, Grenoble, 18–23 August 2002.

Horbaniuc, B., Popescu, A. and Dumitrascu, G. The correlation between the number of fins and the discharge time for a finned heat pipe latent heat storage system. Proceedings of the World Renewable Energy Conference (WREC), p 605 et seq., 1996.

Huang, B.J. et al. Heat-pipe enhanced solar-assisted heat pump water heater. Solar Energy, Vol. 78, No. 3, pp 375–381, 2005.

Jen, T.-C. et al. Use of thermosyphon to cool the cutting tip of a drill. Proceedings of the ASME Heat Transfer/Fluids Engineering Summer Conference, 2004, Vol. 3, pp 963–968, 2004.

Katsuta, M. et al. Design of capillary pumped loop to be boarded on the artificial satellite (USERS) and development of its analytical heat transport model. Proceedings of the 12th International Heat Transfer Conference, Vol. 4, pp 441–446, Grenoble, 18–23 August 2002.

Kempers, R. et al. Effect of fluid loading on the performance of wicked heat pipes. ASME Paper HT-FED2004-56403, 2004.

Kilkis, B.I. Cost optimisation of a hybrid HVAC system with composite radiant panel walls. Appl. Therm. Eng., Vol. 26, pp 10–17, 2006.

Ku, J. et al. Investigation of capillary limit in a loop heat pipe. Proceedings 12th International Heat Transfer Conference, Vol. 4, pp 489–494, Grenoble, 18–23 August 2002.

Larsen, E.H. et al. Investigation of a capillary assisted thermosyphon (CAT) for shipboard electronics cooling. Proceedings of the ASME Heat Transfer/Fluids Engineering Summer Conference, 2004, Vol. 4, pp 485–494, 2004.

Mahfoud, M. and Emadi, D. Application of heat pipe technology in thermal analysis of metals. J. Therm. Anal. Calorim., Vol. 81, No. 1, pp 161–167, 2005.

Makhankov, A. et al. Liquid metal heat pipes for fusion applications. Fusion Eng. Des., Vol. 42, pp 373–379, 1998.

Martinez, F.J.R. et al. Design and experimental study of a mixed energy recovery system, heat pipes and indirect evaporative equipment for air conditioning. Energy Buil., Vol. 35, pp 1021–1030, 2003.

Maydanik, Y.F. et al. Miniature loop hat pipe – a promising means for cooling electronics. IEEE Trans. Components Packaging Technol., Vol. 28, No. 2, pp 290–296, 2005.

Mazumdar, A. and Srivastava, R. Cirus peel gasification using molten sodium heat pipes. Proceedings of the AIChE Annual Meeting, pp 8831–8835, 2005.

Nikkanen, K et al. Pulsating heat pipes for microelectronics cooling. Proceedings of the 2005 AIChE Spring National Meeting, pp 2421–2428, 2005.

Nouri-Borujerdi, A. and Layeghi, M. A review of concentric annular heat pipes. Heat Transf. Eng., Vol. 26, No. 6, pp 45–58, 2005.

Nuwayhid, R.Y. and Hamade, R. Design and testing of a locally made loop-type thermosyphonic heat sink for stove-top thermoelectric generators. Renewable Energy, Vol. 30, pp 1101–1116, 2005.

Prisniakov, K. et al. About the complex influence of vibrations and gravitational fields on serviceability of heat pipes in composition of space-rocket systems. Acta Astronaut., Vol. 55, pp 509–518, 2004.

Razzaque, M.M., Pate, M.B. and Shapiro, H.N. A novel concept of passive shutdown heat removal in advanced nuclear reactors – applications to PRISM and MHTGG. Ann. Nucl. Energy, Vol. 16, No. 9, pp 483–486, 1989.

Rhi, S.H. A cooling system using wickless heat pipes for multi-chip modules: experiment and analysis. KSME J., Vol. 11, No. 4, pp. 208–220, 1997.

Riffat, S.B. and Zhao, X. A novel hybrid heat pipe solar collector/CHP system – Part 1: system design and construction. Renewable Energy, Vol. 29, pp. 2217–2233, 2004.

Rosas, C. et al. Improvement of the cooling process of oil-immersed electrical transformers using heat pipes. IEEE Trans. Power Deliv., Vol. 20, No. 3, pp 1955–1961, 2005.

Sivaraman, B. and Krishna Mohan, N. Experimental analysis of heat pipe solar collector with different L/di ratio of heat pipe. J. Sci. Ind. Res., Vol. 64, No. 9, pp 698–701, 2005.

Song, F. et al. Fluid flow and heat transfer modeling in rotating heat pipes. Proceedings of the 12th International Heat Transfer Conference, Vol. 4, pp 465–470, Grenoble, 18–23 August 2002.

Song, F., Ewing, G. and Ching, C.Y. Effect of inclination on the performance of axial rotating heat pipes. Proceedings of the 13th International Heat Pipe Conference, Shanghai, 2004.

Tan, B.K. et al. Analytical effective length study of a flat plate heat pipe using point source approach. Appl. Therm. Eng., Vol. 25, Nos. 14–15, pp 2272–2284, 2005.

Tanaka, H. et al. A vertical multiple-effect diffusion-type solar still coupled with a heat-pipe solar collector. Desalination, Vol. 160, pp 195–205, 2004.

Thomas, S.K. and Damle, V.C. Fluid flow in axial re-entrant grooves with application to heat pipes. J. Thermophysics Heat Transf., Vol. 189, No. 3, pp 395–405, 2005.

Tu, S.-T., Zhang, H. and Zhou, W.-W. Corrosion failure of high temperature heat pipes. Eng. Fail. Anal., Vol. 6, pp 363–370, 1999.

Yu, Z.-T. et al. Optimal design of the separate type heat pipe heat exchanger. J. Zhejiang Univ. Sci., Vol. 6A (Supplement), pp 23–28, 2005.

APPENDIX 4

A SELECTION OF HEAT-PIPE-RELATED WEB SITES

http://mscweb.gsfc.nasa.gov/545web/Groups/P2/Projects/ (Interesting NASA web site devoted to activities of the Thermal Engineering Branch – not just heat pipes).

http://process-equipment.globalspec.com/SpecSearch/ (The Engineering Search Engine – useful for identifying heat pipe manufacturers – currently early 2005 – lists 21, all in the USA, so not fully comprehensive!).

http://tmrl.mcmaster.ca/ (Gives information on research and related publications). Thermal Management Research Laboratory, McMaster University, Canada.

www.aascworld.com/heatpipespanel.htm (Data on heat pipe panels for spacecraft, constructed by AASC). Applied Aerospace Structures Corp., Stockton, California, USA.

www.apricus-solar.com (Manufacturer of heat-pipe-based solar collectors). Nanjing, China.

www.astro-r.co.jp (Manufacturer). Astro Research Corporation, Kanagawa, Japan.

www.cast.cn/en/ (Research Institute working on cryogenic heat pipes and VCHPs for space applications in China). Chinese Academy of Space Technology, China.

www.commerce.com.tw/products/EN/H/Heat_Pipe.htm (Web site listing heat pipe suppliers in Asia and areas. Lists many large suppliers and a number of minor players that may supply locally). Based in Taiwan.

www.crtech.com (Modelling using CFD etc.). Colorado, USA.

www.deschamps.com (Manufacturer of a range of heat exchangers, including heat-pipe-based units for cabinet cooling). Deschamps Technologies, Natural Bridge Street, Virginia, USA.

www.europeanthermodynamics.com (Heat pipes for electronics cooling). Leicestershire, UK.

www.fujikura.co.uk (Web site of the UK branch of Fujikura. Heat pipe supplier). Chessington, Surrey, UK.

www.furukawa.co.jp/english/ (Manufacturer of a range of heat pipe systems – English sweb site with access to technical articles). The Furukawa Electric Company, Japan.

www.heatsink-guide.com/heatpipes.shtml (Includes information and descriptions of heat sinks using heat pipes).

www.heat-pipes.com (Site of CRS Engineering, manufacturers of a wide range of heat pipes, including units for injection moulding, electronic thermal control, etc.) CRS Engineering Ltd., Alnwick, Northumberland, UK.

www.heatpipe.com (Heat pipe manufacturer). Gainesville, Florida, USA.

www.heatpipeindia.com (Moulding and waste heat recovery applications). Gloden Star Technical Services, Pune, India.

www.hsmarston.co.uk/ (Heat pipe supplier and also makes other compact heat exchangers). H.S. Marston Ltd., Wolverhampton, UK.

www.kellysearch.com/ (A product search engine. Lists 116 heat pipe suppliers (Spring 2005) and is better at identifying international players than Globalspec, but over-emphasises Chinese SMEs).

www.lanl.gov/world/views/news/releases (Site of Los Alamos Laboratory – includes heat pipe data). Los Alamos, New Mexico, USA.

www.mjm-engineering.com (Heat pipes and consultancy on thermal design based upon use of heat pipes). MJM Engineering Co., Naperville, Illinois, USA.

www.norenproducts.com/Heat_Pipe_Product_Guide.html (The heat pipe catalogue of Noren Products). Noren Products, Inc., Menlo Park, California, USA.

www.nottingham.ac.uk/sbe/ (The web site of the School of the Built Environment at Nottingham University, where heat pipes related to renewable energy and other uses are researched). Nottingham University, UK.

www.pipcar.co.uk (Heat pipes for core cooling in injection moulding etc.). Tonbridge, Kent, UK.

www.pr.afrl.af.mil/facilities/pr_east/heat.htm (Data on heat transfer R&D facilities, including heat pipe R&D). Air Force Research Laboratory, USA.

www.silverstonetek.com (Heat pipes for electronics thermal control, including sintered units). SilverStone Technology Co., Ltd.

www.spcoils.co.uk (Manufacturer of heat pipe heat exchangers and dehumidifiers). S & P Coil Products Ltd., Leicester, UK.

www.swales.com (The web site of Swales Aerospace – heat pipe manufacturer, including loop heat pipes). Swales Aerospace, Maryland, USA.

www.thermacore.com (The web site of Thermacore Inc., part of Modine and a major manufacturer and supplier of heat pipes and related systems). Thermacore, Inc., Lancaster, Pennsylvania, USA.

www.transterm.ro/ (The web site of the Romanian heat pipe manufacturer, Transterm). Transterm, 2200 Brasov, Romania.

www.itmo.by/ (The web site of the Luikov Heat & Mass Transfer Institute, where much research on innovative heat pipes is carried out). P. Brovka, Minsk, Belarus.

APPENDIX 5

CONVERSION FACTORS

Physical Quantity		
Mass	1 lb	= 0.4536 kg
Length	1 ft	= 0.3048 m
	1 in	= 0.0254 m
Area	1 ft^2	= 0.0929 m^2
Force	1 lbf	= 4.448 N
Energy	1 Btu	= 1.055 kJ
	1 kWh	= 3.6 MJ
Power	1 hp	= 745.7 W
Pressure	1 lbf/in^2	= 6894.76 N/m^2
	1 bar	= 10^5 N/m^2
	1 atm	= 101.325 kN/m^2
	1 torr	= 133.322 N/m^2
Dynamic Viscosity	1 Poise	= 0.1 Ns/m^2
Kinematic Viscosity	1 stoke	= 10^{-4} m^2/s
Heat Flow	1 Btu/h	= 0.2931 W
Heat Flux	1 Btu/ft^2h	= 3.155 W/m^2
Thermal Conductivity	1 Btu/ft^2h °F/ft	= 1.731 W/m^2 °C/m
Heat Transfer Coefficient	1 Btu/ft^2h °F	= 5.678 W/m^2 °C

NOMENCLATURE

A_c	Circumferential flow area
A_w	Wick cross-sectional area
C_p	Specific heat of vapour, constant pressure
C_v	Specific heat of vapour, constant volume
D	Sphere density in Blake–Kozeny equation
H	Constant in the Ramsey–Shields–Eotvös equation
J	4.18 J/g mechanical equivalent of heat
K	Wick permeability
L	Enthalpy of vapourisation or latent heat of vapourisation
M	Molecular weight
M	Mach number
M	Figure of merit
N	Number of grooves or channels
Nu	Nusselt number
Pr	Prandtl number
P	Pressure
ΔP	Pressure difference
$\Delta P_{c\,max}$	Maximum capillary head
ΔP_l	Pressure drop in the liquid
ΔP_v	Pressure drop in the vapour
ΔP_g	Pressure drop due to gravity
Q	Quantity of heat
R	Radius of curvature of liquid surface
R_o	Universal gas constant $= 8.3 \times 10^3$ J/K kg mol
Re	Reynolds number
Re_r	Radial Reynolds number
Re_b	A bubble Reynolds number
S	Volume flow per second
T	Absolute temperature
T_c	Critical temperature
T_v	Vapour temperature
ΔT_s	Superheat temperature
T_w	Heated surface temperature

V Volume
V_c Volume of condenser
V_R Volume of gas reservoir
W_e Weber number
a Groove width
a Radius of tube
b Constant in the Hagen–Poiseuille Equation
c Velocity of sound
d_a Artery diameter
d_w Wire diameter
f Force
g Acceleration due to gravity
g_c Rohsenhow correlation
h Capillary height, artery height, coefficient of heat transfer
k Boltzmanns Constant $= 1.38 \times 10^{-23}$ J/K
k_w Wick thermal conductivity – k_s solid phase, k_l liquid phase
l Length of heat pipe section defined by subscripts, Section 2.3.4
l_{eff} Effective length of heat pipe
m Mass
m Mass of molecule
m Mass flow
n Number of molecules per unit volume
g Heat flux
r Radius
r Radial co-ordinate
r_e Radius in the evaporator section
r_c Radius in the condensing section
r_H Hydraulic radius
r_v Radius of vapour space
r_w Wick radius
u Radial velocity
v Axial velocity
y Co-ordinate
z Co-ordinate
α Heat transfer coefficient
β Defined as $(1+k_s/k_l)/(1-k_s/k_l)$
δ Constant in Hsu Formula – thermal layer thickness
ε Fractional voidage
θ Contact angle
ϕ Inclination of heat pipe
ϕ_c Function of channel aspect ratio
λ Characteristic dimension of liquid/vapour interface
μ Viscosity
μ_l Dynamic viscosity of liquid

μ_v	Dynamic viscosity of vapour
γ	Ratio of specific heats
ρ	Density
ρ_l	Density of liquid
ρ_v	Density of vapour
σ	σ_{LV} used for surface energy where there is no ambiguity
σ_{SL}	Surface energy between solid and liquid
σ_{LV}	Surface energy between liquid and vapour
σ_{SV}	Surface energy between solid and vapour

Other notations are as defined in the text.

Index